W0017133

MATHEMATICAL PHYSICS ELECTRONIC JOURNAL

Print Version

MATHEMATICAL PHYSICS ELECTRONIC JOURNAL

Print Version

Volume 5, 1999

Chief Editors

J.-P. Eckmann
G. Gallavotti
H. Koch

Volume 6, 2000

Chief Editors

P. Collet
H. Koch
C. E. Wayne

Technical Editors

R. de la Llave
H. Koch
C. Radin

Department of Mathematics
University of Texas at Austin
USA

World Scientific
New Jersey • London • Singapore • Hong Kong

Published by

World Scientific Publishing Co. Pte. Ltd.

P O Box 128, Farrer Road, Singapore 912805

USA office: Suite 1B, 1060 Main Street, River Edge, NJ 07661

UK office: 57 Shelton Street, Covent Garden, London WC2H 9HE

British Library Cataloguing-in-Publication Data
A catalogue record for this book is available from the British Library.

MATHEMATICAL PHYSICS ELECTRONIC JOURNAL
Volumes 5 and 6 (Print Version)

Copyright © 2002 by World Scientific Publishing Co. Pte. Ltd.
The copyright of each article is owned by the author(s).

All rights reserved. This book, or parts thereof, may not be reproduced in any form or by any means, electronic or mechanical, including photocopying, recording or any information storage and retrieval system now known or to be invented, without written permission from the Publisher.

For photocopying of material in this volume, please pay a copying fee through the Copyright Clearance Center, Inc., 222 Rosewood Drive, Danvers, MA 01923, USA. In this case permission to photocopy is not required from the publisher.

ISBN 981-02-4881-4 (pbk)

Printed in Singapore.

Preface

The main goal in the creation of MPEJ was to have a journal that was distributed as widely as possible in the community and that had as its only criterion for acceptance the quality of the papers. In particular, we did not want that considerations of the number of pages to be published every year entered into decisions of acceptance.

These goals are well served by the free electronic distribution of the papers and the electronic supplements available from:

- `http://www.ma.utexas.edu/mpej`
- `http://mpej.unige.ch/mpej`
- `http://www.maia.ub.es/mpej`

and the about 20 mirrors organized by the European Mathematical Information Service (EMIS) whose main server is:

- `http://www.emis.de`

We think that adding a print version of the papers to the electronic distribution will increase the availability and, given that there is no commitment of pages nor any financial incentive for editors or authors, it will allow MPEJ to keep quality as the sole criterion.

We thank World Scientific and, specially Dr. Sen Hu for the organization of this print version.

MPEJ

MATHEMATICAL PHYSICS ELECTRONIC JOURNAL* ISSN 1086-6655

Volume 5, 1999

Chief Editors

J.-P. Eckmann
Université de Genève
Geneva

G. Gallavotti
Università di Roma La Sapienza
Rome

H. Koch
The University of Texas
Austin

Editorial Board

J.E. Avron
Israel Institute of Technology
Haifa

G. Benettin
Università di Padova
Padua

P. Collet
Ecole Polytechnique
Palaiseau

P. Constantin
The University of Chicago
Chicago

J. Feldman
University of British Columbia
Vancouver

R. Kotecký
Charles University
Prague

A. Kupiainen
Helsinki University
Helsinki

R. de la Llave
The University of Texas
Austin

H. Spohn
Universität München
Munich

H. Tasaki
Gakushin University
Tokyo

C.E. Wayne
Boston University
Boston

* For an electronic version of this volume, and for information on subscriptions and other matters, see `http://www.ma.utexas.edu/mpej/` or `http://mpej.unige.ch/mpej/` or send an empty e-mail message to `mpej@math.utexas.edu` for instructions.

Contents
Volume 5 (1999)

M P E J

MATHEMATICAL PHYSICS ELECTRONIC JOURNAL

ISSN 1086-6655
Volume 5, 1999

Paper 1
Received: June 19, 1998, Accepted: Jan 6, 1999
Editor: J. Avron

Lower bounds on wave packet propagation by packing dimensions of spectral measures

I. Guarneri[1,2], H. Schulz-Baldes[3]

[1] International Centre for the Study of Dynamical Systems,
Università di Milano, sede di Como, via Lucini 3,
22100 Como, Italy

[2] Istituto Nazionale di Fisica della Materia, Unitá di
Milano, Via Celoria 16, 20133 Milano, Italy

[3] Technische Universität Berlin, Fachbereich Mathematik,
Straße des 17. Juni 136, 10623 Berlin, Germany

Abstract

We prove that, for any quantum evolution in $\ell^2(\mathbf{Z}^D)$, there exist arbitrarily long time scales on which the qth moment of the position operator increases at least as fast as a power of time given by q/D times the packing dimension of the spectral measure. Packing dimensions of measures and their connections to scaling exponents and box-counting dimensions are also discussed.

1 Introduction.

Let H be a selfadjoint operator in a separable Hilbert space $\mathcal{H}$. Let $|\psi\rangle$ be any vector in $\mathcal{H}$, μ its spectral measure relative to H and $\mathcal{B} \equiv \{|n\rangle\}_{n\in\mathbf{N}}$ a Hilbert basis in $\mathcal{H}$. As *time* t increases, the *wave packet* $e^{-\imath tH}|\psi\rangle$ spreads out over the basis $\mathcal{B}$. In particular, it is known [G1, G2, La, Co] that, for $q > 0$, the qth moment of the probability distribution associated with the Fourier expansion of $e^{-\imath tH}|\psi\rangle$ over the basis $\mathcal{B}$ increases, in time average, at least as fast as a power of time given by q times the Hausdorff dimension of μ. This qualitatively means that propagation is faster the more continuous the spectral measure is, the degree of continuity being measured by the lower pointwise dimension of the spectral measure.

The present work is summarized by the qualitative remark that, at any given time t, the wave packet can only probe continuity of the spectral measure on the level of spectral resolution achieved at time t, hence on a spectral scale of the order of $1/t$. This remark is relevant to the case of non-exactly scaling spectral measures. For such measures, the upper and lower pointwise dimensions do not coincide, meaning that there are arbitrarily small spectral scales at which the measure is somewhere in the spectrum scaling faster than expressed by the Hausdorff dimension. We accordingly obtain that on arbitrarily long time scales a larger lower bound is valid, given by the packing dimension of the spectral measure.

There is at least one abstract example of a zero-Hausdorff dimensional spectral measure with packing dimension equal to 1, which leads to ballistic transport [G3]. Spectral measures of a similar nature are likely to occur also for concrete Schrödinger operators. Such may be the case with the quasi-ballistic dynamics exhibited by the Harper model with Liouville incommensuration [La].

The Hausdorff dimension of a measure is related to its lower pointwise dimensions by the theory of Rogers and Taylor [Ro], whose relevance in the present context was advocated by Last [La]. Our present result calls *upper* pointwise dimensions into play. Apart from [G3], these have not yet found their way in the theory of quantum transport. In the appendix, we therefore elaborate on results by Cutler [Cu] and develop a treatment, in a sense dual to Rogers' and Taylor's, which connects such dimensions to the packing dimension of a measure. Furthermore, we show that these packing dimensions can also be calculated by a box-counting procedure.

In the next two sections we establish preliminaries and notations. In Section 4 we state and prove our main result Theorem 1. The main element of its proof is Proposition 2 which is basically a restatement of existing results. However, we give here an alternative *ab initio* derivation which makes no use of Strichartz theorem, but is based on Proposition 1 combined with Last's argument [La]. A modified version of Proposition 1 allowed to prove also *upper* bounds for a special class of Hamiltonians [GS]. Finally, in Section 5, we transpose the main result to Hamiltonians on a D-dimensional lattice.

A useful discussion with Y. Last is acknowledged.

2 Growth exponents.

For given time T and $\epsilon \in (0,1)$, we define the *minimal carrier* $\overline{n}(\epsilon, T)$ of the wave packet on the basis $\mathcal{B}$ as follows:

$$\overline{n}(\epsilon, T) = \min \left\{ \overline{n} \in \mathbf{N} \;\middle|\; \sum_{n \geq \overline{n}} p_n(T) \leq \epsilon \right\} , \tag{1}$$

where the $p_n(T)$ is the average probability up to time T in the basis state $|n\rangle \in \mathcal{B}$, given by:

$$p_n(T) = \int_0^T \frac{dt}{T} |\langle n|e^{-\imath Ht}|\psi\rangle|^2 .$$

Upper and lower growth exponents of the minimal carriers are defined as:

$$\beta_0^+(\epsilon) = \limsup_{T\to\infty} \frac{\log(\overline{n}(\epsilon,T))}{\log(T)} , \qquad \beta_0^-(\epsilon) = \liminf_{T\to\infty} \frac{\log(\overline{n}(\epsilon,T))}{\log(T)} , \tag{2}$$

We also define $\beta_0^\pm = \lim_{\epsilon\to 0} \beta_0^\pm(\epsilon)$. Let us further introduce the qth moment $M_q(T)$ of the distribution $p_n(T)$ by

$$M_q(T) = \sum_{n\geq 0} n^q \, p_n(T) ,$$

which can be interpreted as the time-average up to time T of the expectation value of the qth moment ($q \neq 0$) of the *position operator* associated with the basis $\mathcal{B}$. The corresponding growth exponents are

$$\beta_q^+(\epsilon) = \limsup_{T\to\infty} \frac{\log(M_q(T))}{q \log(T)} , \qquad \beta_q^-(\epsilon) = \liminf_{T\to\infty} \frac{\log(M_q(T))}{q \log(T)} , \tag{3}$$

We have $\beta_q^+ \leq \beta_0^-$ whenever $q < 0$ and $\beta_q^- \geq \beta_0^+$ whenever $q > 0$. Lower bounds on $\beta_0^\pm$ convey stronger information than lower bounds on $\beta_q^\pm$, $q > 0$. For instance, if H has a pure point spectrum then $\beta_0^\pm = 0$, because minimal carriers remain bounded in time; still, moments may display nonvanishing growth exponents.

The lower growth exponents β_q^- characterize the minimal spreading of the wave packet: for all $T \geq 0$, $M_q(T) \geq C(\delta) T^{q(\beta_q^- - \delta)}$ holds for all $\delta > 0$ with appropriate constants $C(\delta)$. The upper exponents β_q^+ give the fastest possible spreading on sequences of times: for all $\delta > 0$, $M_q(T) \leq C(\delta) T^{q(\beta_q^+ + \delta)}$ holds at all times T, but there exists a diverging sequence $(T_k)_{k\geq 1}$ such that $M_q(T_k) \geq C'(\delta) T_k^{q(\beta_q^+ - \delta)}$ for all $k \geq 1$.

3 Spectral dimensions.

Given a (Borel) probability measure μ on $\mathbf{R}$, we define its lower and upper pointwise dimensions at $E \in \mathrm{supp}(\mu)$ as follows:

$$\underline{d}_\mu(E) = \liminf_{\epsilon\to 0}\frac{\log(\mu([E-\epsilon,E+\epsilon]))}{\log(\epsilon)}\,, \qquad \overline{d}_\mu(E) = \limsup_{\epsilon\to 0}\frac{\log(\mu([E-\epsilon,E+\epsilon]))}{\log(\epsilon)}\,, \tag{4}$$

while for $E \notin \text{supp}(\mu)$, $\underline{d}_\mu(E) = \overline{d}_\mu(E) = \infty$.

The (upper) Hausdorff dimension $\dim_{\mathrm{H}}^{+}(\mu)$ and the (upper) packing dimension $\dim_{\mathrm{P}}^{+}(\mu)$ of the measure μ can then be defined as follows:

$$\dim_{\mathrm{H}}^{+}(\mu) = \mu\text{-}\operatorname*{ess\,sup}_{E\in\mathbf{R}} \underline{d}_\mu(E)\,, \qquad \dim_{\mathrm{P}}^{+}(\mu) = \mu\text{-}\operatorname*{ess\,sup}_{E\in\mathbf{R}} \overline{d}_\mu(E)\,, \tag{5}$$

where, for a real function f, $\mu\text{-}\operatorname{ess\,sup}_{E\in\mathbf{R}} f(E)$ denotes the μ-essential supremum of f, *i.e.* the infimum over all sets Δ of full μ-measure of the quantity $\sup_{E\in\Delta} f(E)$. Although the above definitions are optimally suited to our present purposes, dimensions of Borel measures are more properly defined and discussed in the appendix.

In the rest of this article, we shall use dyadic partitions of the real axis in intervals $I_j^N = ((j-1)2^{-N}, j2^{-N}]$, $j \in \mathbf{Z}$. Of course, any other hierarchic partition could be used. For $E \in \mathbf{R}$, we shall denote $I_{j(E)}^N$ the dyadic interval of the Nth generation to which E belongs. We shall in particular make use of the fact that

$$\underline{d}_\mu(E) = \liminf_{N\to\infty}\frac{\log(\mu(I_{j(E)}^N))}{\log(|I_{j(E)}^N|)} = \liminf_{N\to\infty}\frac{-\log_2(\mu(I_{j(E)}^N))}{N}\,. \tag{6}$$

Similar equalities hold for the upper pointwise dimension $\overline{d}_\mu(E)$, with lim inf replaced by lim sup.

4 Lower bounds on growth exponents.

The following proposition expresses in a quantitative way the fact that the time evolution up to time T does not resolve details of the spectrum on scales smaller than $1/T$ ($\hbar = 1$). Up to time T it is thus possible to work with an approximate Hamiltonian with discrete spectrum the eigenvalues of which have a spacing of order $1/T$.

We use the notation $|\chi_j^N\rangle = \chi_j^N(H)|\psi\rangle$, where χ_j^N is the characteristic function of I_j^N.

Proposition 1 *Given $\epsilon > 0$, we associate to any time T a generation index N so that:*

$$2^{(N-1)} < \frac{T}{\sqrt{\epsilon}} \le 2^N\,. \tag{7}$$

Then for any family of indices $\mathcal{F} \subset \mathbf{N}$ and $\psi \in \mathcal{H}$ with $\|\psi\| \le 1$, the following estimate holds true:

$$\sum_{n\in\mathcal{F}}\int_0^T \frac{dt}{T}|\langle n|e^{-\imath Ht}|\psi\rangle|^2 \le 2\epsilon + \frac{8\pi}{\sqrt{\epsilon}}\sum_{n\in\mathcal{F}}\sum_{j\in\mathbf{Z}}|\langle n|\chi_j^N\rangle|^2\,. \tag{8}$$

Proof. Throughout this proof, we understand that N and T are related to each other via (7). For $0 \leq t \leq T$, we approximate $e^{-\imath Ht}|\psi\rangle$ by:

$$|\psi_T(t)\rangle \;=\; \sum_{j\in\mathbf{Z}} e^{-\imath tE_j^N}|\chi_j^N\rangle \;,$$

where $E_j^N = j2^{-N}$. Then we have:

$$\|\,|\psi_T(t)\rangle - e^{-\imath tH}|\psi\rangle\|^2 \;\leq\; \sum_{j\in\mathbf{Z}}\int_{I_j^N} d\mu(E)\,t^2|E-E_j^N|^2 \;.$$

As $|E-E_j^N| \leq 2^{-N}$ for $E \in I_j^N$, it follows from (7) that the latter expression is less than ϵ as long as $0 \leq t \leq T$. Thus

$$\sum_{n\in\mathcal{F}}\int_0^T \frac{dt}{T}|\langle n|e^{-itH}|\psi\rangle|^2 \;\leq\; 2\,\epsilon + 2\sum_{n\in\mathcal{F}}\int_0^T \frac{dt}{T}|\langle n|\psi_T(t)\rangle|^2 \;. \tag{9}$$

Now the integral can be bounded as follows:

$$\begin{aligned}\int_0^T \frac{dt}{T}|\langle n|\psi_T(t)\rangle|^2 \;&\leq\; \int_0^{\pi 2^{N+1}} \frac{dt}{T}\,|\langle n|\psi_T(t)\rangle|^2 \\ &=\; \sum_{j\in\mathbf{Z}}\sum_{l\in\mathbf{Z}}\langle n|\chi_{I_j^N}\rangle\langle\chi_{I_l^N}|n\rangle\,\frac{1}{T}\int_0^{\pi 2^{N+1}} dt\,e^{-i(E_j^N-E_l^N)t} \;.\end{aligned}$$

The latter integral yields $2^{N+1}\pi\delta_{jl}$. Putting this into (9) and recalling (7), we directly get inequality (8). □

Proposition 2 *For integer N and $\alpha \in (0,1)$, let $A_{N,\alpha}$ be the union of all the dyadic intervals of the Nth dyadic generation which have measure $\mu(I_j^N) < 2^{-N\alpha}$. Set $b_N = \mu(A_{N,\alpha})$. If $b_N > 0$, then the following holds true for T satisfying $b_N 2^{N-1} < 9T \leq b_N 2^N$:*

$$\overline{n}\left(\frac{b_N}{2}, T\right) \;>\; C(\alpha)\,b_N^{3-\alpha}\,T^\alpha \;, \tag{10}$$

where $C(\alpha) > 0$ is only dependent on α.

Proof. Let us define $|\psi_N\rangle = \chi_{A_{N,\alpha}}(H)|\psi\rangle$, so $\|\,|\psi_N\rangle\|^2 = b_N > 0$. We wish to apply Proposition 1, with the following specifications: replace $|\psi\rangle$ by $|\psi_N\rangle$ and choose $\mathcal{F}$ as the integers smaller than a given m, finally set $\epsilon = (b_N/9)^2$. Condition (7), which makes Proposition 1 applicable, then becomes precisely $b_N 2^{N-1} < 9T \leq b_N 2^N$.

The spectral measure of $|\psi_N\rangle$ is $d\mu_N(E) = \chi_{A_{N,\alpha}}(E)d\mu(E)$. Hence the sum on the right-hand side of (8) is restricted to the set $\mathcal{J}_{N,\alpha}$ of indices $j \in \mathbf{Z}$ for which $\mu(I_j^N) < 2^{-N\alpha}$. The sum can be estimated as follows:

$$\sum_{n\in\mathcal{F}}\sum_{j\in\mathcal{J}_{N,\alpha}} |\langle n|\chi_j^N\rangle|^2 \le \sum_{n\in\mathcal{F}}\sum_{j\in\mathcal{J}_{N,\alpha}} \|\chi_j^N\|^2\,\|\chi_j^N(H)|n\rangle\|^2$$
$$= \sum_{n\in\mathcal{F}}\sum_{j\in\mathcal{J}_{N,\alpha}} \mu_N(I_j^N)\|\chi_j^N(H)|n\rangle\|^2$$
$$\le m \max_{j\in\mathcal{J}_{N,\alpha}} \mu_N(I_j^N)$$
$$\le m\,2^{-N\alpha}\,.$$

We choose m as follows:

$$m = \frac{1}{4\pi}\left(\frac{b_N}{9}\right)^3 2^{N\alpha}\,, \tag{11}$$

and denote $P_m = \sum_{n<m} |n\rangle\langle n|$. Substituting all the above in Proposition 1, we obtain

$$\int_0^T \frac{dt}{T}\|P_m(t)\psi_N\|^2 \le \left(\frac{2b_N}{9}\right)^2\,,$$

where $P_m(t) = e^{\imath Ht}P_m e^{-\imath Ht}$. Now we continue as in [La, Theorem 6.1]. Let $|\psi'_N\rangle = |\psi\rangle - |\psi_N\rangle$. Then $|\psi'_N\rangle$ is orthogonal to $|\psi_N\rangle$, and

$$\int_0^T \frac{dt}{T}\|P_m(t)|\psi\rangle\|^2 \le \int_0^T \frac{dt}{T}\|P_m(t)|\psi_N\rangle\|^2 + 2\|\,|\psi'_N\rangle\|\int_0^T \frac{dt}{T}\|P_m(t)|\psi_N\rangle\| + \|\,|\psi'_N\rangle\|^2$$
$$\le \left(\frac{2b_N}{9}\right)^2 + \|\,|\psi'_N\rangle\|^2 + \frac{4b_N}{9}\,.$$

Whence, recalling $\|\,|\psi_N\rangle\|^2 + \|\,|\psi'_N\rangle\|^2 = 1$, we get:

$$\int_0^T \frac{dt}{T}\|(1-P_m(t))|\psi\rangle\|^2 \ge \|\,|\psi_N\rangle\|^2 - \left(\frac{2b_N^2}{9}\right)^2 - \frac{4b_N}{9} \ge \frac{b_N}{2}\,.$$

because $\|\,|\psi_N\rangle\|^2 = b_N < 1$. It therefore follows from the definition of a minimal carrier that $\overline{n}(b_N/2, T) \ge m$. The proof is concluded on recalling the definition of m and ϵ as well as the connection between N and T. □

A link to spectral dimensions is provided by (we also use lim sup and lim inf in their set-theoretic meaning; further we suppress the subscripts $N \to \infty$):

$$\limsup A_{N,\alpha} = \limsup\left\{E\in\mathbf{R}\,|\,\mu(I^N_{j(E)}) < 2^{-N\alpha}\right\}$$
$$= \left\{E\in\mathbf{R}\,\middle|\,\limsup\frac{-\log_2\mu(I^N_{j(E)})}{N} > \alpha\right\}$$
$$= \left\{E\in\mathbf{R}\,|\,\overline{d}_\mu(E) > \alpha\right\} \tag{12}$$

In a completely analogous way, we get

$$\liminf A_{N,\alpha} = \left\{E \in \mathbf{R} \,|\, \underline{d}_\mu(E) > \alpha\right\} \tag{13}$$

At this point, we derive lower bounds. First of all, if $\alpha < \dim_{\mathrm{H}}(\mu)$, then (5) and (13) show that $\liminf \mu(A_{N,\alpha}) \geq \mu(\liminf A_{N,\alpha}) > 0$, so b_N is larger than some b_0 for all N. Inserting such a b_0 in Proposition 2, we immediately get $\overline{n}(b_0/2, T) \geq \mathrm{const}\, T^\alpha$ for all T. This proves the already known lower bound, $\beta_0^- \geq \dim_{\mathrm{H}}(\mu)$. In addition, we can now prove the existence of a sequence of times on which the large packing dimension of the "thin" part of the spectral measure forces the wave packet to travel possibly farther than imposed by the Hausdorff dimension.

Theorem 1 *For all positive q, the upper growth exponent of the q-th moment of the position operator satisfies $\beta_q^+ \geq \dim_{\mathrm{P}}^+(\mu)$.*

Proof. It is obviously sufficient to consider the case $\dim_{\mathrm{P}}^+(\mu) > 0$. Then, if $\alpha < \dim_{\mathrm{P}}(\mu)$, it follows from the definition (5) of the packing dimension and from (12) that $\mu(\limsup A_{N,\alpha}) > 0$. Therefore, from the Borel-Cantelli lemma we get that $\sum_N \mu(A_{N,\alpha}) = \infty$. This in turn implies that there is a sequence of integers $N_k \to \infty$ such that $b_{N_k} = \mu(A_{N_k,\alpha}) > N_k^{-2}$.

In this situation, Proposition 2 says that

$$\overline{n}\left(\frac{b_{N_k}}{2}, T_k\right) > \frac{C(\alpha)}{N_k^{6-2\alpha}}\, T_k^\alpha .$$

In this inequality, $k \in \mathbf{N}$ is arbitrary, $C(\alpha)$ is a numerical factor only depending on α, the sequence of times T_k satisfies $b_{N_k} 2^{N_k - 1} < 9T_k \leq b_{N_k} 2^{N_k}$ for all k, and $b_{N_k} > N_k^{-2}$. From all that it follows that $T_k \to \infty$ as $k \to \infty$ and that $N_k < f^{-1}(T_k)$ where f is the function $f(x) = 2^x/(18\, x^2)$. Hence we get that, for all $k \in \mathbf{N}$,

$$\overline{n}\left(\frac{b_{N_k}}{2}, T_k\right) > \frac{C(\alpha)}{(f^{-1}(T_k))^{6-2\alpha}}\, T_k^\alpha .$$

We denote the right-hand side by C_k. Since $\alpha > 0$, C_k will be eventually larger than 1. From the definition (1) of a minimal carrier, we obtain that, at all large enough k, the total probability supported by basis states $|n\rangle$ with $n > C_k - 1$ is larger than $b_{N_k}/2$. For such k's, the Chebyshev inequality yields:

$$M_q(T_k) \geq \frac{b_{N_k}}{2}\, (C_k - 1)^q .$$

Whence, replacing C_k and using $b_{N_k} > N_k^{-2}$, we get

$$M_q(T_k) > \frac{1}{2(f^{-1}(T_k))^2} \left(\frac{C(\alpha) T_k^\alpha}{(f^{-1}(T_k))^{6-2\alpha}} - 1\right)^q .$$

Since $\lim_{T\to\infty} \log(f^{-1}(T))/\log(T) = 0$, Theorem 1 follows from the definition of the exponent β_q^+. □

Remark 1 The present proof does not allow to bound β_0^+ below by means of $\dim_P^+(\mu)$. In order to do that one needs to know that $\limsup \mu(A_{N,\alpha}) > 0$. We do not know whether that follows from $\dim_P^+(\mu) > \alpha$. Nevertheless we can prove the weaker result given in Proposition 3 below.

For our next result, let us define

$$\mathcal{D}(\mu) = \sup_{\epsilon} \limsup_{N\to\infty} \frac{\log_2(\mathcal{N}(\mu, N, \epsilon))}{N} \tag{14}$$

where, for given N and ϵ, we define $\mathcal{N}(\mu, N, \epsilon)$ as the minimal number of dyadic intervals of generation N which support more than $1-\epsilon$ of the measure μ. It is shown in the appendix that $\mathcal{D}(\mu)$ is bigger than or equal to the box-counting information dimension of μ and smaller than or equal to its packing dimension.

Proposition 3 $\beta_0^+ \geq \mathcal{D}(\mu)$.

Proof. It is enough to show that, if α is less than the expression on the rhs of (14), then $\limsup \mu(A_{N,\alpha}) > 0$. Let us choose ϵ and a sequence N_k so that

$$\alpha < \lim_{k\to\infty} \frac{\log_2(\mathcal{N}(\mu, N_k, \epsilon))}{N_k} . \tag{15}$$

If $\lim \mu(A_{N,\alpha}) = 0$, then the sequence of characteristic functions $\chi_{A_{N_k,\alpha}}$ converges to 0 in μ-measure. So there is a sequence N_{k_j} such that $\chi_{A_{N_{k_j},\alpha}}$ converges to 0 μ-almost everywhere. We can then find a compact K, with $\mu(K) > 1-\epsilon$, so that $\chi_{A_{N_{k_j},\alpha}}$ converges uniformly to 0 in K. Hence, K is eventually a subset of all the $A^c_{N_{k_j},\alpha}$. Thus, for all sufficiently large j, K has a covering by dyadic intervals of the N_{k_j}th generation, everyone of which is not smaller than $2^{-\alpha N_{k_j}}$ in μ-measure. There cannot be more than $(1-\epsilon)2^{\alpha N_{k_j}}$ intervals in those coverings. Therefore, $\mathcal{N}(\mu, N_{k_j}, \epsilon) \leq (1-\epsilon)2^{\alpha N_{k_j}}$, which contradicts (15). □

Remark 2 An explicit albeit abstract illustration of the lower bounds proven above is given in [G3]. There H is multiplication by E in $L^2((0, 2\pi), \mu)$, $|n\rangle = \exp(2\pi \imath n F(E))$, where $F(E) = \mu((0, E))$, and the measure μ has $\dim_H^+(\mu) = 0 = \dim_I^-(\mu)$, $\dim_P^+(\mu) = 1 = \dim_I^+(\mu)$. That measure is constructed following [RJLS]. The motion in this model is ballistic, in the sense that $\beta_q^+ = 1$ for all $q > 0$.

5 Lower bound for covariant lattice Hamiltonians.

Here we transpose the results of the last section to quantum diffusion of dynamics governed by covariant Hamiltonians on the lattice. Because disordered media and quasicrystals can be described by these models, this situation is of particular physical interest. We do not furnish the technical proofs for the various statements in this section because they can be completed along the lines of [SB].

Let the space of disorder or quasicrystaline configurations Ω be a compact and metrizable space on which is given an action T of the group $\mathbf{Z}^D$. We suppose that to each configuration $\omega \in \Omega$ there is a bounded operator $H_\omega : \ell^2(\mathbf{Z}^D) \to \ell^2(\mathbf{Z}^D)$ and that this operator family is strongly continuous in ω and covariant with respect to a projective representation U of the translation group $\mathbf{Z}^D$ on $\ell^2(\mathbf{Z}^D)$, that is

$$U(a)H_\omega U(a)^* \;=\; H_{T^a\omega}\;, \qquad a \in \mathbf{Z}^D\;.$$

Finally we fix an invariant and ergodic probability measure $\mathbf{P}$ on Ω. Let now $\psi_\omega \in \ell^2(\mathbf{Z}^D)$ be a cyclic vector for H_ω and μ_ω its spectral measure. It can then be shown by the same techniques as in [SB] that the packing dimension $\dim_{\mathrm{P}}(\mu_\omega, \Delta)$ in the Borel set $\Delta \subset \mathbf{R}$ (see the appendix for the definition) is $\mathbf{P}$-almost surely constant and thus defines the packing dimension of the local density of states in Δ. It is smaller than or equal to the packing dimension of the density of states $\mathcal{N}$ which is defined to be the disorder average of the spectral measure $\mu_{\omega,|0\rangle}$ of the state $|0\rangle$ localized at the origin, namely we have

$$\dim_{\mathrm{P}}(\mu_\omega, \Delta) \;\leq\; \dim_{\mathrm{P}}(\mathcal{N}, \Delta)\;, \qquad \mathbf{P}\text{-a.s.}\;.$$

Next we define the time averaged moments of the position operator $\vec{X}$ on $\ell^2(\mathbf{Z}^D)$ by

$$\hat{M}_{q,\Delta,\omega}(T) \;=\; \int_0^T \frac{dt}{T} \langle 0|\, \Pi_\omega(\Delta)\, |e^{\imath H_\omega t}\vec{X}e^{-\imath H_\omega t} - \vec{X}|^q\, \Pi_\omega(\Delta)\, |0\rangle\;, \qquad q \neq 0\;,$$

where $\Pi_\omega(\Delta)$ is the spectral projection of H_ω to Δ. The corresponding growth exponents $\hat{\beta}^{\pm}_{q,\omega}(\Delta)$ are defined as in (3). Modification of the proof of Theorem 1 (for fixed H_ω) leads to

$$\dim_{\mathrm{P}}(\mu_\omega, \Delta) \;\leq\; D\,\hat{\beta}^{+}_{q,\omega}(\Delta)\;, \qquad \forall\;\omega \in \Omega\;.$$

Appendix: various dimensions of Borel measures.

The aim of this appendix is to review and extend known results about lower and upper pointwise dimensions, Hausdorff and packing dimensions, as well as fractal and box-counting information dimensions of Borel measures on the real line (the extension to $\mathbf{R}^d$ is immediate). Links between lower pointwise dimensions and Hausdorff dimensions were first established by Rogers and Taylor (see [Ro]). The corresponding theory connecting upper pointwise dimensions and packing dimensions was given by Cutler [Cu], whose results we extend here to a completely dual treatment to that in [Ro, Chapters 3.2 and 3.3]. Fractal dimensions of Borel measures were studied by one of the authors [G2] and we complete here the results given in the latter reference. We also review box-counting information dimensions and establish relations to Hausdorff and packing dimensions. Throughout this appendix, μ and ν are Borel probability measures on $\mathbf{R}$, Δ is a Borel subset of $\mathbf{R}$ and $\gamma \in \mathbf{R}$.

The lower and upper pointwise dimensions of μ at a point $E \in \mathbf{R}$ were defined in equation (4) of Section 3. For various alternative definitions, see [RJLS, SB]. In [SB] it is proven that $\underline{d}_\mu(E) \leq \underline{d}_\nu(E)$ μ-almost surely, by a similar proof $\overline{d}_\mu(E) \leq \overline{d}_\nu(E)$ μ-almost surely. These results imply [SB] that $E \mapsto \underline{d}_\mu(E)$ and $E \mapsto \overline{d}_\mu(E)$ are Borel functions in $L^\infty(\mathbf{R}, \mu)$ taking values in $[0, 1]$ and only depending on the measure class of μ. Let us further introduce as in [Ro] the upper and lower γ-derivative of μ at E by

$$\overline{D}^\gamma_\mu(E) = \limsup_{\epsilon \to 0} \frac{\mu([E-\epsilon, E+\epsilon])}{\epsilon^\gamma}\,, \qquad \underline{D}^\gamma_\mu(E) = \liminf_{\epsilon \to 0} \frac{\mu([E-\epsilon, E+\epsilon])}{\epsilon^\gamma}\,.$$

Links between lower pointwise dimensions and upper γ-derivatives are given by

$$\underline{d}_\mu(E) < \gamma \;\Rightarrow\; \overline{D}^\gamma_\mu(E) = \infty \;\Rightarrow\; \underline{d}_\mu(E) \leq \gamma\,, \tag{16}$$

$$\gamma < \underline{d}_\mu(E) \;\Rightarrow\; \overline{D}^\gamma_\mu(E) = 0 \;\Rightarrow\; \gamma \leq \underline{d}_\mu(E)\,. \tag{17}$$

Analogous relations hold between upper pointwise dimensions and lower γ-derivatives. Next we denote the γ-Hausdorff and the γ-packing measure by M^γ_{H} and M^γ_{P} respectively. The Hausdorff and packing dimension of Δ are denoted by $\dim_{\mathrm{H}}(\Delta)$ and $\dim_{\mathrm{P}}(\Delta)$. As we shall use it in the proofs below, let us recall the definition of the packing measure [TT]. A δ-packing of an arbitrary set $D \subset \mathbf{R}$ is a countable disjoint collection $(B(E_k, r_k))_{k \in \mathbf{N}}$ of closed balls centered at $E_k \in D$ and with radius $r_k \leq \delta/2$. A positive set function is defined by

$$M^{\gamma,\delta}_{\mathrm{P}}(D) = \sup\left\{ \sum_{k \in \mathbf{N}} (2r_k)^\gamma \;\middle|\; (B(E_k, r_k))_{k \in \mathbf{N}} \;\; \delta\text{-packing of } D \right\}. \tag{18}$$

The γ-packing measure is constructed in two steps:

$$\tilde{M}^\gamma_{\mathrm{P}}(D) = \lim_{\delta \to 0} M^{\gamma,\delta}_{\mathrm{P}}(D)\,, \tag{19}$$

$$M^\gamma_{\mathrm{P}}(D) = \inf\left\{ \sum_{n \in \mathbf{N}} \tilde{M}^\gamma_{\mathrm{P}}(\Delta_n) \;\middle|\; \Delta_n \text{ Borel}, \; \bigcup_{n \in \mathbf{N}} \Delta_n = D \right\}, \tag{20}$$

that is, M^γ_{P} is an metric outer measure in the sense of Caratheodory. The corresponding Borel measure is also denoted by M^γ_{P}. The packing dimension $\dim_{\mathrm{P}}(\Delta)$ of a Borel set Δ is defined as the infimum of all γ such that $M^\gamma_{\mathrm{P}}(\Delta) = 0$.

For the case of the upper γ-derivative and the γ-Hausdorff measure (the inequalities on the left hand-side of (21) and (22)), the following theorem summarizes the main technical results of [Ro, Chapter 3.2 and 3.3]. For the case of the lower γ-derivative and the γ-packing dimension, it is strictly speaking new, but the proof uses similar techniques as in [Cu]. Not only the results, but also the proofs show some kind of duality between Hausdorff dimensions and lower pointwise dimensions on one side and packing dimensions and upper pointwise dimensions on the other: for the Hausdorff measure case, the proof of (21) is based on a covering lemma, and (22) follows directly from the definitions; for the packing measure case, the situation is just the converse.

Theorem 2 *For* $\lambda > 0$ *we have*

$$M_{\mathrm{H}}^{\gamma}(\{E \in \Delta \mid \overline{D}_{\mu}^{\gamma}(E) > \lambda\}) \leq \frac{6^{\gamma}}{\lambda}, \qquad M_{\mathrm{P}}^{\gamma}(\{E \in \Delta \mid \underline{D}_{\mu}^{\gamma}(E) > \lambda\}) \leq \frac{2^{\gamma}}{\lambda}, \tag{21}$$

and

$$\mu(\{E \in \Delta \mid \overline{D}_{\mu}^{\gamma}(E) < \lambda\}) \leq 2^{-\gamma}\lambda\, M_{\mathrm{H}}^{\gamma}(\Delta), \quad \mu(\{E \in \Delta \mid \underline{D}_{\mu}^{\gamma}(E) < \lambda\}) \leq 2^{-\gamma}\lambda\, M_{\mathrm{P}}^{\gamma}(\Delta). \tag{22}$$

Proof. The first inequalities in (21) and (22) being proven in [Ro], we here only prove the packing dimension part. Let $R_{\mu}^{\gamma}(\delta,\lambda) = \{E \in \Delta \mid \inf_{\epsilon<\delta} \mu([E-\epsilon, E+\epsilon])\epsilon^{-\gamma} \geq \lambda\}$. Then for any $E \in R_{\mu}^{\gamma}(\delta,\lambda)$, $\mu([E-\epsilon, E+\epsilon]) \geq \lambda\epsilon^{\gamma} \quad \forall\ \epsilon < \delta$. Therefore, if $(B(E_k, r_k))_{k\in\mathbf{N}}$ is a δ-packing of $R_{\mu}^{\gamma}(\delta,\lambda)$, we have $\sum_k (2r_k)^{\gamma} \leq 2^{\gamma}/\lambda$ because the elements of the δ-packing are disjoint. Consequently, $M_{\mathrm{P}}^{\gamma}(R_{\mu}^{\gamma}(\delta,\lambda)) \leq M_{\mathrm{P}}^{\gamma,\delta}(R_{\mu}^{\gamma}(\delta,\lambda)) \leq 2^{\gamma}\mu(\Delta)/\lambda$. Moreover, $\bigcup_{n\in\mathbf{N}} R_{\mu}^{\gamma}(1/n,\lambda) = \{E \in \Delta \mid \underline{D}_{\mu}^{\gamma}(E) > \lambda\}$ so that $M_{\mathrm{P}}^{\gamma}(\{E \in \Delta \mid \underline{D}_{\mu}^{\gamma}(E) > \lambda\}) = \sup_n M_{\mathrm{P}}^{\gamma}(R_{\mu}^{\gamma}(1/n,\lambda)) \leq 2^{\gamma}\mu(\Delta)/\lambda$ due to the σ-additivity of M_{P}^{γ}.

To prove (22), we shall use (as in [Cu]) the fact that any Borel measure on $\mathbf{R}$ possesses the centered Vitali covering property [Be]. A centered Vitali covering of Δ is a set of closed balls containing for any $E \in \Delta$ and $\delta > 0$ some closed ball $B(E,r)$ with $r \leq \delta$. The centered covering property of μ means that every centered Vitali covering of Δ contains a countable set of disjoint balls B_k such that $\mu(\Delta \backslash \bigcup_k B_k) = 0$.

Let then $\Delta_n \subset \{E \in \Delta \mid \underline{D}_{\mu}^{\gamma}(E) < \lambda\}$. For any $E \in \Delta_n$ and $\delta > 0$, there exists $r < \delta$ such that $\mu([E-r, E+r]) \leq \lambda r^{\gamma}$. Hence the set of balls $B(E,r)$ such that $r \leq \delta$, $E \in \Delta_n$ and that $\mu([E-r, E+r]) \leq \lambda r^{\gamma}$ is a centered Vitali covering of Δ_n. Let $(B(E_k, r_k))_{k\in\mathbf{N}}$ be the associated δ-packing satisfying $\mu(\Delta_n \backslash \bigcup_k B(E_k, r_k)) = 0$ as given by the centered Vitali covering property. Then $\mu(\Delta_n) \leq \sum_k \mu(B(E_k, r_k)) \leq \lambda \sum_k r_k^{\gamma}$. As this holds for any $\delta > 0$, we have $\mu(\Delta_n) \leq 2^{-\gamma}\lambda \tilde{M}_{\mathrm{P}}^{\gamma}(\Delta_n)$. As the decomposition $\Delta = \bigcup_n \Delta_n$ in (20) can be chosen disjoint, the result follows. □

Before drawing the for us interesting consequences of Theorem 2, let us introduce some further notations. We define $\underline{d}_{\mu}^{+}(\Delta) = \mu\text{-}\operatorname{ess\,sup}_{E\in\Delta} \underline{d}_{\mu}(E)$ as well as $\underline{d}_{\mu}^{-}(\Delta) = \mu\text{-}\operatorname{ess\,inf}_{E\in\Delta} \underline{d}_{\mu}(E)$, and similarly $\overline{d}_{\mu}^{+}(\Delta)$ and $\overline{d}_{\mu}^{-}(\Delta)$. By the above remarks, these quantities only depend on the measure class of μ [SB].

The upper Hausdorff dimension $\dim_{\mathrm{H}}^{+}(\mu, \Delta)$ of μ in Δ is defined by the infimum of the Hausdorff dimensions of all Borel subsets $\Delta' \subseteq \Delta$ satisfying $\mu(\Delta') = \mu(\Delta)$ [Yo]. The lower Hausdorff dimension $\dim_{\mathrm{H}}^{-}(\mu, \Delta)$ of μ in Δ is defined as the supremum of the α's such that $\Delta' \subset \Delta$, $\dim_H(\Delta') \leq \alpha$ imply $\mu(\Delta') = 0$. The packing dimensions $\dim_{\mathrm{P}}^{\pm}(\mu, \Delta)$ are once more defined similarly. If $\Delta = \mathbf{R}$, we further drop the specification.

For the case of the Hausdorff dimension and lower pointwise dimensions, the first of the following corollaries already appears in [Ro]. A version of Corollary 2 can be found in [Cu]. Corollaries 4 and 5 appear in [G2, Co] and in [G2], respectively, for the case of the Hausdorff dimension; here we give a different proof.

Corollary 1 $\dim_{\rm H}(\Delta) < \underline{d}_\mu^-(\Delta)$ *implies* $\mu(\Delta) = 0$. $\dim_{\rm P}(\Delta) < \overline{d}_\mu^-(\Delta)$ *implies* $\mu(\Delta) = 0$.

Corollary 2 *The following identities hold:*

$$\dim_{\rm H}(\{E \in \Delta \mid \underline{d}_\mu(E) \le \underline{d}_\mu^+(\Delta)\}) = \underline{d}_\mu^+(\Delta) \; ,$$

$$\dim_{\rm P}(\{E \in \Delta \mid \overline{d}_\mu(E) \le \overline{d}_\mu^+(\Delta)\}) = \overline{d}_\mu^+(\Delta) \; .$$

Corollary 3
i) *Let* $\underline{\Delta}^+ = \{E \in \Delta \mid \underline{d}_\mu(E) = \underline{d}_\mu^+(\Delta)\}$. *Either* $\mu(\underline{\Delta}^+) = 0$ *or* $\dim_{\rm H}(\underline{\Delta}^+) = \underline{d}_\mu^+(\Delta)$.
ii) *Let* $\overline{\Delta}^+ = \{E \in \Delta \mid \overline{d}_\mu(E) = \overline{d}_\mu^+(\Delta)\}$. *Either* $\mu(\overline{\Delta}^+) = 0$ *or* $\dim_{\rm P}(\overline{\Delta}^+) = \overline{d}_\mu^+(\Delta)$.

Corollary 4 $\dim_{\rm H}^+(\mu, \Delta) = \underline{d}_\mu^+(\Delta)$ *and* $\dim_{\rm P}^+(\mu, \Delta) = \overline{d}_\mu^+(\Delta)$.

Corollary 5 $\dim_{\rm H}^-(\mu, \Delta) = \underline{d}_\mu^-(\Delta)$ *and* $\dim_{\rm P}^-(\mu, \Delta) = \overline{d}_\mu^-(\Delta)$.

Due to the complete symmetry of the results in Theorem 2, it is sufficient to prove the corollaries for the Hausdorff dimension case.

Proof of Corollary 1. Let γ be such that $\dim_{\rm H}(\Delta) < \gamma < \underline{d}_\mu^-(\Delta)$. Then $M_{\rm H}^\gamma(\Delta) = 0$ and by (17) and (22) one has

$$\mu(\Delta) = \mu(\{E \in \Delta \,|\, \gamma < \underline{d}_\mu(E)\}) \le \mu(\{E \in \Delta \,|\, \overline{D}_\mu^\gamma(E) = 0\}) = 0 \; .$$

□

Proof of Corollary 2. First note that, for any $d > 0$, the set $\Delta_d = \{E \in \Delta \mid \underline{d}_\mu(E) < d\}$ has a Hausdorff dimension less or equal than d: in fact, by equations (16) and (21) one has $M_{\rm H}^d(\Delta_d) \le M_{\rm H}^d(\{E \in \Delta \mid D_\mu^d = \infty\}) = 0$. Now for $n \in \mathbf{N}$ let $d_n = \underline{d}_\mu^+(\Delta) + 1/n$. Then $\Delta_0 = \bigcap_{n\in\mathbf{N}} \Delta_{d_n} = \{E \in \Delta | \underline{d}_\mu(E) \le \underline{d}_\mu^+(\Delta)\}$ satisfies according to the above $\dim_{\rm H}(\Delta_0) \le \underline{d}_\mu^+(\Delta)$. Now suppose there exists a γ such that $\dim_{\rm H}(\Delta_0) < \gamma < \underline{d}_\mu^+(\Delta)$. Then $M_{\rm H}^\gamma(\Delta_0) = 0$ and because of $\mu(\Delta_0) = \mu(\Delta)$, (17) and (22) we have

$$\mu(\{E \in \Delta \,|\, \gamma < \underline{d}_\mu(E)\}) \;=\; \mu(\{E \in \Delta_0 \,|\, \gamma < \underline{d}_\mu(E)\}) \;\le\; \mu(\{E \in \Delta \,|\, \overline{D}_\mu^\gamma(E) = 0\}) \;=\; 0 \; , \tag{23}$$

which is in contradiction to $\gamma < \underline{d}_\mu^+(\Delta)$. Therefore $\dim_{\rm H}(\Delta_0) = \underline{d}_\mu^+(\Delta)$. □

Proof of Corollary 3. If $E \in \underline{\Delta}^+$, then $\overline{D}_\mu^\gamma(E) = 0$ for all $\gamma < \underline{d}_\mu^+(\Delta)$. Hence $\underline{\Delta}^+ \subset \{E \in \underline{\Delta}^+ | \overline{D}_\mu^\gamma(E) < \lambda\}$ for any $\lambda \ge 0$. Now by (22), $\mu(\underline{\Delta}^+) \le 2^{-\gamma}\lambda M_{\rm H}^\gamma(\underline{\Delta}^+)$. Therefore either $\mu(\underline{\Delta}^+) = 0$ or $M_{\rm H}^\gamma(\underline{\Delta}^+) = \infty$ which by Corollary 2 implies $\dim_{\rm H}(\underline{\Delta}^+) = \underline{d}_\mu^+(\Delta)$. □

Proof of Corollary 4. We have $\dim_{\rm H}(\mu, \Delta) \le \dim_{\rm H}(\Delta_0) = \underline{d}_\mu^+(\Delta)$ with Δ_0 as above. Suppose a γ exists such that $\dim_{\rm H}(\mu, \Delta) < \gamma < \underline{d}_\mu^+(\Delta)$. Then there exists a set $\Delta' \subset \Delta$ with $M_{\rm H}^\gamma(\Delta') = 0$ and $\mu(\Delta') = \mu(\Delta)$. As in (23), this is impossible. □

Proof of Corollary 5. Corollary 1 directly implies $\underline{d}_\mu^-(\Delta) \le \dim_H^-(\mu, \Delta)$. Furthermore, by the definition of essential infimum there must exist a set $\Delta' \subseteq \Delta$ with $\mu(\Delta') > 0$

such that $\underline{d}_\mu^+(\Delta') \geq \underline{d}_\mu^-(\Delta)$. Then Corollary 2 implies $\dim_H(\Delta') \geq \underline{d}_\mu^-(\Delta)$. Therefore $\underline{d}_\mu^-(\Delta) \geq \dim_{\overline{H}}(\mu, \Delta)$ by the definition of the latter dimension. □

A measure μ is said to have *exact (Hausdorff) dimension* d in a Borel set Δ if $\dim_{\overline{H}}(\mu,\Delta) = \dim_H^+(\mu,\Delta)$, that is, if its lower pointwise dimension is μ-a.e. equal to d in Δ. In parallel, one can say that μ has *exact (packing) dimension* d' in Δ if its upper pointwise dimension is μ -a.e. equal to d' in Δ. In ref. [G2] an *exactly scaling* measure μ was defined as one for which the upper and lower local dimensions have a common μ-a.e. constant value. For exactly scaling measures the Hausdorff and the packing dimension coincide, but the converse is not true.

In the physical literature, box-counting dimensions have found wide use because of their easy numerical implementation. It is interesting to note that packing dimensions can also be computed by a box-counting procedure. First we recall that the upper and lower box-counting dimension of a compact set K are defined as

$$\dim_{\rm B}^+(K) = \limsup_{\delta\to 0} \frac{\log(N_\delta(K))}{-\log(\delta)}, \qquad \dim_{\rm B}^-(K) = \liminf_{\delta\to 0} \frac{\log(N_\delta(K))}{-\log(\delta)},$$

where $N_\delta(K)$ denotes the minimal number of closed intervals of size δ needed to cover K (or equivalently, by the box-counting theorem [Fa], the number of elements of a grid cover of size δ which overlap K). The upper and lower fractal dimensions of μ in $\Delta \subset \mathbf{R}$ are now defined as in [G2] by

$$\dim_{\rm F}^\pm(\mu,\Delta) = \sup_{\eta>0} \inf_{K\subset\Delta} \{ \dim_{\rm B}^\pm(K) \mid \mu(K) > \mu(\Delta) - \eta \,,\ K \text{ compact} \} .$$

Theorem 3 $\dim_{\rm H}^+(\mu,\Delta) \leq \dim_{\rm F}^-(\mu,\Delta) \leq \dim_{\rm F}^+(\mu,\Delta) = \dim_{\rm P}^+(\mu,\Delta)$.

Proof. To prove the first inequality, we replace in the definition of $\dim_{\rm F}^-(\mu,\Delta)$ the value of $\dim_{\rm F}^-(K)$ by $\dim_{\rm H}(K)$ and call the result $\tilde{d}$. Then there exist a sequence $(K_n)_{n\geq 1}$ of compact subsets of Δ satisfying $\mu(K_n) > \mu(\Delta) - 1/n$ and $\lim_{n\to\infty} \dim_{\rm H}(K_n) = \tilde{d}$. Now $\bigcup_n K_n$ supports μ in Δ and has Hausdorff dimension $\tilde{d}$ such that $\tilde{d} \geq \dim_{\rm H}^+(\mu,\Delta)$. On the other hand, $\tilde{d} \leq \dim_{\rm F}^-(\mu,\Delta)$ because $\dim_{\rm H}(K) \leq \dim_{\rm F}^-(K)$ [Fa].

Next let us recall from [G2] that $\dim_{\rm F}^+(\mu,\Delta) \leq \dim_{\rm P}^+(\mu,\Delta)$, so we only need to prove the converse inequality. Let $\eta_n \in \mathbf{R}$ and $K_n \subset \Delta$, $n \in \mathbf{N}$, be such that $\mu(K_n) > \mu(\Delta) - \eta_n$, $\lim_{n\to\infty} \eta_n = 0$ and $\sup_{n\geq 1} \dim_{\rm B}^+(K_n) = \dim_{\rm F}(\mu,\Delta)$. We set $K_\infty = \bigcup_{n\geq 1} K_n$ so that $\mu(K_\infty) = \mu(\Delta)$. Hence, by definition of $\dim_{\rm P}^+(\mu,\Delta)$ and countable stability of packing dimensions,

$$\dim_{\rm P}^+(\mu,\Delta) \leq \dim_{\rm P}(K_\infty) = \sup_{n\geq 1} \dim_{\rm P}(K_n) \leq \sup_{n\geq 1} \dim_{\rm B}^+(K_n) = \dim_{\rm F}(\mu,\Delta) ,$$

where we used the fact that $\dim_{\rm P}(K) \leq \dim_{\rm B}^+(K)$ for any set $K \subset \mathbf{R}$ [Fa]. □

We finally examine the dimension $\mathcal{D}(\mu)$ introduced in (14) as well as box-counting information dimensions defined next. Given $N \in \mathbf{N}$, the Shannon entropy ("missing information") of the measure μ relative to the partition of the real line in dyadic intervals I_j^N, $j \in \mathbf{Z}$), of the Nth generation is given by:

$$S_N(\mu) \;=\; -\sum_{j\in\mathbf{Z}} \mu(I_j^N)\log_2(\mu(I_j^N)) \;, \tag{24}$$

where by convention those j's for which $\mu(I_j^N) = 0$ do not contribute in the sum. We consider the class $\mathcal{E}$ of measures μ for which $S_1(\mu)$ is finite. Then $S_N(\mu)$ is also finite for all N (see below) and the upper and lower information dimensions of μ are defined by

$$\dim_I^+(\mu) \;=\; \limsup_{N\to\infty}\frac{S_N(\mu)}{N}\;, \qquad \dim_I^-(\mu) \;=\; \liminf_{N\to\infty}\frac{S_N(\mu)}{N}\;.$$

Then we have the following:

Theorem 4 *If $\mu \in \mathcal{E}$, then $\dim_H^-(\mu) \le \dim_I^-(\mu)$ and $\dim_I^+(\mu) \le \mathcal{D}(\mu) \le \dim_P^+(\mu)$.*

Proof. Let $A_{N,\alpha}$ be defined as in Proposition 2. It is immediate that:

$$\dim_I^-(\mu) \;\ge\; \alpha\,\liminf \mu(A_{N,\alpha}) \;\ge\; \alpha\,\mu(\liminf A_{N,\alpha})\;,$$

which is not less than α if $\alpha < \dim_H^-(\mu)$ because of (13) and of Corollary 5. This proves the first inequality. To prove the second one, let $\mathcal{N}(\mu, N, \epsilon)$ be as in equation (14), and let $B_N \subset \mathbf{R}$ be the union of exactly $\mathcal{N}(\mu, N, \epsilon)$ dyadic intervals of the Nth generation, with $\mu(B_N) > 1 - \epsilon$. Let us fix $N_0 \in \mathbf{N}$ so that

$$-\sum_{|j|\ge N_0} \mu(I_j^1)\log_2(\mu(I_j^1)) \;<\; \epsilon\;.$$

We denote $K = (-N_0, N_0]$ and we note that, if ϵ is small enough, then $\mu(K^c) < \epsilon$. Now we split the sum in (24) in three terms $S_i(\mu, N, \epsilon)$, $i = 1,2,3$, which result of summing over different sets J_i of indices j, namely $J_1 = \{j \in \mathbf{Z}|I_j^N \subset K^c\}$, $J_2 = \{j \in \mathbf{Z}|I_j^N \cap B_N \cap K \ne \emptyset\}$ and $J_3 = \{j \in \mathbf{Z}|I_j^N \subset K \setminus B_N\}$. We denote P_i, $i = 1,2,3$, the total measure of the intervals whose label belongs in J_i. From well-known properties of the conditional entropy it follows that

$$S_1(N+1) \;\le\; S_1(N) + P_1\log_2(2) \;\le\; S_1(N) + \epsilon \;\le\; (N+1)\epsilon\;.$$

Again from conditioning we also get that

$$S_2(N) \;\le\; P_2\log_2(\#J_2) - P_2\log_2(P_2) \;\le\; \log_2(\mathcal{N}(\epsilon,\mu,N)) + c\;,$$

where c is the maximum of $-x\log_2 x$ in $(0,1)$. In the same way,

$$S_3(N) \;\le\; P_3\log_2(\#J_3) - P_3\log_2(P_3) \;\le\; \epsilon\log_2(N_0 2^{N+1}) + \epsilon\log_2\left(\frac{1}{\epsilon}\right)\;.$$

Putting all these estimates together, we get

$$\frac{S_N(\mu)}{N} \leq \frac{\log_2(\mathcal{N}(\epsilon,\mu,N))}{N} + 2\epsilon + \mathcal{O}\left(\frac{1}{N}\right) ,$$

whence:

$$\dim_I^+(\mu) \leq 2\epsilon + \mathcal{D}(\mu) \leq 2\epsilon + \dim_F^+(\mu) ,$$

because a Nth generation dyadic covering of a compact K with $\mu(K) > 1-\epsilon$ cannot be obtained with less than $\mathcal{N}(\epsilon,\mu,N)$ intervals, due to the very definition of the latter quantity. As ϵ is arbitrary and $\dim_F^+(\mu) = \dim_P^+(\mu)$, we get the second inequality in the thesis. □

References

[Be] A. S. Besicovitch, *A general form of the covering principle and relative differentiation of additive functions*, Proc. Camb. Phil. Soc. **41**, 103-110 (1945).

[Co] J.-M. Combes, in *Differential Equations with applications to Mathematical Physics*, edited by W.F.Ames, E.M.Harrel and I.V.Herod (Academic, Boston 1993); J.-M. Barbaroux, J.-M. Combes, R. Montcho, *Remarks on the relation between quantum dynamics and fractal spectra*, J. Math. Anal. and Appl. 213, 698-722 (1997).

[Cu] C. D. Cutler, *Measure disintegrations with respect to σ-stable monotone indices and the pointwise representation of packing dimensions*, Supp. Rend. Circ. Mat. Palermo **28**, serie II, 319-339 (1992).

[Fa] K. Falconer, *Fractal Geometry - Mathematical Foundations and Applications*, John Wiley & Sons, 1990.

[G1] I. Guarneri, *Spectral properties of quantum diffusion on discrete lattices*, Europhys. Lett., **10**, 95-100, (1989); *On an estimate concerning quantum diffusion in the presence of a fractal spectrum* , Europhys. Lett., **21**, 729-733, (1993); and I. Guarneri, G. Mantica, *On the asymptotic properties of quantum dynamics in the presence of fractal spectrum*, Ann. Inst. H. Poincaré, **61**, 369-379, (1994).

[G2] I. Guarneri, *Singular continuous spectra and discrete wave packet dynamics*, J. Math. Phys. **37**, 5195-5206, (1996).

[G3] I. Guarneri, *On the Dynamical Meaning of Spectral Dimensions*, to appear in Ann. Inst. H. Poincaré, 1998.

[GM] I. Guarneri, G. Mantica, *Multifractal Energy Spectra and Their Dynamical Implications*, Phys. Rev. Lett. **73**, 3379-3383 (1994).

[GS] I. Guarneri, H. Schulz-Baldes, *Upper bounds for quantum dynamics governed by Jacobi matrices with self-similar spectra*, preprint, Como 1998.

[KKK] R. Ketzmerick, K. Kruse, S. Kraut, T. Geisel, *What determines the spreading of a wave packet?*, Phys. Rev. Lett. **79**, 1959-1962 (1997).

[La] Y. Last, *Quantum dynamics and decomposition of singular continuous spectra*, J. Funct. Analysis **142**, 402-445 (1996).

[RJLS] R. del Rio, S. Jitomirskaya, Y. Last, B. Simon, *Operators with singular continuous spectrum: IV. Hausdorff dimension, rank-one perturbations and localization*, J. d'Analyse Math. **69**, 153-200 (1996).

[Ro] C. A. Rogers, *Hausdorff measures*, (Cambridge University Press, Cambridge, 1970).

[SB] H. Schulz-Baldes, J. Bellissard, *Anomalous transport: a mathematical framework*, Rev. Math. Phys. **10**, 1-46 (1998).

[TT] S. J. Taylor, C. Tricot, *Packing measure and its evaluation for a Brownian path*, Trans. Amer. Math. Soc. **288**, 679-699 (1985); S. J. Taylor, C. Tricot, *The packing measure of rectifiable subsets of the plane*, Math. Proc. Camb. Phil. Soc. **99**, 285-296 (1986).

[Yo] L. S. Young, *Dimension, Entropy, and Lyapunov exponents*, Erg. Th. Dyn. Sys. **2**, 109-124 (1982).

M P E J

MATHEMATICAL PHYSICS ELECTRONIC JOURNAL

ISSN 1086-6655
Volume 5, 1999

Paper 2
Received: Oct 19, 1998, Accepted: Apr 8, 1999
Editor: P. Collet

Eigenvalue Asymptotics for the Dirac Operator in Strong Constant Magnetic Fields

G.D.RAIKOV

Section of Mathematical Physics
Institute of Mathematics and Informatics
Bulgarian Academy of Sciences
P.O.B. 373, 1090 Sofia, Bulgaria
E-mail: gdraikov@omega.bg

Abstract. We consider the three-dimensional Dirac operator H with constant magnetic field and electric potential which decays at infinity. We study the asymptotic behaviour of the discrete spectrum of H as the norm of the magnetic field grows unboundedly.

1 Introduction

Let $H_0(b)$ be the three-dimensional Dirac operator in constant magnetic field $B = (0,0,b)$, $b > 0$. Choosing an appropriate gauge, system of units, and coordinates, we can write

$$H_0(b) = \sum_{j=1,2,3} \alpha_j \Pi_j(b) + \beta$$

where

$$\alpha_j := \begin{pmatrix} 0 & \sigma_j \\ \sigma_j & 0 \end{pmatrix}, \; j = 1,2,3, \; \beta := \begin{pmatrix} I_2 & 0 \\ 0 & -I_2 \end{pmatrix},$$

σ_j, $j = 1,2,3$, are the Pauli matrices

$$\sigma_1 := \begin{pmatrix} 0 & 1 \\ 1 & 0 \end{pmatrix}, \sigma_2 := \begin{pmatrix} 0 & -i \\ i & 0 \end{pmatrix}, \sigma_3 := \begin{pmatrix} 1 & 0 \\ 0 & -1 \end{pmatrix},$$

I_2 is the unit 2×2 matrix, Π_j, $j = 1,2,3$, are the components of the extended momentum

$$\Pi_1 = \Pi_1(b) := -i\frac{\partial}{\partial x} + \frac{by}{2}, \Pi_2 = \Pi_2(b) := -i\frac{\partial}{\partial y} - \frac{bx}{2}, \Pi_3 := -i\frac{\partial}{\partial z},$$

and $X = (x,y,z) \in \mathbf{R}^3$. It is well-known that for each $b \geq 0$ we have

$$\sigma(H_0(b)) = \sigma_{\text{ess}}(H_0(b)) = (-\infty, -1] \cup [1, +\infty). \tag{1.1}$$

Further, let $V : \mathbf{R}^3 \to \mathbf{R}$ be the electric (scalar) potential. We shall say that V is in the class $\mathcal{L}$ if and only if for each $\varepsilon > 0$ it can be written as $V = V_1 + V_2$ with $V_1 \in L^3(\mathbf{R}^3)$, and $\sup_{X \in \mathbf{R}^3} |V_2(X)| \leq \varepsilon$. Throughout the paper we assume $V \in \mathcal{L}$, unless more restrictive assumptions are imposed.
In particular, $V \in \mathcal{L}$ entails the compactness of the operator $VH_0(b)^{-1}$. Set

$$H(b) := H_0(b) + VI_4 = H_0(b) + V$$

where I_4 is the unit 4×4 matrix. Since the operator $VH_0(b)^{-1}$ is compact, we have $\sigma_{\text{ess}}(H(b)) = \sigma_{\text{ess}}(H_0(b))$, and hence (1.1) implies

$$\sigma_{\text{ess}}(H(b)) = (-\infty, -1] \cup [1, +\infty).$$

However, the discrete spectrum of the operator $H(b)$ might be non-empty. The aim of the present paper is to investigate the asymptotic distribution as $b \to \infty$ of the eigenvalues of $H(b)$ lying in the gap $(-1,1)$ of its essential spectrum.

2 Statement of the main result

Let $T = T^*$ be a selfadjoint operator in a Hilbert space. Denote by $P_{\mathcal{I}}(T)$ its spectral projection corresponding to the interval $\mathcal{I} \subset \mathbf{R}$. Set

$$\mathcal{N}(\lambda_1, \lambda_2; T) = \text{rank}\, P_{(\lambda_1,\lambda_2)}(T),\ \lambda_1, \lambda_2 \in \mathbf{R},\ \lambda_1 < \lambda_2,$$

$$N(\lambda; T) := \text{rank}\, P_{(-\infty,\lambda)}(T),\ \lambda \in \mathbf{R},$$

$$n_{\pm}(s; T) := \text{rank}\, P_{(s,+\infty)}(\pm T), s > 0.$$

If T is a linear compact operator which is not necessarily selfadjoint, put

$$n_*(s; T) := \text{rank}\, P_{(s^2,+\infty)}(T^*T), s > 0.$$

In what follows if $X = (x, y, z) \in \mathbf{R}^3$ we shall write occasionally $X = (X_\perp, z)$ where $X_\perp = (x, y)$ are the variables on the plane perpendicular to the magnetic field $B = (0, 0, b)$, while z is the variable along B. Fix $X_\perp \in \mathbf{R}^2$ and set

$$\chi(X_\perp) := \chi_0 + V(X_\perp, .)\, I_2$$

where

$$\chi_0 := \begin{pmatrix} 1 & -i\frac{d}{dz} \\ -i\frac{d}{dz} & -1 \end{pmatrix}.$$

Proposition 2.1 *Let $V \in \mathcal{L}$. Then for almost every $X_\perp \in \mathbf{R}^2$ the operator $\chi(X_\perp)$ is defined as an operator sum selfadjoint in $L^2(\mathbf{R}; \mathbf{C}^2)$. Moreover, for almost every $X_\perp \in \mathbf{R}^2$ the operator $V(X_\perp, .)\chi_0^{-1}$ is compact and, therefore,*

$$\sigma_{\text{ess}}(\chi(X_\perp)) = \sigma_{\text{ess}}(\chi_0) = (-\infty, -1] \cup [1, +\infty).$$

The proof of the proposition is contained in Section 7.
Let λ_1 and λ_2 be real numbers such that $-1 < \lambda_1 < \lambda_2 < 1$. Introduce the magnetic integrated density of states

$$\mathcal{D}(\lambda_1, \lambda_2) = \mathcal{D}_V(\lambda_1, \lambda_2) := \int_{\mathbf{R}^2} \mathcal{N}(\lambda_1, \lambda_2; \chi(X_\perp))\, dX_\perp.$$

Proposition 2.2 *Let $V \in \mathcal{L}$, $\lambda_1, \lambda_2 \in \mathbf{R}$, $-1 < \lambda_1 < \lambda_2 < 1$. Then $\mathcal{D}_V(\lambda_1, \lambda_2) < \infty$.*

The proof of this proposition can also be found in Section 7.
We shall say that a point $\lambda \in (-1, 1)$ is regular if and only if

$$\text{vol}\, \left\{X_\perp \in \mathbf{R}^2 |\, \dim \text{Ker}\, (\chi(X_\perp) - \lambda) \geq 1\right\} = 0.$$

Note that λ_1 (respectively, λ_2) is a regular point if and only if $\lim_{\varepsilon \to 0} \mathcal{D}(\lambda_1 + \varepsilon, \lambda_2) = \mathcal{D}(\lambda_1, \lambda_2)$ (respectively, $\lim_{\varepsilon \to 0} \mathcal{D}(\lambda_1, \lambda_2 + \varepsilon) = \mathcal{D}(\lambda_1, \lambda_2)$).

Theorem 2.1 *Let $V \in \mathcal{L}$, $\lambda_1, \lambda_2 \in \mathbf{R}$, $-1 < \lambda_1 < \lambda_2 < 1$. Assume that the points λ_1 and λ_2 are regular. Then we have*

$$\lim_{b \to \infty} b^{-1} \mathcal{N}(\lambda_1, \lambda_2; H(b)) = \frac{1}{2\pi} \mathcal{D}(\lambda_1, \lambda_2). \tag{2.1}$$

The present paper could be regarded as a supplement to [7] where strong-magnetic-field spectral asymptotics for the Schrödinger and Pauli operators have been considered. The methods applied here are close to the ones used in [7]. However, the Dirac operator $H(b)$ studied in this paper as well as the auxiliary operator $\chi(X_\perp)$ are not semibounded in contrast to the Schrödinger and Pauli operators. This additional difficulty is overcome by the application of a simple but yet non-trivial generalization of the well-known Birman-Schwinger principle.

Various types of spectral properties and, in particular, eigenvalue asymptotics for the Dirac operator with or without magnetic field have been studied in [8], [3], [4], [5], [6]. However, the asymptotic behaviour as $b \to \infty$ of $\mathcal{N}(\lambda_1, \lambda_2; H(b))$ has never been investigated before.

The paper is organized as follows. The next four brief sections contain auxiliary results. A formulation of the Kac-Murdock-Szegö theorem borrowed from [7], can be founded in Section 3. Section 4 is devoted to the generalization of the Birman-Schwinger principle concerning the number of the eigenvalues situated in a gap of the essential spectrum of a selfadjoint operator. In Section 5 we describe certain spectral properties of the unperturbed operator $H_0(b)$. In Section 6 we perform some preliminary estimates. Finally, Propositions 2.1-2.2 are proved in Section 7, and Theorem 2.1 – in Section 8.

3 The Kac-Murdock-Szegö theorem

In this section we follow closely the exposition of [7, Subsection 3.1]. For the reader's convenience, we reproduce a suitable version of the Kac-Murdock-Szegö theorem whose proof can be found in [7, Subsection 3.1].

In the sequel we shall denote by S_∞ the space of linear compact operators acting in a given Hilbert space, and by S_p, $p \in [1, \infty)$, – the Schatten–von Neumann spaces of operators $T \in S_\infty$ for which the norm $\|T\|_p := \left(\mathrm{Tr}\, |T|^p\right)^{1/p}$ is finite.

Moreover, we shall say that the function ν defined on $\mathbf{R} \setminus \{0\}$ is in the class $\mathcal{C}$ if it is non-decreasing on $(-\infty, 0)$ and $(0, \infty)$, non-negative on $(-\infty, 0)$, and non-positive on $(0, \infty)$.

Lemma 3.1 *Let $\{T(b)\}_{b>0}$ be a family of selfadjoint compact operators satisfying the estimate $\|T(b)\| \leq t_0$ with $t_0 > 0$ independent of b. Let $\nu \in \mathcal{C}$. Assume that $\nu(t) = 0$ for $|t| > t_0$. Suppose that there exists a real $p \geq 1$ such that the following three conditions are fulfilled:*

(i) *$T(b) \in S_p$ for each $b > 0$;*

(ii) *the quantity* $\int_{\mathbf{R}\setminus\{0\}} |t|^p \, d\nu(t)$ *is finite;*
(iii) *the limiting relations*

$$\lim_{b\to\infty} b^{-1} \operatorname{Tr} T(b)^l = \int_{\mathbf{R}\setminus\{0\}} t^l \, d\nu(t)$$

hold for each integer $l \geq p$.
Let $t \neq 0$ *be a continuity point of* ν. *Then we have*

$$\lim_{b\to\infty} b^{-1} \, n_-(-t; T(b)) = \nu(t) \quad \text{if} \quad t < 0,$$

$$\lim_{b\to\infty} b^{-1} \, n_+(t; T(b)) = -\nu(t) \quad \text{if} \quad t > 0.$$

Remark. We shall use Lemma 3.1 only with $t < 0$.

4 The generalized Birman-Schwinger principle

One of the versions of the classical Birman-Schwinger principle (cf. [2, Lemma 1.1]) says that if $\mathcal{H}_0 = \mathcal{H}_0^* \geq 0$, $\mathcal{V} = \mathcal{V}^*$, and $|\mathcal{V}|^{1/2}(\mathcal{H}_0 + 1)^{-1/2} \in S_\infty$, then for each $\lambda > 0$ we have

$$N(-\lambda; \mathcal{H}_0 + \mathcal{V}) = n_-(1; (\mathcal{H}_0 + \lambda)^{-1/2}\mathcal{V}(\mathcal{H}_0 + \lambda)^{-1/2}) \tag{4.1}$$

where the sum $\mathcal{H}_0 + \mathcal{V}$ should be understood in the quadratic-forms sense.
Lemma 4.1 below contains a generalization of (4.1) to the case where $\mathcal{H}_0$ is not necessarily semibounded.
Related arguments in the special case where $\mathcal{H}_0$ coincides with the free Dirac operator have already appeared in [2, Section 5]. Much later arguments of this type have been employed in [4] and [5].
Let $\mathcal{H}_0$ be a linear operator selfadjoint in the Hilbert space H. Assume $\lambda_1, \lambda_2 \in \mathbf{R}$, $\lambda_1 < \lambda_2$, $[\lambda_1, \lambda_2] \subset \rho(\mathcal{H}_0)$.
Set $\mathcal{R}(\lambda_1, \lambda_2; \mathcal{H}_0) := ((\mathcal{H}_0 - \lambda_1)(\mathcal{H}_0 - \lambda_2))^{-1/2}$. Since $[\lambda_1, \lambda_2]$ is in the resolvent set of $\mathcal{H}_0$ the operator $(\mathcal{H}_0 - \lambda_1)(\mathcal{H}_0 - \lambda_2)$ is positive-definite, and hence the operator $\mathcal{R}(\lambda_1, \lambda_2; \mathcal{H}_0)$ is well-defined and bounded. Set

$$\mathcal{G}(\lambda_1, \lambda_2; \mathcal{H}_0) := \left(\mathcal{H}_0 - \frac{1}{2}(\lambda_1 + \lambda_2)\right) \mathcal{R}(\lambda_1, \lambda_2; \mathcal{H}_0).$$

Evidently, $\mathcal{G}(\lambda_1, \lambda_2; \mathcal{H}_0)$ is bounded. Further, let $\mathcal{V}$ be a symmetric operator on $D(\mathcal{H}_0)$ such that $\mathcal{V}(\mathcal{H}_0 + i)^{-1} \in S_\infty$, which is equivalent to $\mathcal{V}\mathcal{R}(\lambda_1, \lambda_2; \mathcal{H}_0) \in S_\infty$. Set

$$\mathcal{K}(\lambda_1, \lambda_2; \mathcal{H}_0, \mathcal{V}) :=$$

$$\mathcal{R}(\lambda_1, \lambda_2; \mathcal{H}_0)\mathcal{V}^2\mathcal{R}(\lambda_1, \lambda_2; \mathcal{H}_0) + 2\mathrm{Re}\, \mathcal{G}(\lambda_1, \lambda_2; \mathcal{H}_0)\mathcal{V}\mathcal{R}(\lambda_1, \lambda_2; \mathcal{H}_0).$$

Obviously, $\mathcal{K}(\lambda_1, \lambda_2; \mathcal{H}_0, \mathcal{V}) \in S_\infty$.

Lemma 4.1 *Let $\mathcal{H}_0$ be a linear operator selfadjoint in the Hilbert space* H*,* $\lambda_1, \lambda_2 \in \mathbf{R}$*,* $\lambda_1 < \lambda_2$*, and* $[\lambda_1, \lambda_2] \subset \rho(\mathcal{H}_0)$*. Let $\mathcal{V}$ be a symmetric operator on $D(\mathcal{H}_0)$ such that* $\mathcal{V}(\mathcal{H}_0 + i)^{-1} \in S_\infty$*. Then we have*

$$\mathcal{N}(\lambda_1, \lambda_2; \mathcal{H}_0 + \mathcal{V}) = n_-(1; \mathcal{K}(\lambda_1, \lambda_2; \mathcal{H}_0, \mathcal{V})) \tag{4.2}$$

where the sum $\mathcal{H}_0 + \mathcal{V}$ should be understood in the operator sense.

Proof. Obviously,

$$\mathcal{N}(\lambda_1, \lambda_2; \mathcal{H}_0 + \mathcal{V}) = N\left(\frac{1}{4}(\lambda_1 - \lambda_2)^2; (\mathcal{H}_0 + \mathcal{V} - \frac{1}{2}(\lambda_1 + \lambda_2))^2\right). \tag{4.3}$$

The minimax principle implies that the quantity at the right-hand side of (4.3) is equal to the maximal dimension of the linear subsets of $D(\mathcal{H}_0)$ whose non-zero elements u satisfy the inequality

$$\left\|\mathcal{H}_0 u + \mathcal{V} u - \frac{1}{2}(\lambda_1 + \lambda_2) u\right\|^2 < \frac{1}{4}(\lambda_1 - \lambda_2)^2 \|u\|^2$$

where $\|.\|$ denotes the norm in H. This last inequality can be re-written as

$$\|\sqrt{(\mathcal{H}_0 - \lambda_1)(\mathcal{H}_0 - \lambda_2)} u\|^2 < -\|\mathcal{V} u\|^2 - 2\mathrm{Re}\left\langle \left(\mathcal{H}_0 - \frac{1}{2}(\lambda_1 + \lambda_2)\right) u, \mathcal{V} u \right\rangle \tag{4.4}$$

where $\langle ., . \rangle$ denotes the inner product in H. Note that the operator $\mathcal{R}(\lambda_1, \lambda_2; \mathcal{H}_0)$ maps bijectively H on $D(\mathcal{H}_0)$. Set $u = \mathcal{R}(\lambda_1, \lambda_2; \mathcal{H}_0) w$, $w \in$ H, in (4.4). Hence, (4.4) is equivalent to

$$\begin{aligned} \|w\|^2 < -\|\mathcal{V}\mathcal{R}(\lambda_1, \lambda_2; \mathcal{H}_0) w\|^2 - \\ 2\mathrm{Re}\left\langle \left(\mathcal{H}_0 - \frac{1}{2}(\lambda_1 + \lambda_2)\right) \mathcal{R}(\lambda_1, \lambda_2; \mathcal{H}_0) w, \mathcal{V}\mathcal{R}(\lambda_1, \lambda_2; \mathcal{H}_0) w \right\rangle = \\ - \langle \mathcal{K}(\lambda_1, \lambda_2; \mathcal{H}_0, \mathcal{V}) w, w \rangle . \end{aligned} \tag{4.5}$$

By (4.3), the quantity $\mathcal{N}(\lambda_1, \lambda_2; \mathcal{H}_0 + \mathcal{V})$ coincides with the maximal dimension of the subspaces of H whose non-zero elements w satisfy (4.5). By the minimax principle this maximal dimension equals $n_-(1; \mathcal{K}(\lambda_1, \lambda_2; \mathcal{H}_0, \mathcal{V}))$. Hence, (4.2) is valid. □

In this paper we shall apply Lemma 4.1 in the case $\mathcal{H}_0 = H_0(b)$, $\mathcal{V} = V$, and $[\lambda_1, \lambda_2] \subset (-1, 1)$. Set

$$R(\lambda_1, \lambda_2) \equiv R_b(\lambda_1, \lambda_2) := \mathcal{R}(\lambda_1, \lambda_2; H_0(b)), \tag{4.6}$$

$$G(\lambda_1, \lambda_2) \equiv G_b(\lambda_1, \lambda_2) := \mathcal{G}(\lambda_1, \lambda_2; H_0(b)), \tag{4.7}$$

$$K(\lambda_1, \lambda_2) \equiv K_b(\lambda_1, \lambda_2) := \mathcal{K}(\lambda_1, \lambda_2; H_0(b), V). \tag{4.8}$$

By analogy with (4.6) and (4.7) introduce the operators $\varrho(\lambda_1, \lambda_2)$ and $\gamma(\lambda_1, \lambda_2)$ replacing $H_0(b)$ by χ_0. Similarly, fix $X_\perp \in \mathbf{R}^2$ such that the operator $\chi(X_\perp)$ is well-defined,

and $V(X_\perp,.)\chi_0^{-1} \in S_\infty$, and define the operator $\kappa(X_\perp) \equiv \kappa(\lambda_1,\lambda_2;X_\perp)$ substituting in (4.8) the operator $H_0(b)$ for χ_0, and V for $V(X_\perp,.)$.
Applying (4.2), we obtain

$$\mathcal{N}(\lambda_1,\lambda_2;H(b)) = n_-(1;K_b(\lambda_1,\lambda_2)), \tag{4.9}$$

$$\mathcal{D}(\lambda_1,\lambda_2) = \int_{\mathbf{R}^2} n_-(1;\kappa(\lambda_1,\lambda_2;X_\perp))\,dX_\perp. \tag{4.10}$$

For further references we formulate here a lemma which is closely related to the generalized Birman–Schwinger principle.

Lemma 4.2 *Let the operators $\mathcal{H}_0$ and $\mathcal{V}$ and the numbers $\lambda_1,\ \lambda_2 \in \mathbf{R}$ satisfy the hypotheses of Lemma 4.1. Then the spectrum of $\mathcal{H}_0+\mathcal{V}$ contains at least one of the points λ_1 and λ_2 if and only if the operator $\mathcal{K}(\lambda_1,\lambda_2;\mathcal{H}_0,\mathcal{V})$ has an eigenvalue equal to -1. Moreover,*

$$\sum_{j=1,2} \dim \operatorname{Ker}(\mathcal{H}_0+\mathcal{V}-\lambda_j) = \dim \operatorname{Ker}(\mathcal{K}(\lambda_1,\lambda_2;\mathcal{H}_0,\mathcal{V})+1).$$

Proof. It suffices to note that the equations

$$(\mathcal{H}_0+\mathcal{V}-\lambda_1)(\mathcal{H}_0+\mathcal{V}-\lambda_2)u = 0,\ u \in D(\mathcal{H}_0),$$

and

$$\mathcal{K}(\lambda_1,\lambda_2;\mathcal{H}_0,\mathcal{V})w + w = 0,\ w \in \mathrm{H},$$

are equivalent for $u = \mathcal{R}(\lambda_1,\lambda_2;\mathcal{H}_0)w,\ w \in \mathrm{H}$. □

Corollary 4.1 *Let $\lambda_1,\lambda_2 \in \mathbf{R}$, $-1<\lambda_1<\lambda_2<1$. Then λ_1 and λ_2 are simultaneously regular points if and only if*

$$\operatorname{vol}\left\{X_\perp \in \mathbf{R}^2 | \dim \operatorname{Ker}(\kappa(\lambda_1,\lambda_2;X_\perp)+1) \geq 1\right\} = 0.$$

5 The ground-levels projection

The unperturbed Hamiltonian can be written as

$$H_0(b) = \begin{pmatrix} I_2 & F(b) \\ F(b) & -I_2 \end{pmatrix}$$

where

$$F(b) := \sum_{j=1}^{3} \sigma_j \Pi_j(b) = \begin{pmatrix} \Pi_3 & a(b) \\ a(b)^* & -\Pi_3 \end{pmatrix},$$

$$a(b) := \Pi_1(b) - i\Pi_2(b),\ a(b)^* := \Pi_1(b) + i\Pi_2(b).$$

The commutation relation $[\Pi_1(b), \Pi_2(b)] = ib$ implies

$$a(b)a(b)^* = \Pi_1(b)^2 + \Pi_2(b)^2 - b, \;\; a(b)^*a(b) = \Pi_1(b)^2 + \Pi_2(b)^2 + b.$$

Therefore $F(b)^2$ coincides with the Pauli operator

$$F(b)^2 = \left(\sum_{j=1}^{3} \sigma_j \Pi_j(b)\right)^2 = \begin{pmatrix} \Pi(b)^2 - b & 0 \\ 0 & \Pi(b)^2 + b \end{pmatrix} \tag{5.1}$$

where

$$\Pi(b)^2 := \sum_{j=1,2,3} \Pi_j^2.$$

Moreover,

$$H_0(b)^2 = \begin{pmatrix} F(b)^2 + I_2 & 0 \\ 0 & F(b)^2 + I_2 \end{pmatrix}. \tag{5.2}$$

Define the orthogonal projection p_b by

$$(p_b u)(x,y,z) = \int_{\mathbf{R}^2} \mathcal{P}_b(x,y;x',y')u(x',y',z)\,dx'dy', \;\; u \in L^2(\mathbf{R}^3), \tag{5.3}$$

where

$$\mathcal{P}_b(x,y;x',y') := \frac{b}{2\pi}\exp\left\{-\frac{b}{4}\left[(x-x')^2 + (y-y')^2 + 2i(xy' - yx')\right]\right\}. \tag{5.4}$$

It is essential that $\mathcal{P}_b$ is the integral kernel of the orthogonal projection on $\operatorname{Ker} a(b)^* = \operatorname{Ker} a(b)a(b)^* \subset L^2(\mathbf{R}^2)$. Evidently, p_b commutes with Π_3.
On $L^2(\mathbf{R}^3, \mathbf{C}^4)$ introduce the orthogonal projection

$$P_b := \begin{pmatrix} p_b & 0 & 0 & 0 \\ 0 & 0 & 0 & 0 \\ 0 & 0 & p_b & 0 \\ 0 & 0 & 0 & 0 \end{pmatrix}. \tag{5.5}$$

Obviously, P_b commutes with Π_3 and H_0. Moreover, if $u = (u_1, u_2, u_3, u_4) \in D(H_0)$, we have

$$H_0 P_b u = \begin{pmatrix} 1 & 0 & \Pi_3 & 0 \\ 0 & 0 & 0 & 0 \\ \Pi_3 & 0 & -1 & 0 \\ 0 & 0 & 0 & 0 \end{pmatrix} \begin{pmatrix} p_b u_1 \\ 0 \\ p_b u_3 \\ 0 \end{pmatrix}. \tag{5.6}$$

Put

$$Q_b := \mathrm{Id} - P_b. \tag{5.7}$$

On $\{D(\Pi_3)\}^2 \subset L^2(\mathbf{R}^3; \mathbf{C}^2)$ introduce the operator

$$h_0 := \begin{pmatrix} 1 & \Pi_3 \\ \Pi_3 & -1 \end{pmatrix}. \tag{5.8}$$

Evidently, $\sigma(h_0) = \sigma_{\text{ess}}(h_0) = (-\infty, -1] \cup [1, +\infty)$. Note that if we replace Π_3 by $-i\frac{d}{dz}$ in (5.8), we shall obtain the operator χ_0.
Define the operators r and g substituting $H_0(b)$ for h_0 respectively in (4.6) and (4.7). Taking into account (5.6), we find that the spectral theorem for selfadjoint operators entails the following lemma.

Lemma 5.1 *The restrictions of the operators $H_0(b)$ (respectively, R_b and G_b) on $P_b D(H_0(b))$ (respectively, $P_b L^2(\mathbf{R}^3; \mathbf{C}^4)$) are unitarily equivalent to the restrictions of h_0 (respectively, r and g) on $P_b D(h_0)$ (respectively, $p_b L^2(\mathbf{R}^3; \mathbf{C}^2)$).*

6 Preliminary estimates

Lemma 6.1 *Let $\lambda_1, \lambda_2 \in \mathbf{R}$, $-1 < \lambda_1 < \lambda_2 < 1$. Then the estimates*

$$c_1 \| |H_0(b)|^{-1} u\|^2 \leq \|R_b(\lambda_1, \lambda_2) u\|^2 \leq c_2 \| |H_0(b)|^{-1} u\|^2, \quad \forall u \in L^2(\mathbf{R}^3; \mathbf{C}^4), \tag{6.1}$$

$$c_1 \| |h_0|^{-1} v\|^2 \leq \|r(\lambda_1, \lambda_2) v\|^2 \leq c_2 \| |h_0|^{-1} v\|^2, \quad \forall v \in L^2(\mathbf{R}^3; \mathbf{C}^2), \tag{6.2}$$

$$c_1 \| |\chi_0|^{-1} w\|^2 \leq \|\varrho(\lambda_1, \lambda_2) w\|^2 \leq c_2 \| |\chi_0|^{-1} w\|^2, \quad \forall w \in L^2(\mathbf{R}; \mathbf{C}^2), \tag{6.3}$$

hold for some $c_j(\lambda_1, \lambda_2) > 0$, $j = 1, 2$.

Proof. In order to deduce (6.1), it suffices to note that $\|R_b(\lambda_1, \lambda_2) u\|^2 = \|((H_0(b) - \lambda_1)(H_0(b) - \lambda_2))^{-1/2} u\|^2$, and the quantity $|\lambda^2 (\lambda - \lambda_1)^{-1} (\lambda - \lambda_2)^{-1}|$ is bounded and strictly positive if $\lambda \in \sigma(H_0(b)) = (-\infty, -1] \cup [1, +\infty)$. Estimates (6.2) and (6.3) are completely analogous. □

Let the matrix $M(X) : \mathbf{C}^4 \to \mathbf{C}^4$ be defined for $X \in \mathbf{R}^3$. Denote by $|M(X)|$ the norm of $M(X)$, $X \in \mathbf{R}^3$.

Lemma 6.2 *Let $|M| \in L^p(\mathbf{R}^3)$, $p \geq 2$, $\lambda_1, \lambda_2 \in \mathbf{R}$, $-1 < \lambda_1 < \lambda_2 < 1$. Then the estimate*

$$\|M R_b P_b\|_p^p \leq b c_3 \int_{\mathbf{R}^3} |M(X)|^p \, dX, \quad b > 0, \tag{6.4}$$

holds with c_3 which depends on λ_1 and λ_2, but is independent of b and M.

Proof. Evidently,

$$\|M R_b P_b\| \leq \|M R_b\| \leq c_2 \|M |H_0(b)|^{-1}\| \leq$$
$$c_2 \| \, |H_0(b)|^{-1} \| \| \, |M| \, \|_{L^\infty(\mathbf{R}^3)} = c_2 \| \, |M| \, \|_{L^\infty(\mathbf{R}^3)}. \tag{6.5}$$

On the other hand,

$$\|MR_bP_b\|_2^2 = \|Mp_br\|_2^2 \le c_2^2\| \,|M|\, p_b|h_0|^{-1}\|_2^2 = 2c_2^2\| \,|M|\, p_b(\Pi_3^2+1)^{-1/2}\|_2^2. \quad (6.6)$$

Taking into account (5.3)–(5.4) and

$$((\Pi_3^2+1)^{-1/2}u)(x,y,z) = \frac{1}{2\pi}\int_{\mathbf{R}}\int_{\mathbf{R}} \frac{e^{i(z-z')\zeta}}{(\zeta^2+1)^{1/2}} u(x,y,z')\,d\zeta dz',$$

we get

$$\| \,|M|\, p_b(\Pi_3^2+1)^{-1/2}\|_2^2 =$$

$$\frac{b^2}{(2\pi)^3}\int_{\mathbf{R}^3} |M(x,y,z)|^2 dxdydz \int_{\mathbf{R}^2} e^{-\frac{b}{2}\left((x-x')^2+(y-y')^2\right)}\,dx'dy' \int_{\mathbf{R}} \frac{d\zeta}{\zeta^2+1} =$$

$$= \frac{b}{4\pi}\int_{\mathbf{R}^3} |M(X)|^2\,dX. \quad (6.7)$$

Combining (6.6) with (6.7), we obtain

$$\|MR_bP_b\|_2^2 \le b\frac{c_2^2}{2\pi}\int_{\mathbf{R}^3} |M(X)|^2\,dX. \quad (6.8)$$

Interpolating between (6.5) and (6.8), we find that (6.4) holds with $c_3 = c_2^p/2\pi$. □.

Recall that if $T \in S_p$, $p \ge 1$, then $n_*(\varepsilon; T) \le \varepsilon^{-p}\|T\|_p^p$, $\varepsilon > 0$.

Corollary 6.1 *Under the assumptions of Lemma* 6.2 *the estimate*

$$n_*(\varepsilon; MR_bP_b) \le b\,c_3\varepsilon^{-p}\int_{\mathbf{R}^3} |M(X)|^p\,dX \quad (6.9)$$

holds for every $\varepsilon > 0$ *and* $p \ge 2$.

Lemma 6.3 *Let* $|M| \in L^3(\mathbf{R}^3)$.
(i) *There exists a constant* c_4 *such that for every* $\varepsilon > 0$ *we have*

$$n_*(\varepsilon; MR_bQ_b) \le c_4\varepsilon^{-3}\int_{\mathbf{R}^3} |M(X)|^3\,dX. \quad (6.10)$$

(ii) *Moreover, for every* $\varepsilon > 0$, λ_1, $\lambda_2 \in \mathbf{R}$, $-1 < \lambda_1 < \lambda_2 < 1$, *and* $|M| \in L^3(\mathbf{R}^3)$, *there exists a number* b_0 *such that* $b \ge b_0$ *entails*

$$n_*(\varepsilon; MR_bQ_b) = 0. \quad (6.11)$$

Proof. By Lemma 6.1 we have

$$n_*(\varepsilon; MR_bQ_b) \leq n_*(\varepsilon c_2^{-1}; |M||H_0(b)|^{-1}Q_b), \quad \varepsilon > 0. \tag{6.12}$$

Further, (5.1)-(5.2) entail

$$|H_0(b)|^{-1}Q_b \leq (\Pi^2 + b)^{-1/2}Q_b \leq (\Pi^2 + b)^{-1/2}I_4.$$

Moreover, the operators $|H_0(b)|^{-1}$, Q_b and $(\Pi^2 + b)^{-1/2}I_4$ are pairwise commuting. Therefore, we have

$$n_*(\varepsilon c_2^{-1}; |M||H_0(b)|^{-1}Q_b) \leq 4n_*(\varepsilon c_2^{-1}; |M|(\Pi^2 + b)^{-1/2}). \tag{6.13}$$

The classical Birman-Schwinger principle (see (4.1)) entails

$$n_*(\varepsilon c_2^{-1}; |M|(\Pi^2 + b)^{-1/2}) = N(-b; \Pi^2 - c_2^2\varepsilon^{-2}|M|^2) \leq N(0; \Pi^2 - c_2^2\varepsilon^{-2}|M|^2), \tag{6.14}$$

while the magnetic version of the Cwickel-Lieb-Rozenblioum estimate implies

$$N(0; \Pi^2 - c_2^2\varepsilon^{-2}|M|^2) \leq c_5c_2^3\varepsilon^{-3} \int_{\mathbf{R}^3} |M(X)|^3\, dX \tag{6.15}$$

where c_5 is independent of M and ε (see [1, Theorem 2.15]).
Now, the combination of (6.12)–(6.15) immediately yields (6.10) with $c_4 = c_5c_2^3$.
On the other hand, by the Kato–Simon inequality we have

$$\|\,|M|\,(\Pi^2 + b)^{-1/2}\| \leq \|\,|M|\,(-\Delta + b)^{-1/2}\|. \tag{6.16}$$

Since $|M| \in L^3(\mathbf{R}^3)$, the multiplier by $|M|^2$ is -Δ-form-compact. Therefore

$$\lim_{b\to\infty} \|\,|M|\,(-\Delta + b)^{-1/2}\| = 0. \tag{6.17}$$

Fix $\varepsilon > 0$, and taking into account (6.16)–(6.17), choose b_0 so that $b \geq b_0$ entails

$$\|\,|M|\,(\Pi^2 + b)^{-1/2}\| < \varepsilon c_2^{-1}. \tag{6.18}$$

Now, (6.12) and (6.13) combined with (6.18) imply (6.11). □

Corollary 6.2 *Let $M \in L^3(\mathbf{R}^3)$. Then for every $\varepsilon > 0$ and $b > 0$ we have*

$$n_*(\varepsilon; MR_b) \leq (c_3b + c_4)\left(\frac{\varepsilon}{2}\right)^{-3} \int_{\mathbf{R}^3} |M(X)|^3\, dX. \tag{6.19}$$

Proof. Since $Q_b + P_b = \mathrm{Id}$, we have

$$n_*(\varepsilon; MR_b) = n_*(\varepsilon; MR_bP_b + MR_bQ_b) \leq n_*(\varepsilon/2; MR_bP_b) + n_*(\varepsilon/2; MR_bQ_b).$$

Applying (6.9) with $p = 3$ and (6.10), we get (6.19). □

Remark. In most cases we shall apply Lemmas 6.2 - 6.3 and Corollaries 6.1 - 6.2 with $M = V\,I_4$.

7 Proof of Propositions 2.1 – 2.2

Let $V : \mathbf{R}^3 \to \mathbf{R}$ be a measurable function. Fix $\varepsilon > 0$ and set

$$V_1(X) = V_{1,\varepsilon}(X) = \begin{cases} V(X) \text{ if } |V(X)| > \varepsilon, \\ 0 \text{ otherwise,} \end{cases} \tag{7.1}$$

$$V_2(X) = V_{2,\varepsilon}(X) = V(X) - V_{1,\varepsilon}(X). \tag{7.2}$$

It is easy to check that $V \in \mathcal{L}$ is equivalent to $V_{1,\varepsilon} \in L^3(\mathbf{R}^3)$ for all $\varepsilon > 0$. In what follows if $V \in \mathcal{L}$ and $\varepsilon > 0$ we shall choose the decomposition $V = V_1 + V_2$ with $V_1 \in L^3(\mathbf{R}^3)$ and $\sup_{X \in \mathbf{R}^3} |V_2(X)| \leq \varepsilon$, as in (7.1)–(7.2), and shall call it briefly the *ε-decomposition* of V.

Lemma 7.1 *Let $V \in \mathcal{L}$. Then for almost every $X_\perp \in \mathbf{R}^2$ the operator $V(X_\perp, .)\chi_0^{-1}$ is compact in $L^2(\mathbf{R})$.*

Proof. Fix ε and write the ε-decomposition of V. Choose $X_\perp \in \mathbf{R}^2$ so that

$$\int_{\mathbf{R}} |V_1(X_\perp, z)|^3 \, dz < \infty. \tag{7.3}$$

Evidently, the complement of the set of $X_\perp$ satisfying (7.3), is a null-set. Moreover, (7.3) implies that the operator $V_1(X_\perp, .)\chi_0^{-1}$ is compact.
Now, pick a sequence ε_n such that $\varepsilon_n > 0$, and $\lim_{n\to\infty} \varepsilon_n = 0$. Write the ε_n-decomposition of $V = V_1^{(n)} + V_2^{(n)}$ with $V_j^{(n)} := V_{j,\varepsilon_n}$, $j = 1, 2$. Fix $X_\perp \in \mathbf{R}^2$ such that $\int_{\mathbf{R}} |V_1^{(n)}(X_\perp, z)|^3 \, dz < \infty$ for all n. The complement of such $X_\perp$ is again a null-set being a countable union of null-sets. The operators $V_1^{(n)}(X_\perp, .)\chi_0^{-1}$ are compact, and we have

$$\|V_1(X_\perp, .)\chi_0^{-1} - V_1^{(n)}(X_\perp, .)\chi_0^{-1}\| = \|V_2^{(n)}(X_\perp, .)\chi_0^{-1}\| \leq \varepsilon_n.$$

Since the operator $V(X_\perp, .)\chi_0^{-1}$ can be approximated in norm by compact operators, it is a compact operator itself. □

Remark. Lemma 7.1 entails immediately Proposition 2.1.

Lemma 7.2 *Let $v \in L^p(\mathbf{R})$, $p \geq 2$. Then we have*

$$\|v\varrho\|_p^p \leq c_6 \int_{\mathbf{R}} |v(z)|^p \, dz \tag{7.4}$$

where $c_6 = c_6(p)$ is independent of v.

Proof. Applying Lemma 6.1, we get

$$\|v\varrho\|_p^p \leq c_2^p \|v|\chi_0|^{-1}\|_p^p. \tag{7.5}$$

Evidently,

$$\| v|\chi_0|^{-1}\|_p^p = 4 \left\| v \left(-\frac{d^2}{dz^2} + 1 \right)^{-1/2} \right\|_p^p. \tag{7.6}$$

If $v \in L^\infty(\mathbf{R})$, we have

$$\left\| v \left(-\frac{d^2}{dz^2} + 1 \right)^{-1/2} \right\| \le \sup_{\zeta \in \mathbf{R}} (\zeta^2 + 1)^{-1/2} \, \|v\|_{L^\infty(\mathbf{R})} = \|v\|_{L^\infty(\mathbf{R})}. \tag{7.7}$$

If $v \in L^2(\mathbf{R})$, we have

$$\left\| v \left(-\frac{d^2}{dz^2} + 1 \right)^{-1/2} \right\|_2^2 = \frac{1}{2\pi} \int_{\mathbf{R}} \frac{d\zeta}{\zeta^2 + 1} \, \|v\|^2_{L^2(\mathbf{R})} = \frac{1}{2} \|v\|^2_{L^2(\mathbf{R})}. \tag{7.8}$$

Interpolating between (7.7) and (7.8), and bearing in mind (7.5) and (7.6), we find that (7.4) holds with $c_6 = 2c_2^p$. □

Corollary 7.1 *Let $V \in L^3(\mathbf{R}^3)$. Then the estimate*

$$\int_{\mathbf{R}^2} n_*(\varepsilon; V(X_\perp, .)\varrho) \, dX_\perp \le c_6 \varepsilon^{-3} \int_{\mathbf{R}^3} |V(X)|^3 \, dX \tag{7.9}$$

holds for each $\varepsilon > 0$ with $c_6 = c_6(3)$.

Proof. Fix $X_\perp \in \mathbf{R}^2$ for which $\int_{\mathbf{R}^2} |V(X_\perp, z)|^3 \, dz < \infty$; the complement of the set of such $X_\perp$ is a null set. Applying (7.4), we get

$$n_*(\varepsilon; V(X_\perp, .)\varrho) \le \varepsilon^{-3} \|V(X_\perp, .)\varrho\|_3^3 \le c_6 \varepsilon^{-3} \int_{\mathbf{R}} |V(X_\perp, z)|^3 \, dz.$$

Integrating with respect to $X_\perp \in \mathbf{R}^2$, we get (7.9). □

Corollary 7.2 *Let $V \in L^3(\mathbf{R}^3)$, $-1 < \lambda_1 < \lambda_2 < 1$. Then we have*

$$\mathcal{D}(\lambda_1, \lambda_2) \le c_7 \varepsilon^{-3} \int_{\mathbf{R}^3} |V(X)|^3 \, dX \tag{7.10}$$

with $c_7 = c_7(\lambda_1, \lambda_2) = 2^4 (c_2(\lambda_1, \lambda_2) \|\gamma(\lambda_1, \lambda_2)\|)^3$.

Proof. First, by (4.10) and $\varrho V(X_\perp, .)^2 \varrho \ge 0$, we have

$$\mathcal{D}(\lambda_1, \lambda_2) \le \int_{\mathbf{R}^2} n_-(1; 2\mathrm{Re}\gamma V(X_\perp, .)\varrho) dX_\perp \le \int_{\mathbf{R}^2} n_*(1; 2\|\gamma\| V(X_\perp .,)\varrho) dX_\perp.$$

Further, Corollary 7.1 implies

$$\int_{\mathbf{R}^2} n_*(1; 2\|\gamma\| V(X_\perp, .)\varrho) dX_\perp \le c_6 2^3 \|\gamma\|^3 \int_{\mathbf{R}^3} |V(X)|^3 \, dX. \tag{7.11}$$

Inserting the value of c_6 into (7.11), we obtain (7.10). □

Remark. Proposition 2.2 is implied almost immediately by Corollary 7.2. In order to see that, we assume that $-1 < \lambda_1 < \lambda_2 < 1$, fix $\varepsilon > 0$ such that $\lambda_1 - \varepsilon > -1$ and $\lambda_2 + \varepsilon < 1$, and write the ε-decomposition of V. Then we have

$$\mathcal{D}_V(\lambda_1, \lambda_2) \leq \mathcal{D}_{V_1}(\lambda_1 - \varepsilon, \lambda_2 + \varepsilon),$$

and by (7.10)

$$\mathcal{D}_{V_1}(\lambda_1 - \varepsilon, \lambda_2 + \varepsilon) \leq c_7(\lambda_1 - \varepsilon, \lambda_2 + \varepsilon) \int_{\mathbf{R}^3} |V_1(X)|^3 \, dX.$$

Therefore $\mathcal{D}_V(\lambda_1, \lambda_2) < \infty$.

8 Proof of Theorem 2.1

Let $-1 < \lambda_1 < \lambda_2 < 1$. Introduce the operator

$$k_b \equiv k_b(\lambda_1, \lambda_2) := r(\lambda_1, \lambda_2) p_b V^2 p_b r(\lambda_1, \lambda_2) + 2\mathrm{Re}\, g(\lambda_1, \lambda_2) p_b V p_b r(\lambda_1, \lambda_2)$$

where the operator p_b is defined by (5.3), while the operators $r(\lambda_1, \lambda_2)$ and $g(\lambda_1, \lambda_2)$ are introduced at the end of Section 5.
It is easy to check that if $V \in L^p(\mathbf{R}^3)$ then $V p_b r \in S_p$, $p \geq 2$; hence, $V p_b r$ itself as well as $r p_b V^2 p_b r$, $g p_b V p_b r$ and $r p_b V p_b g$ are Hilbert-Schmidt operators.

Proposition 8.1 *Let* $V \in C_0^\infty(\mathbf{R}^3)$, $-1 < \lambda_1 < \lambda_2 < 1$. *Then the asymptotic relations*

$$\lim_{b \to \infty} b^{-1} \mathrm{Tr}\, k_b(\lambda_1, \lambda_2)^l = \frac{1}{2\pi} \int_{\mathbf{R}^2} \kappa(\lambda_1, \lambda_2; X_\perp)^l \, dX_\perp$$

are valid for every integer $l \geq 2$.

Proof. Throughout the proof the parameters λ_1 and λ_2 are fixed, and we omit them in the notations.
For $l \geq 1$ write

$$k_b^l = \sum_{j=1}^{3^l} k_{j,l}(b), \quad \kappa(X_\perp)^l = \sum_{j=1}^{3^l} \kappa_{j,l}(X_\perp),$$

where the terms $k_{j,l}$ and $\kappa_{j,l}$, $j = 1, \ldots, 3^l$, are defined recurrently:

$$k_{1,1}(b) := r p_b V^2 p_b r, \quad k_{2,1}(b) := g p_b V p_b r, \quad k_{3,1}(b) := r p_b V p_b g,$$

$$\kappa_{1,1}(X_\perp) := \varrho V(X_\perp, .)^2 \varrho, \quad \kappa_{2,1}(X_\perp) := \gamma V(X_\perp, .) \varrho, \quad \kappa_{3,1}(X_\perp) := \varrho V(X_\perp, .) \gamma,$$

$$k_{j,l}(b) = \begin{cases} k_{1,1}(b)\ k_{j,l-1}(b), \ j = 1, \ldots, 3^{l-1}, \\ k_{2,1}(b)\ k_{j,l-1}(b), \ j = 3^{l-1}+1, \ldots, 2.3^{l-1}, \ \ l \geq 2, \\ k_{3,1}(b)\ k_{j,l-1}(b), \ j = 2.3^{l-1}+1, \ldots, 3^l, \end{cases}$$

$$\kappa_{j,l}(X_\perp) = \begin{cases} \kappa_{1,1}(X_\perp)\ \kappa_{j,l-1}(X_\perp), \ j = 1, \ldots, 3^{l-1}, \\ \kappa_{2,1}(X_\perp)\ \kappa_{j,l-1}(X_\perp), \ j = 3^{l-1}+1, \ldots, 2.3^{l-1}, \ \ l \geq 2. \\ \kappa_{3,1}(X_\perp)\ \kappa_{j,l-1}(X_\perp), \ j = 2.3^{l-1}+1, \ldots, 3^l, \end{cases}$$

The operators $k_{j,l}(b)$, $j = 1, \ldots, l$, $l \geq 2$, can be written in the form

$$k_{j,l} = E^-_{j,l} W_{1,j,l} T_{1,j,l} \ldots W_{l-1,j,l} T_{l-1,j,l} W_{l,j,l} E^+_{j,l}$$

where the operators $E^-_{j,l}$ and $E^+_{j,l}$ coincide either with r or with g, the operators $W_{s,j,l}$, $s = 1, \ldots, l$, coincide either with $p_b V^2 p_b$ or with $p_b V p_b$, and the operators $T_{s,j,l}$, $s = 1, \ldots, l-1$, coincide either with r^2, or with gr, or with g^2. Note that among the operators $T_{s,j,l}$, $s = 1, \ldots, l-1$, and $E^+_{j,l} \times E^-_{j,l}$, there are either at least one operator r^2, or at least two operators gr.
Analogously,

$$\kappa_{j,l}(X_\perp) = \epsilon^-_{j,l} \omega_{1,j,l}(X_\perp) \tau_{1,j,l} \ldots \omega_{l-1,j,l}(X_\perp) \tau_{l-1,j,l} \omega_{l,j,l}(X_\perp) \epsilon^+_{j,l}$$

where $\epsilon^\pm_{j,l} = \varrho$ if $E^\pm_{j,l} = r$ and $\epsilon^\pm_{j,l} = \gamma$ if $E^\pm_{j,l} = g$, $\omega_{s,j,l}(X_\perp) = V(X_\perp, .)^2$ if $W_{s,j,l} = p_b V^2 p_b$, and $\omega_{s,j,l}(X_\perp) = V(X_\perp, .)$ if $W_{s,j,l} = p_b V p_b$, $s = 1, \ldots, l$, $\tau_{s,j,l} = \varrho^2$ if $T_{s,j,l} = r^2$, $\tau_{s,j,l} = \varrho\gamma$ if $T_{s,j,l} = rg$, and $\tau_{s,j,l} = \gamma^2$ if $T_{s,j,l} = g^2$, $s = 1, \ldots, l-1$.
Obviously,

$$\mathrm{Tr}\ k^l_b = \sum_{j=1}^{3^l} \mathrm{Tr}\ k_{j,l}(b),$$

$$\int_{\mathbf{R}^2} \mathrm{Tr}\ \kappa(X_\perp)^l dX_\perp = \sum_{j=1}^{3^l} \int_{\mathbf{R}^2} \mathrm{Tr}\ \kappa_{j,l}(X_\perp)\ dX_\perp, \ l \geq 2.$$

Hence, it suffices to prove that

$$\lim_{b \to \infty} b^{-1} \mathrm{Tr}\ k_{j,l}(b) = \frac{1}{2\pi} \int_{\mathbf{R}^2} \mathrm{Tr}\ \kappa_{j,l}(X_\perp)\ dX_\perp, \ j = 1, \ldots, 3^l, \ l \geq 2. \tag{8.1}$$

It is not difficult to show that

$$\mathrm{Tr}\ k_{j,l} = \int_{\mathbf{R}^{3l}} \Pi'^{l}_{s=1} w_{s,j,l}(x_{s+1}, y_{s+1}, \zeta_{s+1} - \zeta_s) \mathcal{P}_b(x_{s+1}, y_{s+1}; x_s, y_s) \times$$

$$\mathrm{Tr}\ \Pi^l_{s=1} t_{s,j,l}(\zeta_s)\ \Pi^l_{s=1} dx_s dy_s d\zeta_s \tag{8.2}$$

where

$$w_{s,j,l}(x, y, \zeta) = \begin{cases} \frac{1}{2\pi} \int_{\mathbf{R}} e^{-iz\zeta} V^2(x, y, z) dz & \text{if} \quad W_{s,j,l} = p_b V^2 p_b, \\ \frac{1}{2\pi} \int_{\mathbf{R}} e^{-iz\zeta} V(x, y, z) dz & \text{if} \quad W_{s,j,l} = p_b V p_b, \end{cases}$$

$\mathcal{P}_b$ is introduced in (5.4), $t_{s,j,l}(\zeta)$, $s = 1,\ldots,l-1$, coincides with the matrix-valued symbol of the operator $T_{s,j,l}$, and $t_{l,j,l}(\zeta)$ is the matrix-valued symbol of the operator $E^+_{j,l} \times E^-_{j,l}$. Moreover, the notation $\Pi'^{l}_{s=1}$ means that in the product of l factors the variables x_{l+1}, y_{l+1}, and ζ_{l+1}, should be set equal respectively to x_1, y_1, and ζ_1. Analogously, we have

$$\mathrm{Tr}\ \kappa_{j,l}(X_\perp) \equiv \mathrm{Tr}\ \kappa_{j,l}(x,y) =$$

$$\int_{\mathbf{R}^l} \Pi'^{l}_{s=1} w_{s,j,l}(x,y,\zeta_{s+1}-\zeta_s) \mathrm{Tr}\ \Pi^l_{s=1} t_{s,j,l}(\zeta_s) \Pi^l_{s=1} d\zeta_s,\ X_\perp \equiv (x,y) \in \mathbf{R}^2. \qquad (8.3)$$

In order to prove (8.1), we insert (5.4) into (8.2), and obtain

$$\mathrm{Tr}\ k_{j,l} = \frac{b^l}{(2\pi)^l} \int_{\mathbf{R}^{3l}} \Pi'^{l}_{s=1} w_{s,j,l}(x_{s+1}, y_{s+1}, \zeta_{s+1}-\zeta_s)\times$$

$$\exp\left\{-\frac{b}{4}\left[(x_{s+1}-x_s)^2 + (y_{s+1}-y_s)^2 + 2i(x_{s+1}y_s - y_{s+1}x_s)\right]\right\}$$

$$\mathrm{Tr}\ \Pi^l_{s=1} t_{s,j,l}(\zeta_s) \Pi^l_{s=1} dx_s dy_s d\zeta_s.$$

Change the variables

$$x_1 = x_1',\ y_1 = y_1',$$

$$x_s = b^{-1/2} x_s' + x_1',\ y_s = b^{-1/2} y_s' + y_1',\ s = 2,\ldots,l.$$

Note that the corresponding Jacobian is equal to b^{1-l}. Thus we get

$$\mathrm{Tr}\ k_{j,l} = \frac{b}{(2\pi)^l} \int_{\mathbf{R}^{3l}} w_{l,j,l}(x_1', y_1', \zeta_1 - \zeta_l) e^{\Phi(x_2',\ldots,x_l',y_2',\ldots,y_l')}$$

$$\Pi^{l-1}_{s=1} w_{s,j,l}(x_1' + b^{-1/2} x_{s+1}', y_1' + b^{-1/2} y_{s+1}', \zeta_{s+1} - \zeta_s)\ \mathrm{Tr}\ \Pi^l_{s=1} t_{s,j,l}(\zeta_s)\ \Pi^l_{s=1} dx_s' dy_s' d\zeta_s,$$

where

$$\Phi(x_2,\ldots,x_l,y_2,\ldots,y_l) :=$$

$$-\frac{1}{4}\left\{{x_2}^2 + {y_2}^2 + {x_l}^2 + {y_l}^2 + \sum_{s=2}^{l-1}\left((x_{s+1}-x_s)^2 + (y_{s+1}-y_s)^2 + 2i(x_{s+1}y_s - y_{s+1}x_s)\right)\right\}.$$

Therefore,

$$\lim_{b\to\infty} b^{-1} \mathrm{Tr}\ k_{j,l}(b) =$$

$$\frac{1}{(2\pi)^l} \int_{\mathbf{R}^{3l}} e^{\Phi(x_2',\ldots,x_l',y_2',\ldots,y_l')} \Pi'^{l}_{s=1} w_{s,j,l}(x_1', y_1', \zeta_{s+1}-\zeta_s) \mathrm{Tr}\ \Pi^l_{s=1} t_{s,j,l}(\zeta_s) \Pi^l_{s=1} dx_s' dy_s' d\zeta_s. \qquad (8.4)$$

Changing the variables in the integral at the right hand side of (8.4)

$$x_1' = x_1,\ y_1' = y_1,$$

$$x_s' = x_s - x_1,\ y_s' = y_s - y_1,\ s = 2,\ldots,l,$$

we get

$$\lim_{b\to\infty} b^{-1}\mathrm{Tr}\, k_{j,l}(b) = \int_{\mathbf{R}^{3l}} \Pi'^{l}_{s=1} w_{s,j,l}(x_1, y_1, \zeta_{s+1} - \zeta_s)$$
$$\mathcal{P}_1(x_{s+1}, y_{s+1}; x_s, y_s)\mathrm{Tr}\, \Pi^l_{s=1} t_{s,j,l}(\zeta_s)\Pi^l_{s=1} dx_s dy_s d\zeta_s. \tag{8.5}$$

Since

$$\int_{\mathbf{R}^2} \mathcal{P}_1(x, y; x'', y'')\mathcal{P}_1(x'', y''; x', y')dx''dy'' = \mathcal{P}_1(x, y; x', y'),\ x, y, x', y' \in \mathbf{R},$$

$$\mathcal{P}_1(x, y; x, y) = (2\pi)^{-1},\ x, y \in \mathbf{R},$$

we find that (8.5) is equivalent to

$$\lim_{b\to\infty} b^{-1}\mathrm{Tr}\, k_{j,l}(b) =$$

$$\frac{1}{2\pi}\int_{\mathbf{R}^{l+2}} \Pi'^{l-1}_{s=1} w_{s,j,l}(x_1, y_1, \zeta_{s+1} - \zeta_s)\mathrm{Tr}\, \Pi^l_{s=1} t_{s,j,l}(\zeta_s)\, \Pi^l_{s=1} d\zeta_s\, dx_1 dy_1 =$$
$$\frac{1}{2\pi}\int_{\mathbf{R}^2} \mathrm{Tr}\, \kappa_{j,l}(X_\perp)\, dX_\perp,\ j = 1, \ldots, l,\ l \geq 2,$$

(see (8.3)), which is identical to (8.1). □

Set

$$\nu(s) := \begin{cases} \frac{1}{2\pi}\int_{\mathbf{R}^2} n_-(-s; \kappa(X_\perp))\, dX_\perp,\ s < 0, \\ -\frac{1}{2\pi}\int_{\mathbf{R}^2} n_+(s; \kappa(X_\perp))\, dX_\perp,\ s > 0, \end{cases} \tag{8.6}$$

Note that $s \neq 0$ is a continuity point of ν if and only if

$$\mathrm{vol}\,\left\{X_\perp \in \mathbf{R}^2 | \dim \mathrm{Ker}\, (\kappa(X_\perp) - s) \geq 1\right\} = 0.$$

Corollary 8.1 *Let $t < 0$ be a continuity point of ν. Then we have*

$$\lim_{b\to\infty} b^{-1} n_-(-t; k_b) = \nu(t).$$

The corollary follows immediately from Proposition 8.1 and Lemma 3.1 with $T(b) = k_b$, ν defined as in (8.6), and $t_0 = \|r\|^2\|V\|^2_{L^\infty(\mathbf{R})} + 2\|g\|\|r\|\|V\|_{L^\infty(\mathbf{R})}$.

Proposition 8.2 *Let $V \in C_0^\infty(\mathbf{R}^3)$. Assume that $t < 0$ is a continuity point of ν. Then we have*

$$\lim_{b\to\infty} b^{-1} n_-(-t; K_b) = \nu(t). \tag{8.7}$$

Proof. Evidently

$$n_-(-t; K_b) \geq n_-(-t; P_b K_b P_b) = n_-(-t; k_b).$$

Applying Corollary 8.1, we get

$$\liminf_{b\to\infty} b^{-1}n_-(-t;K_b) \geq \liminf_{b\to\infty} b^{-1}n_-(-t;k_b) = \lim_{b\to\infty} b^{-1}n_-(-t;k_b) = \nu(t). \tag{8.8}$$

On the other hand we have

$$K_b = P_bK_bP_b + Q_bK_bQ_b + 2\text{Re } P_bK_bQ_b =$$

$$P_bK_bP_b + Q_bK_bQ_b + 2\text{Re } P_bR_bV^2R_bQ_b + 2\text{Re } P_bG_bVR_bQ_b + 2\text{Re } P_bR_bVG_bQ_b.$$

Note that

$$R_bVG_b = G_bVR_b + R_bJR_b$$

where

$$J := [V, H_0(b) - \frac{1}{2}(\lambda_1+\lambda_2)] = [V, H_0(b)] = i\left(\frac{\partial V}{\partial x}\alpha_1 + \frac{\partial V}{\partial y}\alpha_2 + \frac{\partial V}{\partial z}\alpha_3\right).$$

It is essential that J is independent of b.
Apply the estimates

$$K_b = P_bK_bP_b + Q_bK_bQ_b + 2\text{Re } P_bR_bV^2R_bQ_b + 4\text{Re } P_bG_bVR_bQ_b + 2\text{Re } P_bR_bJR_bQ_b \geq$$

$$P_bK_bP_b + Q_bK_bQ_b - \varepsilon P_bR_bV^2R_bP_b - \varepsilon^{-1}Q_bR_bV^2R_bQ_b -$$

$$2\varepsilon P_bG_b^2P_b - 2\varepsilon^{-1}Q_bR_bV^2R_bQ_b - \varepsilon P_bR_b^2P_b - \varepsilon^{-1}Q_bR_bJ^*JR_bQ_b \geq$$

$$P_b(K_b - \varepsilon(R_bV^2R_b + 2G_b^2 + R_b^2))P_b - Q_b(\varepsilon G_b^2 + \varepsilon^{-1}R_b(4V^2 + J^*J)R_b)Q_b, \ \varepsilon > 0. \tag{8.9}$$

Now fix $\mu \in (0, -t)$, and choose ε so small that we have $\varepsilon(3\|G_b\|^2 + \|R_b\|^2) \leq \mu/3$; hence $\varepsilon\|2P_bG_b^2P_b + P_bR_b^2P_b + Q_bG_b^2Q_b\| \leq \mu/3$. Then (8.9) entails

$$n_-(-t;K_b) \leq n_-(-t-\mu; P_bK_bP_b) +$$

$$n_+(\mu/3; \varepsilon P_bR_bV^2R_bP_b) + n_+(\mu\varepsilon/3; Q_bR_b(4V^2 + J^*J)R_bQ_b). \tag{8.10}$$

Lemma 6.3(ii) combined with the estimate

$$n_+(2\delta^2; Q_bR_b(4V^2 + J^*J)R_bQ_b) \leq n_*(\delta/2; VR_bQ_b) + n_*(\delta; JR_bQ_b), \quad \delta > 0,$$

implies that the quantity $n_+(\mu\varepsilon/3; Q_bR_b(4V^2+J^*J)R_bQ_b)$ vanishes for b large enough. Hence, (8.10) entails

$$\limsup_{b\to\infty} b^{-1}n_-(-t;K_b) \leq \limsup_{b\to\infty} b^{-1}n_-(-t-\mu; P_bK_bP_b) +$$

$$\limsup_{b\to\infty} b^{-1}n_+(\mu/3; \varepsilon P_bR_bV^2R_bP_b), \ \mu \in (0,-t), \ \varepsilon > 0. \tag{8.11}$$

Corollary 6.1 combined with the estimate $n_+(\delta^2; P_bR_bV^2R_bP_b) \le n_*(\delta; VR_bP_b)$, $\delta > 0$, implies

$$\limsup_{b\to\infty} b^{-1}n_+(\mu/3; \varepsilon P_bR_bV^2R_bP_b) \le c_3\left(\frac{3\varepsilon}{\mu}\right)^{3/2}\int_{\mathbf{R}^3}|V|^3\,dX.$$

Letting $\varepsilon \downarrow 0$, we find that (8.11) entails

$$\limsup_{b\to\infty} b^{-1}n_-(-t; K_b) \le \limsup_{b\to\infty} b^{-1}n_-(-t-\mu; P_bK_bP_b),\ \mu \in (0, -t). \tag{8.12}$$

Now choose a sequence $\{\mu_l\}_{l\ge1}$ such that $\mu_l \in (0,-t)$, $\lim_{l\to\infty}\mu_l = 0$, and all the points $-t-\mu_l$ are continuity points of ν. Then Corollary 8.1 implies

$$\limsup_{b\to\infty} b^{-1}n_-(-t-\mu_l; P_bK_bP_b) = \lim_{b\to\infty} b^{-1}n_-(-t-\mu_l; k_b) = \nu(t+\mu_l). \tag{8.13}$$

Letting $l \to \infty$, we find that (8.11)–(8.13) entail

$$\limsup_{b\to\infty} b^{-1}n_-(-t; K_b) \le \nu(t). \tag{8.14}$$

The combination of (8.8) and (8.14) immediately yields (8.7). □

Proposition 8.3 *Let $V \in L^3(\mathbf{R}^3)$. Assume that $t < 0$ is a continuity point of ν. Then* (8.7) *remains valid.*

Proof. Pick a sequence $\{\delta_l\}_{l\ge1}$, $\lim_{l\to\infty}\delta_l = 0$, and write $V = V_0 + V_1$, where $V_0 = V_{0,l} \in C_0^\infty(\mathbf{R}^3)$, $V_1 = V_{1,l} \in L^3(\mathbf{R}^3)$, and $\|V_{1,l}\|_{L^3(\mathbf{R}^3)} \le \delta_l$.
Introduce the operators $K_{b,0}$, $k_{b,0}$ and $\kappa_0(X_\perp)$, replacing V by $V_{0,l}$. Analogously, define the function $\nu_l \equiv \nu_{0,l}$ substituting κ for κ_0. Choose the sequence $\{\varepsilon_r\}$ such that $0 < \varepsilon_r < \min\{1, -t/2\}$, $\lim_{r\to\infty}\varepsilon_r = 0$, and the points $-t\pm\varepsilon_r$ are continuity points of all functions $\nu_{0,l}$. Evidently,

$$K_b \ge (1-\varepsilon_r^2)K_{b,0} + (1-\varepsilon_r^{-2})R_bV_1^2R_b + 2\mathrm{Re}G_bV_1R_b + 2\varepsilon_r^2\mathrm{Re}G_bV_0R_b,$$

$$K_b \le (1+\varepsilon_r^2)K_{b,0} + (1+\varepsilon_r^{-2})R_bV_1^2R_b + 2\mathrm{Re}G_bV_1R_b - 2\varepsilon_r^2\mathrm{Re}G_bV_0R_b,$$

and, hence,

$$n_-(-t; K_b) \le n_-(-t-\varepsilon_r; K_{b,0}) + n_-(\varepsilon_r/3; (1-\varepsilon_r^{-2})R_bV_1^2R_b) +$$

$$n_-(\varepsilon_r/3; 2\mathrm{Re}\,G_bV_0R_b) + n_-(\varepsilon_r/3; 2\varepsilon_r^2\mathrm{Re}\,G_bV_1R_b),$$

$$n_-(-t; K_b) \ge n_-(-t+\varepsilon_r; K_{b,0}) - n_+(\varepsilon_r/3; (1+\varepsilon_r^{-2})R_bV_1^2R_b) -$$

$$n_+(\varepsilon_r/3; 2\mathrm{Re}G_bV_1R_b) - n_-(\varepsilon_r/3; 2\varepsilon_r^2\mathrm{Re}G_bV_0R_b).$$

Utilizing Corollary 6.2 and Proposition 8.2, we get

$$\limsup_{b\to\infty} b^{-1}n_+(-t; K_b) \le \nu_{0,l}(t+\varepsilon_r) + c_8\delta_l^3 + c_9\varepsilon_r^3\int_{\mathbf{R}^3}|V_0|^3\,dX, \tag{8.15}$$

$$\liminf_{b\to\infty} b^{-1}n_+(-t;K_b) \geq \nu_{0,l}(t-\varepsilon_r) - c_8\delta_l^3 - c_9\varepsilon_r^3\int_{\mathbf{R}^3}|V_0|^3\,dX, \tag{8.16}$$

where c_8 depends on ε_r but is independent on δ_l, while c_9 is independent of both ε_r and δ_l.
Similarly, using Corollaries 7.1 and 7.2, we obtain the estimates

$$\nu_{0,l}(t+\varepsilon_r) \leq \nu(t+2\varepsilon_r) + c_8'\delta_l^3 + c_9'\varepsilon_r^3\int_{\mathbf{R}^3}|V_0|^3\,dX, \tag{8.17}$$

$$\nu_{0,l}(t-\varepsilon_r) \geq \nu(t-2\varepsilon_r) - c_8'\delta_l^3 - c_9'\varepsilon_r^3\int_{\mathbf{R}^3}|V_0|^3\,dX, \tag{8.18}$$

where c_8' depends on ε_r but is independent on δ_l, while c_9' is independent of both ε_r and δ_l.
Letting at first $l \to \infty$ (hence, $\delta_l \downarrow 0$), and then $r \to \infty$ (hence, $\varepsilon_r \downarrow 0$), in (8.15) - (8.18), and taking into account that t is a continuity point of ν, we obtain (8.7). □

Using Lemma 4.1, Corollary 4.1, and Proposition 8.3 with $t = -1$, we deduce the following corollary.

Corollary 8.2 *Let the hypotheses of Theorem* 2.1 *hold. Assume in addition* $V \in L^3(\mathbf{R}^3)$. *Then* (2.1) *holds.*

In order to complete the proof of Theorem 2.1 it remains to show that we can approximate $V \in \mathcal{L}$ by $V \in L^3(\mathbf{R}^3)$.
Let $\lambda_1, \lambda_2 \in \mathbf{R}^3$, $-1 < \lambda_1 < \lambda_2 < 1$, be regular points. Fix $\varepsilon > 0$ such that $\lambda_1 - 3\varepsilon > -1$, $\lambda_2 + 3\varepsilon < 1$, $\lambda_1 + 3\varepsilon < \lambda_2 - 3\varepsilon$, and write the ε-decomposition of V. Evidently,

$$\mathcal{N}(\lambda_1+\varepsilon,\lambda_2-\varepsilon;H_0(b)+V_1) \leq \mathcal{N}(\lambda_1,\lambda_2;H(b)) \leq \mathcal{N}(\lambda_1-\varepsilon,\lambda_2+\varepsilon;H_0(b)+V_1). \tag{8.19}$$

Now chose the numbers $\varepsilon_j \in [0,\varepsilon)$ such that

$$\mathrm{vol}\left\{X_\perp \in \mathbf{R}^2 |\, \dim \mathrm{Ker}\ (\chi_0 + V_1(X_\perp,.) - (\lambda_j \pm \varepsilon \pm \varepsilon_j)) \geq 1\right\} = 0, \quad j = 1,2.$$

Applying Corollary 8.2, we deduce from (8.19) the following estimates

$$\begin{gathered}\limsup_{b\to\infty} b^{-1}\mathcal{N}(\lambda_1,\lambda_2;H(b)) \leq \limsup_{b\to\infty} b^{-1}\mathcal{N}(\lambda_1-\varepsilon,\lambda_2+\varepsilon;H_0(b)+V_1) \leq \\ \limsup_{b\to\infty} b^{-1}\mathcal{N}(\lambda_1-\varepsilon-\varepsilon_1,\lambda_2+\varepsilon+\varepsilon_2;H_0(b)+V_1) = \\ \mathcal{D}_{V_1}(\lambda_1-\varepsilon-\varepsilon_1,\lambda_2+\varepsilon+\varepsilon_2) \leq \mathcal{D}_{V_1}(\lambda_1-2\varepsilon,\lambda_2+2\varepsilon), \end{gathered} \tag{8.20}$$

$$\begin{gathered}\liminf_{b\to\infty} b^{-1}\mathcal{N}(\lambda_1,\lambda_2;H(b)) \geq \liminf_{b\to\infty} b^{-1}\mathcal{N}(\lambda_1+\varepsilon,\lambda_2-\varepsilon;H_0(b)+V_1) \geq \\ \liminf_{b\to\infty} b^{-1}\mathcal{N}(\lambda_1+\varepsilon+\varepsilon_1,\lambda_2-\varepsilon-\varepsilon_2;H_0(b)+V_1) = \end{gathered}$$

$$\mathcal{D}_{V_1}(\lambda_1+\varepsilon+\varepsilon_1,\lambda_2-\varepsilon-\varepsilon_2)\le\mathcal{D}_{V_1}(\lambda_1+2\varepsilon,\lambda_2-2\varepsilon). \tag{8.21}$$

Finally, note the obvious inequalities

$$\mathcal{D}_{V_1}(\lambda_1-2\varepsilon,\lambda_2+2\varepsilon)\le\mathcal{D}_{V}(\lambda_1-3\varepsilon,\lambda_2+3\varepsilon),$$

$$\mathcal{D}_{V_1}(\lambda_1+2\varepsilon,\lambda_2-2\varepsilon)\ge\mathcal{D}_{V}(\lambda_1+3\varepsilon,\lambda_2-3\varepsilon). \tag{8.22}$$

Putting together (8.19)–(8.22), we get

$$\limsup_{b\to\infty} b^{-1}\mathcal{N}(\lambda_1,\lambda_2;H(b))\le\mathcal{D}_{V}(\lambda_1-3\varepsilon,\lambda_2+3\varepsilon), \tag{8.23}$$

$$\liminf_{b\to\infty} b^{-1}\mathcal{N}(\lambda_1,\lambda_2;H(b))\ge\mathcal{D}_{V}(\lambda_1+3\varepsilon,\lambda_2-3\varepsilon). \tag{8.24}$$

Letting $\varepsilon\downarrow 0$ in (8.23)–(8.24), and bearing in mind that λ_1 and λ_2 are regular points, we come to (2.1).

Acknowledgements

This work has been done during author's visit to the University of Regensburg in the summer of 1998 as a DAAD Research Fellow. The financial support of the DAAD is gratefully acknowledged.
The author thanks Prof.H.Siedentop for his warm hospitality, and for several illuminating discussions on the spectral theory of the Dirac operator.
The author was partially supported by the Bulgarian Science Foundation under Grant MM 612/96.

References

[1] J.Avron, I.Herbst, B.Simon, *Schrödinger operators with magnetic fields. I. General interactions*, Duke. Math. J. **45** (1978), 847-883.

[2] M.Š.Birman, *On the spectrum of singular boundary value problems*, Mat. Sbornik **55** (1961) 125-174 (Russian); Engl. transl. in Amer. Math. Soc. Transl., (2) **53** (1966), 23-80.

[3] M.Sh.Birman, A.Laptev, *Discrete spectrum of the perturbed Dirac operator*, Ark.Mat. **32** (1994), 13-32.

[4] C.Cancelier, P.Lévy-Bruhl, J.Nourrigat, *Remarks on the spectrum of the Dirac operator*, Acta Appl.Math. **45** (1996), 349-364.

[5] W.D.Evans, R.T.Lewis, H.Siedentop, J.P.Solovej, *Counting eigenvalues using coherent states with an application to Dirac and Schrödinger operators in the semi-classical limit*, Ark.Mat. **34** (1996), 265-283.

[6] V.Ya.Ivrii, *Microlocal Analysis and Precise Spectral Asymptotics*, Springer Monographs in Mathematics, Springer-Verlag, Berlin- Heidelberg, 1998.

[7] G.D.RAIKOV, *Eigenvalue asymptotics for the Schrödinger operator in strong constant magnetic fields*, Commun. P.D.E. **23** (1998), 1583-1620.

[8] H.TAMURA, *The asymptotic distribution of discrete eigenvalues for Dirac systems*, J.Fac.Sci. Univ. Tokyo Sect. IA Math. **23** (1976), 167-197.

[9] B.THALLER, *The Dirac Equation*, Texts and Monographs in Physics, Springer-Verlag, Berlin-Heidelberg-New York, 1992.

MPEJ

MATHEMATICAL PHYSICS ELECTRONIC JOURNAL

ISSN 1086-6655
Volume 5, 1999

Paper 3
Received: Mar 24, 1999, Revised: Jun 23, 1999, Accepted: Jun 24, 1999
Editor: J. Avron

Propagating Edge States for a Magnetic Hamiltonian

Stephan De Bièvre

UFR de Math. et UMR AGAT – Université Lille 1 (USTL)
59655 Villeneuve d'Ascq Cedex, France – e-mail: debievre@gat.univ-lille1.fr

Joseph V. Pulé

Department of Math. Physics – National University of Ireland, Dublin (UCD)
Belfield, Dublin 4, Ireland – e-mail: Joe.Pule@ucd.ie

Abstract

We study the quantum motion of a charged particle in a half plane, subject to a perpendicular constant magnetic field B and to an arbitrary weak impurity potential W_B (i.e. $||W_B||_\infty < \delta B$, for some δ small enough). We show that there exist states propagating with a speed of size $B^{1/2}$ along the edge, no matter how fast W_B fluctuates. As a consequence, the spectrum of the Hamiltonian is purely absolutely continuous in a spectral interval of size γB ($0 < \gamma < 1$) between the Landau levels of the system without edge or potential, so that the corresponding eigenstates are extended. This then provides a rigorous proof of a phenomenon pointed out by Halperin in his work on the quantum Hall effect.

1 Introduction

It is well known that a classical charged particle, constrained to a plane and subjected to a perpendicular magnetic field will move along physical boundaries when those are present. In the case of a particle moving in a half plane ($x > 0, y \in \mathbb{R}$), it is easy to see that the circular trajectories that are at a distance less than $\sqrt{E}/B$ from the edge will bounce of it in such a way that the particle speeds alongside the edge with a velocity of the order of $\sqrt{E}$, where E denotes the energy of the particle. If, on the other hand, the centre of the trajectory is too far from the edge, it will not affect the motion of the particle.

If, as one would expect, this picture is to carry over to the quantum mechanical situation, then an initial state localized close to the edge in a region of size $B^{-1/2}$ – an *edge state* – should move ballistically along the edge with a speed of order $\sqrt{B}$: here we used that the lowest Landau level, in absence of the edge, is of order B. On the other hand, although states further away from the edge – *bulk states* – should, due to the uncertainty principle, not remain completely localized in the y-direction, as in the classical case, they should nevertheless move much more slowly than the edge states. This picture has long been known to be correct, but as a preparation for the case when an impurity potential is present, we give a precise statement of the above properties in Corollary 2.1. We consider the Hamiltonian

$$H_0 = \frac{1}{2}p_x^2 + \frac{1}{2}(p_y - Bx)^2, \tag{1.1}$$

with a Dirichlet boundary condition at $x = 0$. Corresponding to each Landau band, we introduce the notion of H_0-invariant *edge* and *bulk* spaces, with the following properties. The y-component of the velocity, given by $i[H_0, Y]$, is of order $\sqrt{B}$ on an edge space, whereas it is exponentially small in B on a bulk space (Corollary 2.1). Furthermore, states belonging to the edge spaces are negligibly small at distances much larger than the magnetic length scale $1/\sqrt{B}$, reflecting the intuitively clear fact that the presence of the edge makes itself felt only in a region of size $1/\sqrt{B}$ from the edge. In this sense the edge states are quasi one-dimensional. The eigenfunctions of the restriction of H_0 to the edge spaces are extended along the entire edge.

The existence of non-localized current-carrying quasi one-dimensional edge states plays a role in certain theories of the quantum Hall effect [1] (see [2], [3] and [4] for further details). It is therefore of importance to understand if such states exist in systems exhibiting a quantized Hall resistance. This is argued to be the case in [1], in the case when the full Hamiltonian is obtained by adding a weak impurity potential to H_0. In other words, such potentials are not supposed to destroy the edge states existing in the free case. A very simple and rigorous proof of this statement is given in the present paper. A weak potential is a potential $W_B \in L^\infty(\mathbb{R}_+ \times \mathbb{R})$ satisfying $\delta_B \equiv ||W_B||_\infty < \frac{1}{2}B$. Since the distance between successive Landau levels equals B, such a potential can not close the gaps between the Landau levels of the infinite system without a edge, even though its size can be of order B: in this sense it is weak. It can however fluctuate arbitrarily fast, and in particular on the magnetic length scale, which is of order $1/\sqrt{B}$ (about $50-100$ Angstrom in realistic situations): this is important since, as explained in [2], the weak impurity potential is created by impurities at a distance of order $1/\sqrt{B}$ or less of the layer and can vary rapidly on this length scale. As a typical form for W_B we can keep in mind a potential of the type

$$W_B = \delta B \sum_{i \in \mathbb{Z}^+ \times \mathbb{Z}} u_i(B^\alpha(\vec{x} - \frac{i}{B^\beta}))$$

for some compactly supported site-potentials u_i and exponents $\alpha \geq 0$ and $\beta \geq 1/2$.

For weak potentials, we show that, in a spectral interval of size B between the Landau levels, there are no bound states and that the speed in the y-direction is still of order $\sqrt{B}$. As a consequence, we obtain that in the same spectral interval, the spectrum is absolutely continuous, implying the corresponding eigenstates are extended.

The results described above in the case when no impurity potential is present have been known for a long time and can be obtained by studying explicitly the spectrum and eigenfunctions of H_0, since it is an explicitly solvable Hamiltonian. Such an approach would however not easily extend to the case when an impurity potential is added. Instead, we show in Proposition 2.1 that the magnitude of the speed in the y-direction is strictly positive on the spectral subspaces corresponding to suitable spectral intervals between Landau levels. Such a positive commutator estimate is then shown to be stable under perturbations in section 3, yielding the main results via the virial theorem and the Mourre theory of positive commutators. (Theorem 3.1).

The idea that positive commutator methods and the virial theorem can be used to obtain information about magnetic Hamiltonians in the presence of boundaries was first proposed in [5]. They consider a model with a *soft* edge, modeled by a positive potential V, supported on the negative axis and steeply rising from 0, and prove the absence of eigenvalues in certain regions between the Landau levels in this case. The conjugate operator used in this approach is the quantum observable C_y corresponding to the y-coordinate of the centre of the classical circular orbit: $C_y = y - (p_x/B)$. Classically this is indeed a monotonic function of time for orbits close to the edge, since the Poisson bracket $\{C_y, H\} = \frac{1}{B}(\partial_x V + \partial_x W) < 0$ in that region, provided the impurity potential W has a small enough derivative.

In the present paper, we deal with the problem with a hard edge, as described before. We use the y-coordinate itself as a conjugate operator, proving that the speed in the y-direction, $i[H, Y]$, is strictly negative on edge states. This is marginally surprising since it is not true classically, but it turns out to be extremely simple to understand in terms of the band structure of the free Hamiltonian H_0. Using y also has the important advantage of not introducing derivatives of the potential in the commutator, as is the case when using C_y, and therefore eliminating the need to control their size. In addition, it renders the interpretation of the results in terms of propagation along the edge more transparent. On the down side, it is not obvious the present method will adapt itself easily to cases where the edge is not straight.

Let us point out that we could treat the soft edge in the same way. It seems however that this model does not lend itself to an analysis of the high field regime, which is important for the quantum Hall effect. In that case, the particles will, even in the lowest Landau level, penetrate deeply into the region $x < 0$, so that there is an effective edge around those values of x where $V(x) \sim B$, where V is the edge potential. The high field behaviour of the speed, for example, will then depend crucially on the precise shape of the edge, and this is not satisfactory. We will therefore not deal any further with the soft edge in the following.

The results of [5] on the soft edge have recently been extended [6] to a proof of absence of singular continuous spectrum in suitable intervals between the Landau levels, using the same conjugate operator as in [5] to prove a positive commutator estimate. While finishing the present work, we learned that those results were further extended, still using C_y as a conjugate operator, to the case of the *hard* edge in [7]. A result comparable to our Theorem 3.1 is proven there, but under the additional assumptions that both the first and second derivatives of the impurity potential are small, so that rapid fluctuations in the impurity potential are no longer allowed. In addition, our proof is technically considerably less complicated partially because, in the model with a hard edge, the operator C_y is symmetric but not self-adjoint, leading to complications in applying the Mourre theory of positive commutators.

2 The free Hamiltonian: edge and bulk spaces

To study H_0 in (1.1), we first use the translational invariance in the y-direction to write

$$H_0 = \int_{\mathbb{R}}^{\oplus} dk\ H(k), \quad \text{with} \quad H(k) = -\frac{1}{2}\frac{d^2}{dx^2} + \frac{1}{2}(k - Bx)^2, \quad x > 0, \tag{2.1}$$

acting on $L^2(\mathbb{R}_+ \times \mathbb{R}, dxdk)$, k being the Fourier transform variable conjugate to y. We recall that H_0 is essentially self-adjoint on the space of functions $\varphi \in C_0^\infty(\overline{\mathbb{R}_+} \times \mathbb{R})$

vanishing on the boundary [8].

The spectrum of $H(k)$ consists of isolated non-degenerate eigenvalues $E_n(k)$, $n \in \mathbb{N}_0$, with *normalized* eigenfunctions $\varphi_n(x,k)$. We will write $\mathcal{H}_n$ for the nth *band space*, namely the space consisting of vectors of the form $f(k)\varphi_n(x,k)$, $f \in L^2(\mathbb{R}, dk)$. This is an H_0-invariant subspace of $L^2(\mathbb{R}_+ \times \mathbb{R}, dxdk)$; we shall on occasion view it as a subspace of $L^2(\mathbb{R}_+ \times \mathbb{R}, dxdy)$ as well, with the same notation. To understand the behaviour of the $E_n(k)$ and the $\varphi_n(x,k)$, and in particular their dependence on B, we introduce the following scaling:

$$\tilde{x} = \sqrt{B}x,\ \tilde{y} = \sqrt{B}y,\ H_0 = B\tilde{H}_0,\ \tilde{H}_0 = -\frac{1}{2}\frac{\partial^2}{\partial \tilde{x}^2} + \frac{1}{2}(\frac{1}{i}\frac{\partial}{\partial \tilde{y}} - \tilde{x})^2. \tag{2.2}$$

Note that, strictly speaking, H_0 is unitarily equivalent to $B\tilde{H}_0$, not equal to it, but since the unitary transformation is just the rescaling of the variables, we allow ourselves this slight abuse of notation. Again

$$\tilde{H}_0 = \int_{\mathbb{R}}^{\oplus} d\kappa\ \tilde{H}(\kappa),$$

with

$$\tilde{H}(\kappa) = -\frac{1}{2}\frac{d^2}{d\tilde{x}^2} + \frac{1}{2}(\kappa - \tilde{x})^2, \qquad \tilde{x} > 0. \tag{2.3}$$

Here κ is the Fourier transform variable conjugate to $\tilde{y}$, so that $ky = \kappa\tilde{y}$ and hence $k = \sqrt{B}\kappa$. The spectrum of $\tilde{H}(\kappa)$ consists of isolated eigenvalues $\alpha_n(\kappa)$. The normalized eigenfunctions of $\tilde{H}(\kappa)$, at each fixed κ, $\tilde{\varphi}_n(\cdot,\kappa)$ are given by

$$\tilde{\varphi}_n(\tilde{x},\kappa) = C_n D_{\alpha_n(\kappa)-1/2}(\sqrt{2}(\tilde{x} - \kappa)), \tag{2.4}$$

where $D_{\alpha-1/2}$ is the Whittaker function ([9] p686) with parameter α and $\alpha_n(\kappa)$ is determined by the boundary condition

$$D_{\alpha_n(\kappa)-1/2}(-\sqrt{2}\kappa) = 0. \tag{2.5}$$

One can check that the eigenvalues $\alpha_n(\kappa)$ are smooth functions of κ. The other properties of the $\alpha_n(\kappa)$ that we shall be needing are collected in the following Lemma:

Lemma 2.1 *(i)* $\alpha_n(0) = 2n + 3/2, \qquad \alpha_n'(0) = -\dfrac{(2n+2)!}{n!(n+1)!\pi^{1/2}2^{2n}},$

(ii) $\alpha_{n+1}(\kappa) - \alpha_n(\kappa) > 1,$

(iii) $\alpha_n'(\kappa) = -\frac{1}{2}|\tilde{\varphi}_n'(0,\kappa)|^2 < 0,$

(iv) $\alpha_n(\kappa) > (n+\frac{1}{2})$ *and there exist* $C_n > 0$ *so that, for all* $\kappa \geq 0$,

$$\alpha_n(\kappa) - (n+\frac{1}{2}) \leq C_n \exp -\frac{1}{4}(\kappa - \sqrt{n})^2.$$

(v) For all $n \in \mathbb{N}_0$ *and all* $\epsilon > 0$, *there exist positive constants* X_n, $K_{n,\epsilon}, C_{n,\epsilon}$ *so that, for all* $\kappa \geq K_{n,\epsilon}$ *and all* $\tilde{x} \in [0, X_n]$,

$$|\tilde{\varphi}_n(\tilde{x},\kappa)|^2 \leq C_{n,\epsilon} \exp\left\{-\frac{1}{2}(1-\epsilon)(\tilde{x}-\kappa)^2\right\},$$

and

$$|\alpha_n'(\kappa)| \leq C_{n,\epsilon} \exp\left\{-\frac{1}{2}(1-\epsilon)\kappa^2\right\}.$$

(vi) For all $\kappa < 0$, $|\alpha_n'(\kappa)| > |\kappa|$ *and* $\lim_{\kappa\to-\infty} \alpha_n(\kappa) = \infty$.

The proof of this lemma, which uses only standard techniques of Schrödinger operator theory, is postponed to Section 4. Numerical computation of the $\alpha_n(\kappa)$ indicates that they are convex functions of κ (see Figure 1). We have not been able to prove this result and shall not use it. If it is true, the statements of our results below can be simplified somewhat.

It is clear from the above that the spectrum of $\tilde{H}_0$ and hence of H_0 is absolutely continuous and fills the entire half-axis from 1/2 to infinity. The bands $E_n(k)$ can now be written

$$E_n(k) = B\alpha_n(\frac{k}{\sqrt{B}}). \tag{2.6}$$

Writing $\varphi_n(x,k)$ for the normalized eigenfunctions of $H(k)$ (at each fixed k), we have

$$\varphi_n(x,k) = B^{1/4}\tilde{\varphi}_n(\sqrt{B}x, \frac{k}{\sqrt{B}}). \tag{2.7}$$

These simple observations will now allow us to define within $\mathcal{H}_n$ *edge spaces* and *bulk spaces* as follows. We define, for each $\sigma > 0, \gamma > 0$:

$$\begin{aligned}\mathcal{H}_{n,e}(\sigma,\gamma) &\cong L^2(]-\infty,\sigma B^\gamma], dk) \subset \mathcal{H}_n, &(2.8)\\ \mathcal{H}_{n,b}(\sigma,\gamma) &\cong L^2([\sigma B^\gamma,\infty), dk) \subset \mathcal{H}_n, &(2.9)\\ \mathcal{H}_n &= \mathcal{H}_{n,e} \oplus \mathcal{H}_{n,b}. &(2.10)\end{aligned}$$

Note that these spaces are H_0 invariant. We will call $\mathcal{H}_{n,e}(\sigma,\gamma)$ an edge space for all $\gamma \leq 1/2$ and $\mathcal{H}_{n,b}(\sigma,\gamma)$ a bulk space for all $\gamma > 1/2$. For a different approach to the definition of bulk and edge spaces, in the case of a bounded geometry, we refer to [10].

To understand those definitions, recall first that a standard stationary phase argument shows that $-\partial_k E_n(k_0) = -\sqrt{B}\alpha'_n(k_0/\sqrt{B})$ is the group speed in the y-direction of a wave packet $f(k)\varphi_n(x,k)$ with the support of f close to k_0. If k_0 is inside an interval $(-\infty, k_B]$ where k_B is of order $\sqrt{B}$ or smaller, the wave packet belongs to the edge space $\mathcal{H}_{n,e}(\sigma, 1/2)$ and it follows from Lemma 2.1 that such a wave packet speeds along the edge in the y direction with a velocity of order $\sqrt{B}$. In addition, it follows from standard exponential estimates on the eigenfunctions $\tilde{\varphi}_n$ (as in the proof of Lemma 2.1) that in this case the wave packet is exponentially small for x much bigger than $1/\sqrt{B}$. If, on the other hand, k_0 belongs to an interval of the form $[k_B, \infty[$ with k_B of order $B^\gamma, \gamma > \frac{1}{2}$, then the group velocity is exponentially small in B (see Lemma 2.1 (iv)). In addition, if $f(k)\varphi_n(x,k) \in \mathcal{H}_{n,b}(\sigma,\gamma)$, with $\gamma > \frac{1}{2}$, then Lemma 2.1(iv) immediately implies that

$$\int_0^{\frac{1}{\sqrt{B}}} \int_{-\infty}^{\infty} |f(k)\varphi_n(x,k)|^2 dk dx \leq C_{n,\epsilon} \exp -(1-\epsilon)(\sigma^2 B^{2\gamma-1} - 1),$$

so that the wave packet is exponentially small in the region $0 \leq x \leq \frac{1}{\sqrt{B}}$ close to the edge. We note also that the spectrum of H_0 restricted to a bulk space $\mathcal{H}_{n,b}(\sigma,\gamma)$ is an exponentially small interval (in B) just above the nth Landau level (Lemma 2.1 (iii)), that we will refer to as the bulk spectrum. The spectrum of H_0 restricted to an edge space $\mathcal{H}_{n,e}(\sigma, 1/2)$ – *the edge spectrum* – is on the other hand of the form $[B(n+\frac{1}{2}+c_\sigma), \infty)$. In particular, it fills up an interval of size B below the $(n+1)$th Landau level, including the latter.

To give a formulation of the above statements that is at once more precise and does not use the band structure of the Hamiltonian H_0, so that it has a chance to pass to the perturbed Hamiltonian, we now turn to the statement and proof of a *positive commutator estimate.* We will show that the speed $V_y = i[H_0, Y]$, where the operator Y is multiplication by y, is strictly negative away from the Landau levels. This is the content of the following proposition, which will be generalized to the perturbed Hamiltonian in the next section.

Let $L_n = (n+1/2, n+3/2]$ be the nth Landau band when $B = 1$.

Proposition 2.1 *Let $\Delta \subset L_n$ be a closed interval with $|\Delta| < 1$. Let*

$$\nu_-(\Delta) = \inf_{\{n',\kappa \mid \alpha_{n'}(\kappa)\in\Delta\}} |\alpha'_{n'}(\kappa)| > 0 \tag{2.11}$$

and

$$\nu_+(\Delta) = \sup_{\{n',\kappa \mid \alpha_{n'}(\kappa)\in\Delta\}} |\alpha'_{n'}(\kappa)| > 0. \tag{2.12}$$

If $\tilde{Y}$ is multiplication by $\tilde{y}$ and $\tilde{P}_0(\Delta)$ is the spectral projection of $\tilde{H}_0$ onto Δ, then

$$\nu_-(\Delta)\tilde{P}_0(\Delta) \leq \tilde{P}_0(\Delta)i[\tilde{Y},\tilde{H}_0]\tilde{P}_0(\Delta) \leq \nu_+(\Delta)\tilde{P}_0(\Delta), \tag{2.13}$$

and consequently, for ψ with $||\psi|| = 1$ in the range of $\tilde{P}_0(\Delta)$

$$-\nu_+(\Delta)t \leq \langle\psi_t, \tilde{Y}\psi_t\rangle - \langle\psi_0, \tilde{Y}\psi_0\rangle \leq -\nu_-(\Delta)t. \tag{2.14}$$

Proof: It follows from Lemma 2.1 that $\alpha_{n'}^{-1}(\Delta) \cap \alpha_{n''}^{-1}(\Delta) = \emptyset$ if $n' \neq n''$, since $|\Delta| < 1$. Thus for any ψ we can write

$$\tilde{P}_0(\Delta)\psi(\tilde{x},\kappa) = \sum_{n'=0}^{n} \psi_{n'}(\tilde{x},\kappa), \tag{2.15}$$

where

$$\psi_{n'}(\tilde{x},\kappa) = \beta_{n'}(\kappa)\mathbf{1}_{\alpha_{n'}^{-1}(\Delta)}(\kappa)\tilde{\varphi}_{n'}(\tilde{x},\kappa) \tag{2.16}$$

and

$$\beta_{n'}(\kappa) = \int_{\mathbb{R}_+} \overline{\psi(\tilde{x},\kappa)}\;\tilde{\varphi}_{n'}(\tilde{x},\kappa)dx. \tag{2.17}$$

Since $i[\tilde{Y},\tilde{H}_0] = \tilde{x} - p_{\tilde{y}}$, it is clear that $\langle\psi_{n'}, i[\tilde{Y},\tilde{H}_0]\psi_{n''}\rangle = 0$ if $n' \neq n''$ since the supports of $\psi_{n'}$ and of $\psi_{n''}$ are disjoint in the κ variable. On the other hand,

$$\langle\psi_{n'}, i[\tilde{Y},\tilde{H}_0]\psi_{n'}\rangle = \int_{\alpha_{n'}^{-1}(\Delta)} d\kappa\; |\beta_{n'}(\kappa)|^2 \int_{\mathbb{R}_+} d\tilde{x}(\tilde{x}-\kappa)|\varphi_{n'}(\tilde{x},\kappa)|^2, \tag{2.18}$$

and, by the Feynman-Hellman theorem

$$\int_{\mathbb{R}_+} d\tilde{x}(\tilde{x}-\kappa)|\tilde{\varphi}_{n'}(\tilde{x},\kappa)|^2 = -\alpha_{n'}'(\kappa) = |\alpha_{n'}'(\kappa)|.$$

The proposition is now immediate.

□

Using the scaling behaviour in B we now have the following Corollary:

Corollary 2.1
(i) Let $n \in \mathbb{N}$ be fixed and let $\Delta \subset ((n+\frac{1}{2})B, (n+\frac{3}{2})B]$ be a closed interval with $|\Delta| < B$. Then

$$\sqrt{B}\nu_-(B^{-1}\Delta)P_0(\Delta) \leq P_0(\Delta)i[Y,H_0]P_0(\Delta) \leq \sqrt{B}\nu_+(B^{-1}\Delta)P_0(\Delta). \tag{2.19}$$

where $B^{-1}\Delta = \{E/B \mid E \in \Delta\}$.

(ii) For all $n \in \mathbb{N}$, for all $\sigma > 0$ there exists a constant $C_{n,\sigma} > 0$ so that for all $\psi \in \mathcal{H}_{n,e}(\sigma, 1/2)$ and for all B

$$\langle \psi, i[Y, H_0]\psi \rangle \geq \sqrt{B} \inf_{\kappa \leq \sigma} |\alpha_n'(\kappa)| \, ||\psi||^2 \; > C_{n,\sigma}\sqrt{B}||\psi||^2. \tag{2.20}$$

(iii) Let $\epsilon > 0$. Then for all $n \in \mathbb{N}_0$, for all $\sigma > 0$ there exists a constant $C_{n,\sigma,\epsilon} > 0$ so that for all B and for all $\psi \in \mathcal{H}_{n,b}(\sigma, 1/2 + \epsilon)$

$$\langle \psi, i[Y, H_0]\psi \rangle \leq \sqrt{B} \sup_{\kappa \geq \sigma B^\epsilon} |\alpha_n'(\kappa)| \, ||\psi||^2 \; < C_{n,\sigma,\epsilon}\sqrt{B} \exp\left\{-\frac{1}{2}(1-\epsilon)\sigma^2 B^{2\epsilon}\right\} ||\psi||^2. \tag{2.21}$$

Proof: This is now an immediate consequence of Lemma 2.1 and of the proof of Proposition 2.1.

□

Remark 2.1 *Parts (ii) and (iii) of the corollary state that the speed in the y direction is at least of order $\sqrt{B}$ for any edge state and at most of order $\exp -B^{2\epsilon}$ for any bulk state.*

3 Adding a weak impurity potential

We now consider the Hamiltonian

$$H = H_0 + W_B$$

where $W_B \in L^\infty(\mathbb{R}_+ \times \mathbb{R}, dxdy)$ is a real potential satisfying $||W_B||_\infty \leq AB$ where $A < \infty$ is independent of B. Let

$$\tilde{H} = \tilde{H}_0 + \tilde{W}_B; \tilde{W}_B(\tilde{x}, \tilde{y}) = B^{-1} W_B(\frac{\tilde{x}}{\sqrt{B}}, \frac{\tilde{y}}{\sqrt{B}})$$

and let $\tilde{P}(\cdot)$ denote the spectral family of $\tilde{H}$.

Our main result is then the following theorem, which should be compared to Proposition 2.1. For $\lambda < 1$, let $L_n^\lambda = (n + 1/2 + \lambda, n + 3/2]$. Let

$$\nu(n, \lambda) = \nu_-(L_n^\lambda) = \inf\{|\alpha_{n'}'(\kappa)| \mid n' \leq n, \; n+1/2+\lambda < \alpha_{n'}(\kappa) \leq n+3/2\} > 0. \tag{3.1}$$

Theorem 3.1 *Let $n \in \mathbb{N}$ be fixed. Let λ, $\lambda' > 0$ with $\lambda + \lambda' < 1$ and let $L_n^{\lambda,\lambda'} = (n + 1/2 + \lambda, n + 3/2 - \lambda')$. There exists $\delta(n, \lambda, \lambda') > 0$ such that if $||W_B||_\infty < \delta(n, \lambda, \lambda')B$ and $\epsilon < \delta(n, \lambda, \lambda')$, then, for all $\alpha \in L_n^{\lambda,\lambda'}$, for the interval $\Delta \equiv (\alpha - \epsilon, \alpha + \epsilon)$,*

$$\tilde{P}(\Delta)i[\tilde{Y}, \tilde{H}]\tilde{P}(\Delta) \geq \frac{1}{2}\nu(n, \lambda/2)\tilde{P}(\Delta). \tag{3.2}$$

Consequently if $||W_B||_\infty < \delta(n, \lambda, \lambda')B$, then

$$\sigma_{\text{sing}}(\tilde{H}) \cap L_n^{\lambda,\lambda'} = \emptyset. \tag{3.3}$$

Clearly we can give a scaled up version of this theorem:

Corollary 3.1 *Let $n \in \mathbb{N}$ be fixed and let λ, $\lambda' > 0$ with $\lambda + \lambda' < 1$. There exists $\delta(n, \lambda, \lambda') > 0$ such that if $||W_B||_\infty < \delta(n, \lambda, \lambda')B$, then*

$$\sigma_{\text{sing}}(H) \cap (B(n + 1/2 + \lambda), B(n + 3/2 - \lambda')) = \emptyset. \tag{3.4}$$

It is useful to have the following variant of Theorem 3.1. Here we fix a bound on $||W_B||_\infty/B$ and give the dependence on this bound of the endpoints a,b of the interval $(a, b) \subset L_n$, such that (a, b) contains only absolutely continuous spectrum.

Theorem 3.2 *Let $n \in \mathbb{N}$ be fixed. Suppose that $||W_B||_\infty < \delta B$ where $\delta < 1/2$. Let λ_n, $\lambda'_n \in (0, 1/2)$ be such that $\lambda_n \nu(n, \lambda_n/2)^2 > 2^9(n + 2)\delta$ and $\lambda'_n \nu(n, 1/4)^2 > 2^9(n + 2)\delta$ then for all $\alpha \in L_n^{\lambda_n,\lambda'_n}$, there exists an interval Δ containing α such that*

$$\tilde{P}(\Delta)i[\tilde{Y}, \tilde{H}]\tilde{P}(\Delta) \geq \frac{1}{2}\nu(n, \lambda_n/2)\tilde{P}(\Delta). \tag{3.5}$$

Therefore

$$\sigma_{\text{sing}}(\tilde{H}) \cap L_n^{\lambda_n,\lambda'_n} = \emptyset. \tag{3.6}$$

Note that, given δ, no λ_n and λ'_n satisfying the conditions of the theorem might exist. Nevertheless, it is clear that for sufficiently small δ, the above results guarantee the existence of an interval of absolutely continuous spectrum between the Landau levels. The scaled up version of this theorem is then:

Corollary 3.2 *Under the conditions of Theorem 3.2,*

$$\sigma_{\text{sing}}(H) \cap (B(n + 1/2 + \lambda_n), B(n + 3/2 - \lambda'_n)) = \emptyset. \tag{3.7}$$

Proof of Theorem 3.1: Note first that $i[\tilde{Y}, \tilde{H}_0] = i[\tilde{Y}, \tilde{H}]$, so that the result would follow from Proposition 2.1 if we could replace $\tilde{P}(\Delta)$ by $\tilde{P}_0(\Delta)$. This can indeed be achieved with a few tricks and at not too high a cost, provided one replaces the interval Δ by an auxiliary one Δ', that is larger but for which $\nu(\Delta')$ is not too small. Let $\sigma \equiv \min(\lambda, \lambda')/4$ and let Δ' be the interval $[\alpha - \sigma, \alpha + \sigma] \subset L_n^{\lambda/2, \lambda'/2}$. Let Δ be the interval $[\alpha - \epsilon, \alpha + \epsilon]$, where $\epsilon \leq \sigma$. Let $\psi \in \tilde{P}(\Delta)\mathcal{H}$. Then, recalling that $||\tilde{W}_B||_\infty \leq A$,

$$||(\tilde{H}_0 - \alpha)\psi|| \leq ||(\tilde{H} - \alpha)\tilde{P}(\Delta)\psi|| + A||\psi|| \leq (\epsilon + A)||\psi||.$$

Hence

$$||\tilde{P}_0(\Delta'^c)\psi|| \leq ||\frac{1}{\tilde{H}_0 - \alpha}\tilde{P}_0(\Delta'^c)||\ ||(\tilde{H}_0 - \alpha)\psi|| \leq \sigma^{-1}(\epsilon + A)||\psi||, \tag{3.8}$$

since $\min\{|\lambda - \alpha| \mid \lambda \in \Delta'^c\} \geq \sigma$. Clearly

$$\begin{aligned} i\langle\psi, [\tilde{Y}, \tilde{H}]\psi\rangle \ &\geq\ i\langle\tilde{P}_0(\Delta')\psi, [\tilde{Y}, \tilde{H}_0]\tilde{P}_0(\Delta')\psi\rangle \\ &\quad -2||[\tilde{Y}, \tilde{H}_0]\tilde{P}_0(\Delta'^c)\psi||\ ||\psi||. \end{aligned} \tag{3.9}$$

The required positivity will come from the first term, so we only have to control the last one. We find

$$\begin{aligned} ||[\tilde{Y}, \tilde{H}_0]\tilde{P}_0(\Delta'^c)\psi|| \ &\leq\ 2\langle\tilde{P}_0(\Delta'^c)\psi, \tilde{H}_0\tilde{P}_0(\Delta'^c)\psi\rangle^{1/2} \\ &\leq\ 2||\tilde{H}_0\tilde{P}_0(\Delta'^c)\psi||^{1/2}||\tilde{P}_0(\Delta'^c)\psi||^{1/2}. \end{aligned}$$

But

$$\begin{aligned} ||\tilde{H}_0\tilde{P}_0(\Delta'^c)\psi||^{1/2} \ &\leq\ (||\tilde{H}\psi|| + A||\psi||)^{1/2} \\ &\leq\ (n + 3/2 + A)^{1/2}||\psi||^{1/2} \\ &\leq\ (n+2)^{1/2}||\psi||^{1/2}, \end{aligned}$$

if $A \leq 1/2$. Therefore

$$||[\tilde{Y}, \tilde{H}_0]\tilde{P}_0(\Delta'^c)\psi|| \leq 2(n+2)^{1/2}\sigma^{-1/2}(\epsilon + A)^{1/2}||\psi||.$$

Inserting this into (3.9) yields

$$\begin{aligned} i\langle\psi, [\tilde{Y}, \tilde{H}]\psi\rangle &\geq i\langle\tilde{P}_0(\Delta')\psi, [\tilde{Y}, \tilde{H}_0]\tilde{P}_0(\Delta')\psi\rangle \\ &\quad -4(n+2)^{1/2}\sigma^{-1/2}(\epsilon + A)^{1/2}||\psi||^2. \end{aligned} \tag{3.10}$$

On the other hand, since $|\Delta'| = 2\sigma < 1$, Proposition 2.1 states that

$$i\langle\tilde{P}_0(\Delta')\psi, [\tilde{Y}, \tilde{H}_0]\tilde{P}_0(\Delta')\psi\rangle \geq \nu(\Delta')||\tilde{P}_0(\Delta')\psi||^2 \geq \nu(n, \lambda/2)||\tilde{P}_0(\Delta')\psi||^2,$$

where $\nu(n, \lambda/2)$ is defined in (3.1). Inserting this into (3.10) and using (3.8) together with the observation that $||\psi||^2 = ||\tilde{P}_0(\Delta')\psi||^2 + ||\tilde{P}_0(\Delta'^c)\psi||^2$, yields

$$i\langle\psi, [\tilde{Y}, \tilde{H}]\psi\rangle \geq \nu(n, \frac{\lambda}{2})[1 - \left(\frac{(\epsilon + A)^2}{\sigma^2} + \frac{4(n+2)^{1/2}(\epsilon + A)^{1/2}}{\sigma^{1/2}\nu(n, \frac{\lambda}{2})}\right)]||\psi||^2. \tag{3.11}$$

Let $\delta(n, \lambda, \lambda') = \min\left(\frac{\sigma\nu(n,\lambda/2)^2}{2^9(n+2)}, \frac{\sigma}{4}, \frac{1}{2}\right)$. Then if $A < \delta$ and $\epsilon < \delta$, one has that $4(n+2)^{1/2}\sigma^{-1/2}(\epsilon + A)^{1/2} < \frac{1}{4}\nu(n, \lambda/2)$ and $((\epsilon + A)/\sigma)^2 \leq 1/4$, so that the first statement in the theorem follows. To prove (3.3) it is now sufficient to use (3.2) and to apply the Mourre theory of positive commutators. For a textbook treatment, we refer to [11]; see also [12] for a concise review of the domain questions involved. The latter are trivial in the present case. Indeed, the commutator $[H_0, Y] = [H, Y]$ is obviously relatively H_0 bounded, the domain of the Hamiltonian is invariant under the unitary group $\exp isY$ and the second commutator $[[H_0, Y], Y]$ is bounded.

□

4 Proof of Lemma 2.1

(i) This follows from a computation using standard properties of the Hermite polynomials.

(ii) The Whittaker functions satisfy the recurrence relations ([9] p688):

$$-D'_{\alpha+1/2}(x) - \frac{1}{2}xD_{\alpha+1/2}(x) = -(\alpha + 1/2)D_{\alpha-1/2}(x) \tag{4.1}$$

$$-D'_{\alpha-3/2}(x) + \frac{1}{2}xD_{\alpha-3/2}(x) = D_{\alpha-1/2}(x). \tag{4.2}$$

Consider $H_e(\kappa)$ given by the same expression as for $\tilde{H}(\kappa)$ in (2.3) but with elastic boundary condition

$$\psi'(0) = \lambda\psi(0), \quad 0 \leq \lambda < \infty;$$

instead of the Dirichlet boundary condition ($\lambda = \infty$). Let $\beta_n(\kappa)$, $n = 0, 1, 2, \ldots$ be the eigenvalues of $H_e(\kappa)$. Then standard arguments (see [14], Chapter 1, section 3) show that

$$\beta_0(\kappa) < \alpha_0(\kappa) < \beta_1(\kappa) < \alpha_1(\kappa) < \beta_2(\kappa) < \ldots \tag{4.3}$$

If $\kappa \geq 0$, put $\lambda = \kappa$. Then from (4.1) it follows that the eigenfunctions of $H_e(\kappa)$ are $D_0(\sqrt{2}(\tilde{x} - \kappa))$, $D_{\alpha_{n-1}(\kappa)+1/2}(\sqrt{2}(\tilde{x} - \kappa))$, $n = 1, 2, \ldots$ with eigenvalues $\beta_0(\kappa) = 1/2$, $\beta_n(\kappa) = \alpha_{n-1}(\kappa) + 1$, $n = 1, 2, \ldots$ Thus

$$\frac{1}{2} < \alpha_0 < \alpha_0 + 1 < \alpha_1 < \alpha_1 + 1 < \ldots$$

If $\kappa \le 0$ put $\lambda = -\kappa$. In this case it follows from (4.2) that the eigenfunctions of $H_e(\kappa)$ are $D_{\alpha_n(\kappa)-3/2}(\sqrt{2}(\tilde{x}-\kappa))$, $n = 0, 1, \ldots$ with eigenvalues $\beta_n(\kappa) = \alpha_n(\kappa) - 1$. Therefore

$$\alpha_0 - 1 < \alpha_0 < \alpha_1 - 1 < \alpha_1 < \alpha_2 - 1 < \ldots.$$

(iii) We put $V_\kappa(\tilde{x}) = \frac{1}{2}(\tilde{x}-\kappa)^2$ and use the Feynman-Hellman formula to write (see [13])

$$\begin{aligned}\alpha_n'(\kappa) &= -\int_0^\infty V_\kappa'(\tilde{x})\tilde{\varphi}_n^2(\tilde{x},\kappa)\,d\tilde{x}\\ &= 2\int_0^\infty V_\kappa(\tilde{x})\tilde{\varphi}_n(\tilde{x},\kappa)\tilde{\varphi}_n'(\tilde{x},\kappa)\,d\tilde{x}\\ &= \int_0^\infty \varphi_n''(\tilde{x},\kappa)\varphi_n'(\tilde{x},\kappa)\,d\tilde{x} + 2\alpha_n(\kappa)\int_0^\infty \varphi_n(\tilde{x},\kappa)\varphi_n'(\tilde{x},\kappa)\,d\tilde{x},\end{aligned}$$

from which the result follows. Note that by uniqueness $\varphi_n'(0)$ cannot be zero.

(iv) Here we will use a perturbative argument, treating the Dirichlet boundary condition at 0 as a perturbation. We note first that, by the min-max principle, $\alpha_n(\kappa) > n + \frac{1}{2}$. Now, let h_n denote the nth Hermite function and let $h_{n,\kappa}(x) = h_n(x-\kappa)$. Let θ be a smooth function such that $\theta(x) = 0$ for $x \le 0$ and $\theta(x) = 1$ for $x \ge 1$ We compute

$$(\tilde{H}(\kappa) - (n+\frac{1}{2}))\theta h_{n,\kappa} = [\tilde{H}(\kappa), \theta]h_{n,\kappa} = \frac{1}{2}(-\theta'' - 2i\theta' p)h_{n,\kappa}.$$

Now, since the supports of θ' and θ'' are contained in $[0,1]$, and since

$$||\theta' p h_n||^2 = \langle h_{n,\kappa}, [p, \theta'^2] p h_{n,\kappa}\rangle + \langle \theta'^2 h_{n,\kappa}, p^2 h_{n,\kappa}\rangle,$$

one easily concludes there exists a constant C_n so that

$$||(\tilde{H}(\kappa) - (n+\frac{1}{2}))\theta h_{n,\kappa}|| \le C_n ||\mathbf{1}_{[0,1]} h_{n,\kappa}||^{\frac{1}{2}},$$

where $\mathbf{1}_{[0,1]}$ denotes the characteristic function of $[0,1]$. Standard properties of the Hermite functions then imply that, for κ large enough

$$||(\tilde{H}(\kappa) - (n+\frac{1}{2}))\theta h_{n,\kappa}|| \le C_n \exp -\frac{1}{4}(\kappa - \sqrt{n})^2.$$

This shows that, for n fixed,

$$\mathrm{dist}(\sigma(\tilde{H}(\kappa)), (n+\frac{1}{2})) \le 2C_n \exp -\frac{1}{4}(\kappa - \sqrt{n})^2.$$

For $n = 0$, $|\alpha_0(\kappa) - \frac{1}{2}| = \mathrm{dist}(\sigma(\tilde{H}(\kappa)), \frac{1}{2}) \le 2C_0 \exp -\frac{1}{4}(\kappa)^2$, since $\alpha_1(\kappa) > 3/2$, and (iv) then follows by induction on n.

(v) This only involves a rather straightforward application of the standard method for proving exponential decay estimates on eigenfunctions in a classically forbidden region (see, for example [15, 8]). With $V_\kappa(\tilde{x}) \equiv \frac{1}{2}(\tilde{x}-\kappa)^2$ as before, we first define, for all $\kappa > 0$ large enough, $0 < x_n(\kappa) < \kappa$ by $V_\kappa(x_n(\kappa)) = \alpha_n(\kappa) + 1$. Clearly, for all $0 \leq \tilde{x} \leq x_n(\kappa)$, $\tilde{\varphi}_n(\tilde{x},\kappa)$ and $\tilde{\varphi}_n''(\tilde{x},\kappa)$ have the same sign, which we can assume to be strictly positive. Also, on the same region $\varphi_n'(\tilde{x},\kappa) > 0$. As a result, for any $a \in [0, x_n(\kappa) - 2]$, one has

$$|\tilde{\varphi}_n(a,\kappa)|^2 \leq \int_a^{a+1} |\tilde{\varphi}_n(y,\kappa)|^2 dy.$$

Let, for $0 \leq \tilde{x} \leq x_n(\kappa)$,

$$f_n(\tilde{x},\kappa) = \int_{\tilde{x}}^{x_n(\kappa)} \sqrt{2(V_\kappa(y) - \alpha_n(\kappa) - 1)}\,dy.$$

Note that

$$\frac{1}{2} f_n'(\tilde{x},\kappa)^2 - (V_\kappa(\tilde{x}) - \alpha_n(\kappa) - 1) = 0.$$

We introduce $\eta_n(\tilde{x},\kappa)$, a smooth characteristic function of the interval $[0, x_n(\kappa) - 1]$, with supp $\eta_n' \subset [x_n(\kappa) - 1, x_n(\kappa) - 1/2]$. Then

$$\begin{aligned} \int_a^{a+1} |\tilde{\varphi}_n(y,\kappa)|^2 dy &\leq \exp -2f_n(a+1,\kappa) \int_a^{a+1} \exp 2f_n(y,\kappa) |\tilde{\varphi}_n(y,\kappa)|^2 dy \\ &\leq \exp -2f_n(a+1,\kappa) \langle \psi_n, (V_\kappa - \alpha_n(\kappa) - \frac{1}{2} f'^2_n) \psi_n \rangle, \end{aligned} \tag{4.4}$$

where $\psi_n = \eta_n (\exp f_n) \tilde{\varphi}_n$. A simple computation shows

$$(\exp f_n)(\tilde{H} - \alpha_n)(\exp -f_n)\psi_n = \frac{\tilde{p}^2}{2}\psi_n + (V_\kappa - \alpha_n - \frac{1}{2} f'^2_n)\psi_n + \frac{1}{2}(f_n' \frac{d}{d\tilde{x}} + \frac{d}{d\tilde{x}} f_n')\psi_n,$$

so that

$$\mathrm{Re}\langle \psi_n, (\exp f_n)(\tilde{H} - \alpha_n)(\exp -f_n)\psi_n \rangle \geq \langle \psi_n, (V_\kappa - \alpha_n - \frac{1}{2} f'^2_n)\psi_n \rangle.$$

On the other hand, using the definition of ψ_n and $(\tilde{H} - \alpha_n)\tilde{\varphi}_n = 0$, one has

$$\mathrm{Re}\langle \psi_n, (\exp f_n)(\tilde{H} - \alpha_n)(\exp -f_n)\psi_n \rangle = \mathrm{Re}\langle \tilde{\varphi}_n, (\exp 2f_n)\eta_n \frac{1}{2}[\tilde{p}^2, \eta_n]\tilde{\varphi}_n \rangle.$$

Consequently,

$$|\tilde{\varphi}_n(a,\kappa)|^2 \leq (\exp -2f_n(a+1,\kappa)) \mathrm{Re}\langle \tilde{\varphi}_n, W_n \tilde{\varphi}_n \rangle,$$

where

$$W_n = (\exp 2f_n)\ \eta_n \frac{1}{2}[\tilde{p}^2, \eta_n] = -\frac{1}{2}(\exp 2f_n)(\eta_n\eta_n'' + 2i\eta_n\eta_n'\tilde{p}),$$

so that

$$\mathrm{Re}W_n = -\frac{1}{2}\exp 2f_n(\eta_n\eta_n'' - 2f_n'\eta_n\eta_n' - (\eta_n\eta_n')').$$

It follows from the support properties of η_n' and the definition of f_n that there exists a constant C_n so that for all κ, one has $|\mathrm{Re}W_n| \leq C_n$. As a result, a simple computation shows that, for all ϵ, there exists constants $C_{n,\epsilon}, K_{n,\epsilon}$ so that for all $\kappa > K_{n,\epsilon}$, and for all $0 \leq \tilde{x} < x_n(\kappa) - 2$

$$|\tilde{\varphi}_n(\tilde{x}, \kappa)|^2 \leq C_{n,\epsilon} \exp\left\{-\frac{1}{2}(1-\epsilon)(\kappa - \tilde{x})^2\right\}.$$

This proves the first statement of (v). To prove the second estimate, it is clear from part (iii) that we need to prove the above estimate holds for $|\tilde{\varphi}_n'(\tilde{x}, \kappa)|^2$ as well. If χ is a smooth characteristic function of the interval $[0, 1]$ with support in $[0, 2]$, one has, using the eigenvalue equation and two partial integrations that

$$\begin{aligned} |\tilde{\varphi}_n'(0, \kappa)|^2 &\leq \int_0^1 |\tilde{\varphi}_n'(\tilde{x}, \kappa)|^2 d\tilde{x} \\ &\leq \int_0^\infty \chi(\tilde{x})\tilde{\varphi}_n'(\tilde{x}, \kappa)^2 d\tilde{x} \leq C \int_0^1 |\tilde{\varphi}_n(\tilde{x}, \kappa)|^2 d\tilde{x}, \end{aligned}$$

from which the result follows. The last part of the Lemma is an immediate consequence of the Feynman-Hellman formula.

□

Acknowledgements: The authors thank the referee for suggesting the proof of Lemma 2.1, (ii) and J.Fröhlich, G.M.Graf and J.Walcher for bringing their results to their attention prior to publication. They would also like to thank Forbairt, the Royal Irish Academy, the CNRS and the Ministère des Affaires Etrangères for their financial support.

References

[1] B.I. Halperin, Phys. Rev. **B38**, 2185-2190 (1982).

[2] S. M. Girvin and R.E. Prange, *The quantum Hall effect*, Editors, Springer Verlag, 1987.

[3] X. G. Wen, Phys. Rev. B **43**, 11025 (1991).

[4] J. Fröhlich and U. M. Studer, Rev. Mod. Phys **65**, 733 (1993).

[5] N. Macris, P.A. Martin and J.V. Pulé, *On edge states in semi-infinite quantum Hall systems*, preprint 1998 (to appear in J. Phys. A).

[6] J. Fröhlich, G.M. Graf, and J. Walcher, private communication, 1998.

[7] J. Fröhlich, G.M. Graf, and J. Walcher, *On the extended nature of edge states of Quantum Hall Hamiltonians*, preprint march 1999, math-ph/9903014.

[8] B. Helffer, *Semi-classical analysis for the Schrödinger operator and applications*, Lecture Notes in Mathematics 1336, Springer-Verlag, 1988.

[9] M. Abramowitz and I.A. Stegun: *Handbook of Mathematical Functions*, Dover Publications - New York, 1965.

[10] E. Akkermans, J.E. Avron, R. Narevich and R. Seiler, *Boundary conditions for bulk and edge states in quantum Hall systems*, preprint 1998.

[11] W. Amrein, A. Boutet de Monvel and V. Georgescu, *C_0-groups, Commutator Methods and Spectral Theory of N-Body Hamiltonians*, Birkhäuser, Basel-Boston-Berlin, 1996.

[12] V. Georgescu and C. Gérard, *On the virial theorem in quantum mechanics*, preprint 1998 mp-arc 98-744.

[13] M. Dauge and B. Helffer, J. Diff. Eqns. 104, 2, 243-262 (1993).

[14] B.M. Levitan and I.S. Sargsjan, *Introduction to spectral theory*, AMS, Providence RI, 1975.

[15] S. Agmon, *Lectures on exponential decay of eigenfunctions of second order elliptic equations*, Princeton University Press, 1982.

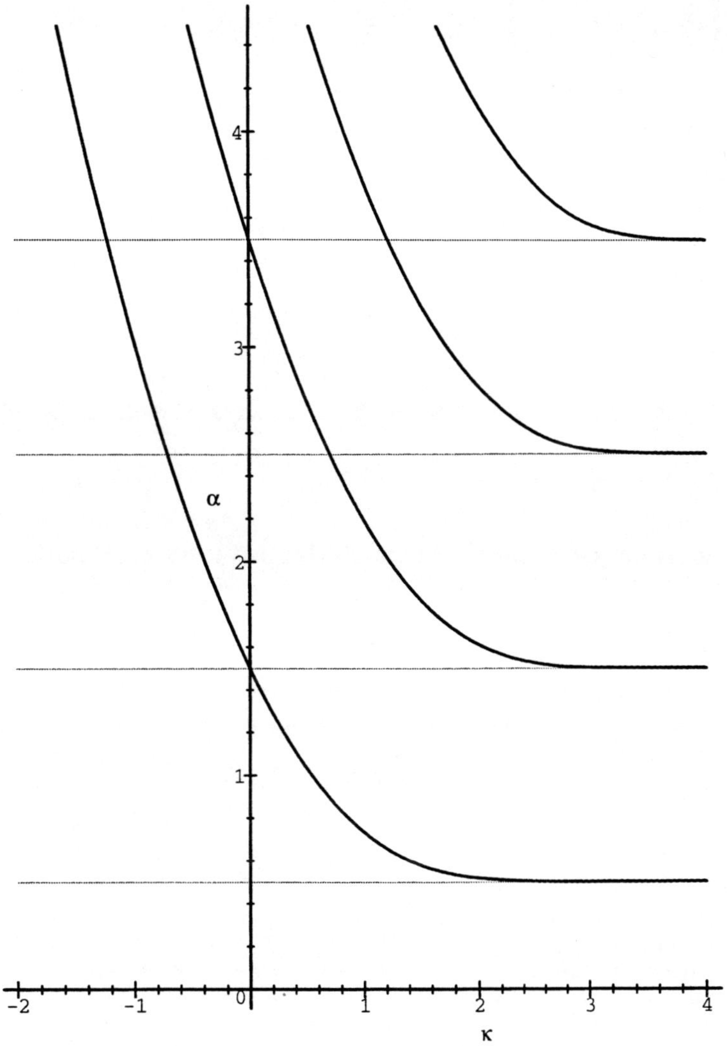

Figure 1: $\alpha_n(\kappa)$ for $n = 0, 1, 2, 3$

M P E J

MATHEMATICAL PHYSICS ELECTRONIC JOURNAL

ISSN 1086-6655
Volume 5, 1999

Paper 4
Received: Jul 7, 1999, Accepted: Sep 29, 1999
Editor: G. Gallavotti

On a conjecture for the critical behaviour of KAM tori

Federico Bonetto* **and Guido Gentile**†

* Mathematics Department, Rutgers University, New Brunswick, 08903 NJ
† Dipartimento di Matematica, Università di Roma Tre, Roma, I-00146

ABSTRACT. *At the light of recent results in literature we review a conjecture formulated in* Math. Phys. Electron. J. **1** (1995), paper 5, 1–13, *about the mechanism of breakdown of invariant sets in KAM problems and the identification of the dominant terms in the perturbative expansion of the conjugating function. We show that some arguments developed therein can be carried out further only in some particular directions, so limiting a possible future research program, and that the mechanism of break down of invariant tori has to be more complicated than as conjectured in the quoted paper.*

1. Introduction

1.1. *The state of the art of [GGM].* In [GGM] a conjecture about the mechanism of breakdown of KAM invariant tori is proposed. Roughly it is based on the following idea (we refer to [GGM] for a more detailed and technical exposition and also for the introduction of the notions used in the following analysis).

If $\mathbb{T} = \mathbb{R}/2\pi\mathbb{Z}$, let $\mathbb{T}^\ell$ be the ℓ-dimensional torus. Consider a Hamiltonian system

$$\mathcal{H} = \underline{\omega}_0 \cdot \underline{A} + \frac{1}{2}\underline{A} \cdot J^{-1}\underline{A} + \varepsilon f(\underline{\alpha}) , \tag{1.1}$$

where $(\underline{\alpha}, \underline{A}) \in \mathbb{T}^\ell \times \mathbb{R}^\ell$ are conjugate variables, J is the matrix of momenta of inertia, $\cdot$ denotes the inner product in $\mathbb{R}^\ell$, $f(\underline{\alpha})$ is a trigonometric polynomial in the angle variables and ε is a parameter.

For concreteness one can suppose that $\mathcal{H}$ describes the Escande-Doveil pendulum, [ED]: $\ell = 2$ and $f(\underline{\alpha}) = a\cos\alpha_1 + b\cos(\alpha_1 - \alpha_2)$, with $(a, b) \in \mathbb{R}^2$.

The solutions of the equations of motion describing the invariant tori for the system (1.1) with Diophantine rotation vector $\underline{\omega}$ can be parameterized as

$$\underline{\alpha} = \underline{\psi} + \underline{h}(\underline{\psi}; \varepsilon) , \qquad \underline{A} = \underline{A}_0 + \underline{H}(\underline{\psi}; \varepsilon) , \tag{1.2}$$

where $\underline{\psi} \in \mathbb{T}^\ell$, $\underline{A}_0 = J(\underline{\omega} - \underline{\omega}_0)$ and $\underline{h}, \underline{H}$, for ε small enough, are analytic functions, whose series expansions admit a graph representation in terms of trees; we refer to [GGM] for details and definitions.

One can consider also trees in which no resonances (see [GGM], §4) are allowed to appear but the perturbative parameter ε is replaced by $\eta_\varepsilon \equiv \varepsilon(1 - \sigma_\varepsilon)^{-1}$, with a suitable (matrix) *form factor* σ_ε taking into account the resummation of all resonances, [GM], and denote by $\underline{h}^*, \underline{H}^*$ the functions so obtained; the series expansion in η will be called *resummed series.*

If ρ is the radius of convergence of the series defining $\underline{h}, \underline{H}$ in (1.2) and $\varepsilon_c \in \mathbb{R}^+$ is the (positive) critical value at which the tori break down, one has $\varepsilon_c \geq \rho$; in general the analyticity domain in ε of the series defining $\underline{h}, \underline{H}$ is not a circle and it can happen that $\varepsilon_c > \rho$.

One can imagine that the following scenario arises: the singular behaviour of $\underline{h}(\underline{\psi}; \varepsilon), \underline{H}(\underline{\psi}; \varepsilon)$ as functions of ε is the same of that of $\underline{h}^*(\underline{\psi}; \eta_\varepsilon), \underline{H}^*(\underline{\psi}; \eta_\varepsilon)$ and while, for $\varepsilon \to \varepsilon_c^-$, σ_ε is still finite, $\eta = \eta_\varepsilon$ goes out the convergence domain (in η) of the series for $\underline{h}^*, \underline{H}^*$. If moreover the analyticity domain in η of $\underline{h}^*, \underline{H}^*$ turns out to be a circle this would mean that η is a more natural parameter for the invariant tori.

If this really happens then the behaviour of the series near the critical value is determined by the trees without resonances, for which the perturbative series would still be meaningful for ε near ε_c.

Under such an assumption a universal behaviour of the series near the critical value is proposed in [GGM]. Predictions of the value of the critical exponent δ are made, conjecturing that only some simple classes of trees are relevant: the linear trees presenting the largest possible number of small divisors (see [GGM], §5).

Considering (for concreteness purposes) a rotation vector $\underline{\omega} = (r, 1)$, where r is the golden mean (see §3.1 below), then one was led to expect that, by denoting by $\{q_n\}$ the denominators of the convergents defined by the continuous fraction expansion for r, a suitable function $Z(\Lambda_{q_n})$, defined in equation (5.2) of [GGM] and recalled in §3.1 below, satisfied the asymptotics

$$|Z(\Lambda_{q_n})| \approx \frac{C^{q_n}}{q_n^\delta} , \qquad C = C(\eta, f) , \tag{1.3}$$

for some positive constant $C(\eta, f)$ depending on ηf.

1.2. *About the standard and semistandard maps.* If true, the above mechanism should work also for the standard map, which is the dynamical system generated by the iteration of the area-preserving map of the cylinder to itself

$$\begin{cases} x' = x + y + \varepsilon \sin x \ , \\ y' = y + \varepsilon \sin x \ , \end{cases} \tag{1.4}$$

where $(x, y) \in \mathbb{T} \times \mathbb{R}$. At least it is generally assumed that area-preserving maps and Hamiltonian flows share the same critical behaviour: in particular they are expected to have the same critical exponent δ; see also [GGM], [CGJ] and comments therein about [M1], [M2].

The rôle of the functions $\underline{h}, \underline{H}$ is now played by two scalar functions u, v (see, for instance, [BG1]) such that

$$x = \alpha + u(\alpha, \varepsilon) \ , \qquad y = 2\pi\omega + v(\alpha, \varepsilon) \ , \tag{1.5}$$

with ω a fixed rotation number and u, v analytic in their arguments [1]. In terms of $\alpha \in \mathbb{T}$, the dynamics is a trivial rotation: $\alpha \to \alpha' = \alpha + 2\pi\omega$.

Analogously to the Hamiltonian case (1.1), one can define two functions u^*, v^*, obtained by considering only the values of the trees contributing to u, v without resonances (and replacing ε with η, where $\eta \equiv \eta_\varepsilon = \varepsilon(1 - \sigma_\varepsilon)^{-1}$ takes into account the resummation of resonances; note that now σ_ε is a scalar).

For the standard map, let us denote by $\rho(\omega)$ and $\varepsilon_c(\omega)$, respectively, the radius of convergence (in ε) and the critical value for fixed rotation number ω.

Consider also the semistandard map, introduced by Chirikov, [C],

$$\begin{cases} x' = x + y + \varepsilon' \exp ix \ , \\ y' = y + \varepsilon' \exp ix \ , \qquad \varepsilon' = \varepsilon/2 \ , \end{cases} \tag{1.6}$$

and call u_0, v_0 the corresponding conjugating functions; let us denote by $\rho_0(\omega)$ the radius of convergence (in ε') for the semistandard map[2].

1.3. *The implications of [D1].* Given a rotation number $\omega \in (0,1)$, let $B(\omega)$ be the function

$$B(\omega) = \sum_{n=0}^{\infty} \frac{\log q_{n+1}}{q_n} \ , \tag{1.7}$$

where $\{q_n\}$ are the convergents defined by the continued fraction expansion for the rotation number $\omega = [a_1, a_2, a_3, \ldots]$, so that $q_n = a_n q_{n-1} + q_{n-2}$, for $n \geq 1$, and $q_{-1} = 0$, $q_0 = 1$. The function (1.7) is related to the Bryuno function introduced by Yoccoz, [Y].

Consider the series defining $u^* \equiv u^*(\alpha, \eta)$. Of course one can write

$$u^*(\alpha, \eta) = \sum_{k=1}^{\infty} \eta^k \left(\sum_{\nu=-k}^{k} u_\nu^{*(k)} e^{i\nu\alpha} \right) \equiv \sum_{k=1}^{\infty} \eta^k u^{*(k)}(\alpha) \ . \tag{1.8}$$

The radius of convergence $R(\omega)$ of a series like (1.8) can be expressed as

$$R^{-1}(\omega) = \sup_{\alpha \in [0, 2\pi]} \limsup_{k \to \infty} \left| u^{*(k)}(\alpha) \right|^{1/k} = \limsup_{k \to \infty} \max_{|\nu| \leq k} \left| u_\nu^{*(k)} \right|^{1/k} \ , \tag{1.9}$$

as the equivalence between the two definitions has been shown in [D1].

On the other hand the coefficient in (1.8) with $\nu = k$ is the same for both the functions u and u^* (and one has $u_k^{(k)} = u_k^{*(k)} = u_{0k}^{(k)}$). Moreover it is easy to see that one has

$$\left| u_{0k}^{(k)} \right| \geq B_1^k e^{2B(\omega)k} \tag{1.10}$$

---

1 The functions u, v are trivially related: $v(\alpha,\varepsilon) = u(\alpha,\varepsilon) - u(\alpha - 2\pi\omega, \varepsilon)$.

2 We could define also the functions u_0^*, v_0^* and the critical value $\varepsilon_{0c}(\omega)$ for the semistandard map, but one has $u_0 = u_0^*, v_0 = v_0^*$, as the semistandard map has no resonances, and $\varepsilon_{0c}(\omega) = \rho_0(\omega)$: to deduce the latter property simply note that the semistandard map is invariant under rotation of ε' in the complex plane.

for k large enough and some constant B_1; again see [D1].

This implies that

$$\limsup_{k\to\infty} \max_{|\nu|\le k} \left| u_\nu^{*(k)} \right|^{1/k} \ge B_1 e^{2B(\omega)} , \qquad (1.11)$$

so that $R(\omega) \le B_1^{-1} e^{-2B(\omega)}$, *i.e.* the radius of convergence in η of the function u^* at best should be of order of $\rho_0(\omega)$ [3].

1.4. *The implications of [D2].* One has $\rho_0(\omega) = C_1(\omega)e^{-2B(\omega)}$ with $C_1(\omega)$ satisfying the bound $C_1^{-1} < C_1(\omega) < C_1$, uniformly in ω, for a suitable constant C_1, [D1]. Also for the standard map one has the same dependence of the radius of convergence $\rho(\omega)$ on the function $B(\omega)$, with a possibly different function $C_1(\omega)$, always admitting a bound from below and from above: the bound from above is proven in [D1], by using the argument recalled in §1.3, and the one from below in [BG2].

In [D2] it is proven that for rotation numbers [4] $\omega = \gamma_n \equiv [n, n, n, \ldots]$, with n large enough, one has $C_2/n > \varepsilon_c(\gamma_n) > C_2^{-1}/n$, for some constant C_2; as $B(\omega) > \log n$, for $\omega = \gamma_n$, then $\varepsilon_c(\gamma_n) > C_2^{-1} n^{-1}$ and $\rho(\gamma_n) < C_1 n^{-2}$, (see [D2], end of section 5). Therefore one can choose n so large that $\rho(\gamma_n)/\varepsilon_c(\gamma_n)$ is smaller than any prefixed quantity.

Note that the difference between the radius of convergence and the critical value becomes relevant only for some rotation numbers (like the noble numbers γ_n, with n large enough). For instance for γ_1, according to numerical simulations, the analyticity domain appears (very) slightly stretched along the imaginary axis so that not only the two quantities are comparable, but even $\varepsilon_c(\gamma_1) = \rho(\gamma_1)$, [FL].

2. Discussion

2.1. *About Equation (1.3).* If the dominant contributions to u^*, v^* were given by summing only the values of trees defining the functions $Z(\Lambda_{q_n})$, as conjectured in [GGM], then the standard map and the semistandard map should have the same critical behaviour. In fact in terms of trees the semistandard map admits the same graph representation of the standard map, with the only (remarkable) difference that the mode labels have all the same signs; then the paths with the maximal number of small divisors in $Z(\Lambda_{q_n})$ are the same for both the standard map and the semistandard map.

This means that, accepting the above conjecture, one ought to expect that the radius of convergence (in ε) of the functions u_0, v_0 for the semistandard map and the radius of convergence (in η) of the functions u^*, v^* for the standard map should be equal to each other and, in particular, of size of $\rho_0(\omega)$.

Furthermore the results listed in §1.3 imply that the functions u^*, v^* can not have a radius of convergence larger than that of the functions u_0, v_0, also considering all trees contributing to them and not only the ones defining the function $Z(\Lambda_{q_n})$.

In other words the semistandard map should capture all the critical behaviour of the standard map: if so there would be contradiction with the results existing in literature. As a matter of fact we shall see in §3 that the behaviour (1.3) conjectured for $Z(\Lambda_{q_n})$ does not hold.

2.2. *About the behaviour of σ_ε near ε_c.* The idea that the critical behaviour can be studied through the series obtained be neglecting the resonances (that is by resumming them and defining a new parameter $\eta \equiv \eta_\varepsilon$, as briefly recalled in §1.1 and in §1.2) is presented in [GGM]. One could also interpret the numerical results of [CGJ], [CGJK] as supporting such an idea: the fact that the KAM iteration represents the good transformation to look at also far from

[3] In the same way one finds that the radius of convergence in ε of the function u at best should be of order of $\rho_0(\omega)$.

[4] With the notations in §1.1, one has $r{=}\gamma_1$.

the KAM analyticity domain could suggest that a perturbative approach to the study of the breakdown of KAM invariant tori in some sense should be possible: this was the idea underlying the analysis performed in [GGM]. Anyway much work has still to be done in this direction; see also §4 below.

From the results listed in §1 the following picture emerges. For ε small enough $|\sigma_\varepsilon| < R|\varepsilon|$ for some constant R; see the theorem in [GGM], §4. When ε grows along the real axis toward the critical value, the possibility to have σ_ε still finite for $\varepsilon = \varepsilon_c$ is consistent with a perturbative approach if one of the following two cases arise:
(1) if σ_ε is finite and smooth in ε for $\varepsilon = \varepsilon_c$, then it can happen that $|\eta_c| \equiv |\varepsilon_c(1-\sigma_{\varepsilon_c})^{-1}|$ becomes equal to the radius of convergence $R(\omega)$ of the functions u^*, v^*;
(2) if σ_ε is finite and singular in ε for $\varepsilon = \varepsilon_c$, while $\eta_c = \varepsilon_c(1-\sigma_{\varepsilon_c})^{-1}$ is smaller than the radius of convergence $R(\omega)$ of the functions u^*, v^*, then the singularity of the tori shows up through the singular dependence of σ_ε on ε at the critical value.

In both cases the perturbative series for u^*, v^* can be used also near the critical value, because for any $|\varepsilon| < \varepsilon_c$ one has that η is smaller than the radius of convergence of the series.

In the case of $\omega = \gamma_n$, in principle two possibilities can be envisaged for the behaviour of σ_ε for ε near ε_c:
(i) either $|\sigma_\varepsilon| < 1$,
(ii) or else σ_ε negative and $|\sigma_\varepsilon| \geq 1$ [5].

Consider $\omega = \gamma_n$, with n large enough, and assume the results in [D2]. Then, if the analyticity domain in η was a circle with radius $R(\omega)$ and, for $\varepsilon \to \varepsilon_c^-(\omega)$, $|\sigma_\varepsilon| < 1$ and $\eta \to R(\omega)$, one should have $|\eta| = |\varepsilon(1-\sigma_\varepsilon)^{-1}| \to R(\omega) \approx \rho_0(\omega)$, hence $|1-\sigma_\varepsilon| \to |1-\sigma_{\varepsilon_c}| \approx \varepsilon_c(\omega)/\rho_0(\omega) > C_0 n$, with $C_0^{-1} = C_1 C_2$, which is incompatible with $|\sigma_\varepsilon| < 1$. So only the possibility (ii) above could be consistent both with [D2] and with the scenario proposed in [GGM], for $\omega = \gamma_n$.

More generally, for $\omega = \gamma_n$, with n large enough, if one had $\eta \to \eta_c \leq R(\omega)$ for $\varepsilon \to \varepsilon_c^-(\omega)$, then $|\sigma_\varepsilon|$ should become at least of order $e^{B(\omega)}$ for $\varepsilon \to \varepsilon_c^-(\omega)$, *i.e.* $|\sigma_\varepsilon| \to |\sigma_{\varepsilon_c}| \geq C_3(\omega) e^{B(\omega)}$, for some function $C_3(\omega)$ bounded uniformly in ω: in fact in this way one would have $|\eta_c| \equiv |\varepsilon_c(1-\sigma_{\varepsilon_c})^{-1}| \approx C_3 \varepsilon_c(\omega) e^{-B(\omega)} \approx C_4 \rho_0(\omega) = R(\omega)$, for some constants C_3, C_4, in case (1) and $|\eta_c| < C_4 \rho_0(\omega) = R(\omega)$ in case (2).

3. Analytic results

3.1. *The definition of* $Z(\Lambda_{q_n})$. If $r = (\sqrt{5}-1)/2$ is the *golden mean*, let us call $\{p_n/q_n\}$ the convergents of the continuous fraction expansion for r, where $\{p_n\}$ is the *Fibonacci sequence* defined by $p_{n+1} = p_n + p_{n-1}$ with $p_{-1} = 1$ and $p_0 = 0$, so that $p_n = q_{n+1}$ with $q_{-1} = 0$ and $q_0 = 1$.

The sequence of numbers $Z_n \equiv Z(\Lambda_{q_n})$ is defined in [GGM], equation (5.2), as sum of the values of a suitable class of trees which can be described by the family Λ_{q_n} of self-avoiding walks on $\mathbb{Z}^2$ starting at $(0,0)$, ending at $(q_n, -p_n)$ and contained in the strip $0 < x < q_n$, except for the left extreme points.

Then the numbers Z_n can be approximately defined by the recursive relation

$$Z_{n+1} = Z_n Z_{n-1} \left(\frac{\varepsilon_{n-1}}{\varepsilon_{n+1}} \right)^2 , \tag{3.1}$$

where, for consistency, we fix $Z_{-1} = r^2$ and $Z_0 = r^{-2}$, [GGM]. Set also $\varepsilon_n = q_n r - p_n$; then $\varepsilon_{-1} = -1$ and $\varepsilon_n = q_n r - q_{n-1}$.

[5] Note that σ_ε positive and $|\sigma_\varepsilon| \geq 1$ is not possible as the analyticity of σ_ε in ε for small ε would imply the existence of a value $\bar{\varepsilon}$ such that $|1-\sigma_{\bar{\varepsilon}}|=0$.

Instead of studying (3.1) define $\lambda_n = \log Z_n$, so that

$$\lambda_{n+1} = \lambda_n + \lambda_{n-1} + 2\left(\log \varepsilon_{n-1} - \log \varepsilon_{n+1}\right) , \tag{3.2}$$

with $\lambda_{-1} = 2\log r$ and $\lambda_0 = -2\log r$.

3.2. *Against the (1.3).* All the following identities can be easily proven by induction using the fact that r is the positive solution of $r^2 + r - 1 = 0$.

First note that $\varepsilon_n = (-1)^n r^{n+1}$. It follows that

$$\lambda_{n+1} = \lambda_n + \lambda_{n-1} - 4\log r . \tag{3.3}$$

which satisfies $\lambda_n = \mu_n + s_n$, where

$$\begin{cases} s_n = -4\log r \sum_{i=0}^{n-1} q_{n-1-i} , & \text{for } n \geq 1 , \\ s_0 = s_{-1} = 0 , \end{cases} \tag{3.4}$$

and μ_n is defined by $\mu_{n+1} = \mu_n + \mu_{n-1}$, with $\mu_{-1} = 2\log r$ and $\mu_0 = -2\log r$. It is now immediate that $\mu_n = -(2\log r)\, q_{n-2}$, for $n \geq 1$, so that we have

$$\lambda_n = -2\log r \left(q_{n-2} + 2\sum_{i=0}^{n-1} q_i \right) , \tag{3.5}$$

for $n \geq 1$.

Let now use that

$$q_n = \frac{r^{-(n+1)} - (-1)^{n+1} r^{n+1}}{r + r^{-1}} . \tag{3.6}$$

Inserting this expression in (3.4) we get, for $n \geq 1$,

$$\begin{aligned} \lambda_n &= \frac{2\log r^{-1}}{r + r^{-1}} \left(r^{-(n-1)} + 2\sum_{i=0}^{n} r^{-i} - (-1)^{n-1} r^{n-1} - 2\sum_{i=0}^{n} (-1)^i r^i \right) = \\ &= 2\log r^{-1} \left(r^{-(n+2)} - 2 + (-1)^{n+2} r^{n+2} \right) , \end{aligned} \tag{3.7}$$

which can be written as

$$\lambda_n = 2\log r^{-1} \left[(r + r^{-1}) q_{n+1} - 2 + 2(-1)^{n+2} r^{n+2} \right] , \tag{3.8}$$

which means

$$Z_n = r^4 r^{-2(r+r^{-1}) q_{n+1}} r^{-4(-1)^{n+2} r^{n+2}} . \tag{3.9}$$

This implies, using that

$$r^{n+1} = \frac{1}{r + r^{-1}} q_n^{-1} \left[1 + (-1)^n r^{2(n+1)} \right] , \tag{3.10}$$

the following asymptotics:

$$|Z(\Lambda_{q_n})| \approx K C^{q_{n+1}} \left(1 + (-1)^n d q_{n+1}^{-1} \right) , \tag{3.11}$$

with $K = r^{-4}$, $C = r^{-2(r+r^{-1})}$ and $d = 4(r + r^{-1})^{-1} \log r^{-1}$. This is clearly in contradiction with (1.3).

4. Conclusions

4.1. *Dominant contributions.* Our attention to [GGM] has been called back by the recent papers [CGJ], [CGJK], where the breakdown of KAM invariant tori (for two-dimensional Hamiltonian systems) is numerically studied through a renormalization group scheme, and by the results in [D2].

Even if it is not true that the terms defining $Z(\Lambda_{q_n})$ are the only relevant ones (as the results in §3 show), one can argue that the real dominant contributions are given by the trees having the mode labels accumulating near the resonant line (*i.e.* such that the small divisors are really as small as possible), but not necessarily belonging to the class described by Λ_{q_n}.

In fact let us compare the tree values for the Escande-Doveil pendulum with the ones for the standard map and accept that they have the same singular behaviour at the critical value (as it is generally believed, [ED]): the small divisors are respectively $(i\underline{\omega}\cdot\underline{\nu})^2$, $\underline{\nu}=(\nu_1,\nu_2)\in\mathbb{Z}^2$, and $2[\cos(2\pi\omega\nu)-1]$, $\nu\in\mathbb{Z}$. So one can note that they have the same smallness problem, with the only difference that for the Escande-Doveil pendulum one can have also small divisors which are not small at all (when $\underline{\nu}$ is nearly parallel to $\underline{\omega}$), while for the standard map one can approximate the quantity $2[\cos(2\pi\omega\nu)-1]$ with something of the form $(i(\omega\nu_1+\nu_2))^2$, with $\nu_1=\nu$ and ν_2 such that $|\omega\nu_1+\nu_2|\le 1$: in other words one can expect that the tree expansion for the standard map is very similar to that of the Escande-Doveil pendulum, but it does contain only the trees with the momenta directed along the resonant line. Of course this imply neither that in the case of linear trees the most dominant contributions are the ones with the mode labels having all the same signs nor that only the linear trees are relevant. As a matter of fact the analysis performed in §3 shows that making a so restrictive assumption leads to wrong results.

4.2. *Resummed series: smooth form factors.* In conclusion an overall behaviour like (1.3) could be still possible in principle, even if a larger class of trees ought to be taken into account.

Let us consider $\omega=\gamma_n$, with n large enough and use the results in §2.2. If for $\varepsilon\to\varepsilon_c$ one had $|\sigma_\varepsilon|<1$, then we have seen that the analyticity domain in η of u^*,v^* can not be a circle: rather, defining $\eta_c(\omega)$ as the equivalent of ε_c for the functions u^*,v^*, the ratio $\eta_c(\omega)/R(\omega)$ would be of the same order of the ratio $\varepsilon_c(\omega)/\rho(\omega)$, and no simplification could arise in considering the resummed series u^*,v^* instead of the original ones u,v. In other words η would not be a "natural" parameter.

Of course in the case of the standard map, deep cancelation mechanisms should intervene in order to enlarge the analyticity domain along the real axis of the series for u^*,v^* from $R(\omega)\approx\rho_0(\omega)$ to a quantity $\eta_c(\omega)$ of the same order of $\varepsilon_c(\omega)$. Analogous considerations could be made for the functions $\underline{h}^*,\underline{H}^*$ in the case of Hamiltonian flows.

On the contrary if, for $\varepsilon\to\varepsilon_c$, one had case (1), possibility (ii) – see §2.2 –, so that $\sigma_\varepsilon\to C_2(\omega)e^{B(\omega)}$, the convergence domain in η for u^*,v^* could still be a circle. Of course, if this is the case, the study of the factor form σ_ε could turn out to be a very difficult task: in fact it could become a nonperturbative problem, as the full dependence on ε would be required for $\varepsilon\to\varepsilon_c$.

A perturbative analysis could still be possible for some rotation number, for instance for $\omega=\gamma_1$, when the radius of convergence and the critical value are expected to be equal (see comments at the end of §1.4); anyway we have no evidence of this, so we leave it as an open problem.

4.3. *Resummed series: singular form factors.* Suppose now that at the critical value σ_ε is finite and singular. One can consider the resummed series for the form factor σ_ε, by expressing it as series in η, *i.e.* $\sigma_\varepsilon=F(\eta)=F(\varepsilon(1-\sigma_\varepsilon)^{-1})$, where the function F admits a graph representation in terms of trees without resonances and with ε replaced with η_ε. If for $\varepsilon=\varepsilon_c$ one has that η is inside (enough) the domain in which the perturbative series for the function F converges, then one can try to truncate the series to the first orders, so obtaining an (approximate) implicit equation for $\sigma\equiv\sigma_\varepsilon$: the solution should be singular at the value $\varepsilon=\varepsilon_c$. So the possibility of a perturbative study of the breakdown of KAM invariant curves has not to be excluded [6].

Further investigation (also from a numerical point of view) in this direction would be highly

[6] Note that for $\omega=\gamma_n$, with n large enough, one can still define the function $F(\eta)$: the problem is that for ε near to the critical value no truncation of the series would be meaningful.

profitable. Also a comparison between the values of the involved quantities for the standard map and the semistandard map would be very enlightening.

Acknowledgments. We want to thank the ESI in Vienna, where part of this work has been done, for hospitality. We also thank V. Mastropietro for interesting discussions and in particular G. Gallavotti for continuous enlightening suggestions and critical comments.

Appendix A1. Numerical analysis

A1.1. *Motivations.* Although the results of §3 are conclusive we report here on some numerical simulations that suggested us those results. We tried to fit λ_n with a slightly more general relation than (1.3), *i.e.* $\lambda_n = k+\delta\log q_n + cq_n$. The problem of this fit is that $q_n \simeq r^{-n}$,so that we have to compute all the involved quantities with a very high accuracy: if not the constant k and the "linear" term δ will be completely lost. We think that the way we did can be of some interest for the reader.

We used the standard Unix command *bc* that is able to execute computation written in a C-like programming language that operates in fixed point notation with number of arbitrary dimension and an arbitrary but prefixed precision. We compute c_i, k_i, δ_i has the best square fit of (3.1) using the value of λ_m with $m = (i-1)100$ to $m = i100$. The only problem is to choose the precision at which the computation are done in such a way to have a desired precision in the value of c_i, k_i, δ_i. This is what we will discuss in the next subsection.

A1.2. *About the precision.* Let p be our chosen precision, *i.e.* we make all operation with p significant digits after the point. This mean that at every elementary operation creates an error $O(10^{-p})$.

If we call θ_i the error due to this round off at step i it's easy to see that the accumulated error on λ_n is $\sum_{i=0}^{n-1}\theta_i' q_{n-1-i} = O(10^{-p}r^n)$. To compute the best quadratic fit we have to solve the equation $\mathbf{A}\vec{\chi} = \vec{v}$, where $\mathbf{A}$ is the symmetric matrix $\mathbf{A} = \sum_{i=0}^{n}\vec{Q}_n\otimes\vec{Q}_n$, with $\vec{Q}_n = (q_n, \log(q_n), 1)$, $\vec{v} = \sum_{i=0}^{n}\vec{Q}_n\lambda_n$ and $\vec{\chi} = (c,\delta,k)$. This mean that $\mathbf{A}_{1,1} \approx r^{-2n}$, while $\mathbf{A}_{1,2}, \mathbf{A}_{1,3} \approx r^{-n}$ and all the other entries $\mathbf{A}_{i,j}$ of $\mathbf{A}$, with $i \geq j$, are of order 1.

Observing that $(\mathbf{A}^{-1})_{i,j} = \mathbf{A}^{i,j}/\det\mathbf{A}$ and that $\det\mathbf{A} \approx r^{-2n} \pm 10^{-p}r^{-2n}$, we get

$$\begin{cases} (\mathbf{A}^{-1})_{1,1} \approx r^{2n} \pm 10^{-p}\ , \\ (\mathbf{A}^{-1})_{1,3}, (\mathbf{A}^{-1})_{1,2} \approx r^n \pm 10^{-p}r^{-n}\ , \\ (\mathbf{A}^{-1})_{2,2}\ ,\ (\mathbf{A}^{-1})_{2,3}\ ,\ (\mathbf{A}^{-1})_{3,3} \approx 1 \pm 10^{-p}r^{-2n}\ . \end{cases} \tag{4.1}$$

The above estimates with the fact that $\vec{v} \approx (r^{-2n}\pm 10^{-p}r^{-2n}, r^{-n}\pm 10^{-p}r^{-n}, r^{-n}\pm 10^{-p}r^{-n})$ implies an error on χ of order $10^{-p}r^{-2n}$.

A1.3. *Numerical results.* The numerical results that we get are summarized in the next table, which, in our opinion, requires no comments. According to our error analysis, we fixed the precision to $p = 2000\log_{10}\omega + 20$ in such a way that the reported numbers should be reliable within 20 digits. The agreement with the analytical evaluation reported in §2 is perfect.

n	c	δ	k
100	3.482081477708027334	-.000312407255699678	-1.915302823598781109
200	3.482081477708027334	-.000000000000000000	-1.924847300238413790
⋮	⋮	⋮	⋮
1000	3.482081477708027334	.000000000000000000	-1.924847300238413790

Table 1. Numerical results. The small deviations for i=1 is due to the fact that our fit for λ_n contains q_n instead of q_{n+1} (see §A1.1 for motivations). Values of n of the form n=100i, $i\in\mathbb{N}$, should be meant to be reported in the table, but no change in the results has been observed starting from n=200: so we can say that numerically the limit is reached for such a value of n.

References

[BG1] A. Berretti, G. Gentile: Scaling properties for the radius of convergence of a Lindstedt series: the standard map, *J. Math. Pure Appl.* **78** (1999), 159–176.

[BG2] A. Berretti, G. Gentile: Bryuno function and the standard map, preprint.

[C] V. Chirikov: A universal instability of many-dimensional oscillator systems, *Phys. Rep.* **52** (1979), 263–379.

[CGJ] C. Chandre, M. Govin, H.R. Jauslin: Kolmogorov-Arnold-Moser renormalization-group approach to the breakup of invariant tori in Hamiltonian systems, *Phys. Rev. E* **57** (1998), (2), 1536–1543.

[CGJK] C. Chandre, M. Govin, H.R. Jauslin, H. Koch: Universality of the breakup of invariant tori in Hamiltonian flows, *Phys. Rev. E* **57** (1998), (6), 6612–6617.

[D1] A.M. Davie: The critical function for the semistandard map, *Nonlinearity* **7** (1994), 219–229.

[D2] A.M. Davie: Renormalization for analytic area-preserving maps, unpublished.

[ED] D.F. Escande, F. Doveil: Renormalization method for computing the threshold of the large-scale stochastic instability in two degrees of freedom Hamiltonian systems, *J. Stat. Phys.* **26** (1981), 257–284.

[FL] C. Falcolini, R. de la Llave: Numerical calculation of domains of analyticity for perturbation theories with the presence of small divisors, *J. Stat. Phys.* **67** (1992), 645–666.

[GGM] G. Gallavotti, G. Gentile, V. Mastropietro: Field Theory and KAM tori, *Math. Phys. Electron. J.* **1** (1995), paper 5, 1–13.

[GM] G. Gentile, V. Mastropietro: Tree expansion and multiscale decomposition for KAM tori, *Nonlinearity* **8** (1995), 1–20.

[M1] R.S. MacKay: A renormalization approach to invariant circles in area-preserving maps, *Physica D* **7** (1983), 283–300.

[M2] R.S. MacKay: *Renormalization in area-preserving maps*, World Scientific, London (1993).

[Y] J.-C. Yoccoz: Théoreme de Siegel, nombres de Brjuno and pôlinomes quadratiques, *Astérisque* **231**, 3–88 (1995).

MPEJ

MATHEMATICAL PHYSICS ELECTRONIC JOURNAL

ISSN 1086-6655
Volume 5, 1999

Paper 5
Received: Jan 14, 1999, Revised: May 12, 1999, Accepted: Nov 18, 1999
Editor: P. Collet

LOCAL PERTURBATIONS OF ENERGY AND KAC'S RETURN TIME THEOREM

Y. LACROIX

Brest

To Gérard Rauzy on the occasion of his sixtieth birthday.

Abstract. We introduce the notion of local perturbations for normalized energies and study their effect on the level of equilibrium measures. Using coupling technics and Kac's return time theorem, we obtain some $\bar{d}$-estimates for the equilibrium measures. These reveal stability of certain energies under local perturbations. They also show how some weak-$\star$ convergence of equilibrium may be obtained in absence of $\| \|_\infty$-accuracy of the energies.

1. INTRODUCTION

This paper concerns Statistical Mechanics - Thermodynamic Formalism (see [9] for basics). However it entirely translates to Probability Theory, where it concerns chains with complete connections [2,3]. There the log of the local transitions for the chain is the normalized energy for Thermodynamics; they describe microscopical interactions for a system with many particles.

Given a transition function g, the pre-cited theories associate to it equilibrium measures. These are stationary and describe the macroscopical aspect of the system after a long time.

1991 *Mathematics Subject Classification.* 28D05, 60J10, 60G10, 82B30.

Key words and phrases. Equilibrium state, g-measure, perturbation, $\bar{d}$-distance.

The author acknowledges support from IML-CNRS in Marseille for attending to the conference "*From crystal to chaos*", Marseille, July 1998, and from both ECOS and DIM (Santiago), where the manuscript was brought to its present stage.

We shall restrict to the case of uniqueness of the equilibrium measure. For such a case to hold there is a relatively important literature - also studying the Ergodic Theoretic properties of the single measure - see e.g. [4]. For a given g we denote by μ_g the uniquely associated equilibrium measure (by hypothesis).

A natural question is the study of the sensitivity, or stability, of the system, for a given interaction g, under perturbation of g. Let $\tilde{g}$ be another local transition function. It always is a perturbation of g. Then the perturbation theory asks about the behavior of $\mu_g - \mu_{\tilde{g}}$ when $\tilde{g}$ is close to g. That is *what is the effect of microscopical perturbations on the macroscopical aspect of the system at equilibrium* ?

Now the meaning of "*$\tilde{g}$ is close to g*" has to be specified. Usually $\tilde{g}$ is thought of as being close to g when $\| g - \tilde{g} \|_\infty$ is small - which is relevant to the Perturbation Theory of Markov chains also : this is $\| \, \|_\infty$ **perturbation theory**.

In this note we introduce and study **local perturbation theory** : we think of $\tilde{g}$ *being close to g if $\mu_{\tilde{g}}(\{g \neq \tilde{g}\})$ is small* (cf. the first statement of Theorem 1).

For classical $\| \, \|_\infty$-perturbation the reader will find in [9] -using [11]- the corresponding basic stability results, and in [2,3] more quantitative aspects. Essentially under the hypothesis of uniqueness of equilibrium it follows that $\mu_{\tilde{g}} \underset{\text{weak-}\star}{\longrightarrow} \mu_g$ as $\| \tilde{g} - g \|_\infty \rightarrow 0$. Convergence in the $\bar{d}$-metric under additional regularity assumptions on g are detailed in [2,3].

However the technics developed to prove this stability under classical perturbation do not apply to the case of local perturbations. The reader will find in Example 3 of the following section a very simple example where this is illustrated.

We have proved in Theorem 1 the stability under local perturbations for certain energies. The proof required the introduction of new ingredients. The main novelty is the use of Kac's return time theorem [5,10], to overcome the absence of $\| \, \|_\infty$-accuracy for $\tilde{g} - g$. Otherwise we use now standard -still powerful- coupling technics [1,2,3,4], [12]. Let us mention that in [7] another technic -of algebraic nature- is developed to prove stability.

We stress that our result also proves that there is no hope to detect empirically strong local variations of the law of a process, producing a time series - whence to determine its law.

The paper is organized as follows. Section 2 introduces the basic notations, presents in Theorem 1 the main result. In Example 3 we choose the simplest case to show how our stability result applies while the classical perturbation results do not.

In Section 3 we briefly sketch our proof. The final Section contains the details of the different steps used to prove Theorem 1.

The author would like to thank M. Babillot for her interest, and providing references [2,3]; also thanks to J. Buzzi for encouragements, and the anonymous referee for making constructive comments about a very preliminary draft of this paper.

2. Notations and statement of result

Our results are presented within the simplest framework as generalizations to larger alphabets and/or sub-shifts of finite type are somewhat easy to guess, but would induce loss of clarity in the exposition.

2.1 : the shift, $\mathcal{G}$-functions, g-measures, and ergodicity.

Let $X = \{0,1\}^{\mathbb{N}}$, and $S : X \to X$ its usual one-sided shift map. This is a covering transformation in the sense of Keane [6]. He proves for such that invariant Borel probability measures and so-called $\mathcal{G}$-functions are intimately related. Now from the Thermodynamical view-point the log of $\mathcal{G}$-functions are normalized energies.

To get something working usual is to restrict generality and define the set of $\mathcal{G}$-functions as follows : we let X have product topology and

$$\mathcal{G} = \{g : X \to]0,1[: g \text{ continuous and for any } x \in X, \sum_{y \in S^{-1}x} g(y) = 1\}.$$

It is standard that such gs are local transitions for chains with complete connections - see e.g. [2].

Let $M(X)$ denote the set of Borel probability measures on X. Let $\mathcal{C}(X)$ denote the set of real valued continuous maps on X, endowed with $\| \|_\infty$. A $g \in \mathcal{G}$ defines a transfer operator $\mathcal{L}_g$ acting continuously, linearly, positively, and contracting $\| \|_\infty$ on $\mathcal{C}(X)$:

$$\mathcal{L}_g f(x) = \sum_{y \in S^{-1}x} g(y) f(y), \quad x \in X.$$

Since moreover $\mathcal{L}_g 1 = 1$, its dual acts on $M(X)$: now $M(X)$ is compact convex whence the Schauder - Tychonov fixed point theorem yields that it has a fixed point.

Such a μ is an *g-measure or equilibrium measure* : since $\mathcal{L}_g(f \circ S) = f$, it must be a S-invariant one. From the Statistical Mechanical view-point, the μ describes macroscopical evolution while the g does so for the microscopical one.

When several such μ exist we speak of phase transition while otherwise we shall agree to say that g is ergodic. If it is, we denote the only μ by μ_g. This will be the case by hypothesis.

2.2 : Variations, and (pseudo)-distances for measures.

For $g, \tilde{g}$ ergodic, if $g \neq \tilde{g}$, then μ_g and $\mu_{\tilde{g}}$ are singular, hence the variation norm is bad to measure $\mu_g - \mu_{\tilde{g}}$. From the macroscopical point of view, measures of cylinders make sense : let

$$w = (w_0, \ldots, w_{|w|-1}) \in \{0,1\}^{|w|}$$

be a word of length $|w|$. Let $[w]$ be the set of $x \in X$ that have their first $|w|$ coordinates equal to those of w.

Then for given $m \geq 1$, we have two probability vectors :

$$\pi_m(g) = (\mu_g([w]))_{|w|=m},$$

and $\pi_m(\tilde{g})$ is defined similarly. Then we can measure

$$\bar{d}_m(\mu_g, \mu_{\tilde{g}}) = \| \pi_m(g) - \pi_m(\tilde{g}) \|_1 = \sum_{|w|=m} |\mu_g([w]) - \mu_{\tilde{g}}([w])|.$$

Ornstein [8] proved fruitful the $\bar{d}$ distance :

$$\bar{d}(\mu_g, \mu_{\tilde{g}}) = \inf\{\nu([(0,1)] \cup [(1,0)]) : \nu \in J(\mu_g, \mu_{\tilde{g}})\},$$

where $J(\mu_g, \mu_{\tilde{g}})$ is the set of joinings between μ_g and $\mu_{\tilde{g}}$, that is the $S \times S$-invariant probability measures on $X \times X$ that go for the first natural projection to μ_g and for the second to $\mu_{\tilde{g}}$.

We shall evaluate both using $\bar{d}_m$ and $\bar{d}$: it holds that $\bar{d}_m \leq m\bar{d}$ (cf. Lemma 5).

Conditions ensuring ergodicity require more than continuity. One is that of summable variations. We will need variations later : put

$$var_m(g) = \max_{|w|=m} \sup_{x,y\in[w]} |g(x) - g(y)|.$$

2.3 : Statement of results.

We assume $0 < \lambda < 1/2$ is such that $g, \tilde{g} > \lambda$. We let μ_λ denote the Bernoulli measure $\mathcal{B}(2\lambda, 1-2\lambda)$ on X. We denote by

$$X_p = \{z \in X : 1 \leq l \leq p-1 \Rightarrow z[l, l+m-1] \neq 0^m\}.$$

We let $m \geq 1$, we let E be an at most countable index set, and for each $i \in E$, p_i denotes an integer and $F_i \subset \{0,1\}^{m+p_i}$. We assume in the second statement of Theorem 1 below that

$$\Delta := \{g \neq \tilde{g}\} \subset \cup_i \cup_{v\in F_i} [v].$$

Theorem 1. *For any $K \geq m+1$,*

$$\max\{\bar{d}_m(\mu_g, \mu_{\tilde{g}}), \bar{d}(\mu_g, \mu_{\tilde{g}})\}$$
$$\leq 2\left[(K-1)\left(var_{m+1}(g) + (1-2\lambda)\mu_{\tilde{g}}(\Delta)\right) + \textstyle\sum_{k>K}(k-K)\mu_\lambda(X_k)\right].$$

As a consequence, with $K = um+1$, $u \geq 1$,

$$\max\{\bar{d}_m(\mu_g, \mu_{\tilde{g}}), \bar{d}(\mu_g, \mu_{\tilde{g}})\}$$
$$\leq 2m\left[u\left(var_{m+1}(g) + (1-2\lambda)(\textstyle\sum_i \#F_i(1-\lambda)^{m+p_i}\right) + m\frac{(1-(2\lambda)^m)^{u-1}}{(2\lambda)^{2m}}\right].$$

Comments 2. *In the first estimate of Theorem 1, we see a quantitative meaning of "$\tilde{g}$ is a small local perturbation of g" : that is $\mu_{\tilde{g}}(\{\tilde{g} \neq g\})$ is small.*

Under the assumption of the second statement ($\{\tilde{g} \neq g\} \subset \cup_{i \in E} \cup_{v \in F_i} [v]$), we have another illustration of this : enough is that $\sum_i \# F_i (1-\lambda)^{m+p_i}$ be small.

To deduce stability under local perturbations -for certain energies- from Theorem 1, second statement, take, for example, a g such that $var_{m+1}(g) = 0$ (that is g depends only on the first $m+1$ coordinates). Then pick a very large u first to make $m^2 u \frac{(1-(2\lambda)^m)^{u-1}}{(2\lambda)^{2m}}$ small ; then take $\tilde{g}$ so that $um\mu_{\tilde{g}}(\{\tilde{g} \neq g\})$ is small also, and conclude that $\bar{d}(\mu, \mu_{\tilde{g}})$ is small (Example 3 below falls into that case).

We emphasize here that the produced bounds require the sequence $(var_m(g))_{m \geq 1}$ to decrease very rapidly to 0. This is due to the nature of the proof : our result produces universal bounds for ergodic g satisfying $\lambda \leq g \leq 1-\lambda$. Now as revealed in [7] a proof technic keeping more information on g yields better bounds.

Now we produce the simplest example we found that illustrates how Theorem 1 even gives some weak-$\star$ convergence of equilibrium, and for which classical theory does not apply because $\| \tilde{g} - g \|_\infty$ remains bigger than or equal to 1/4.

Example 3. *Consider $\Pi : x \mapsto \sum_{i \geq 0} \frac{x_i}{2^{i+1}} \bmod 1$. This is the factor map to the transformation $M : x \mapsto 2x \bmod 1$. By [6] the same notions of $\mathcal{G}$-functions and equilibrium measures can be developed for M on the torus. Call $\mathcal{G}_M$ the corresponding set. Then any $g \in \mathcal{G}_T$ is such that $g \circ \Pi \in \mathcal{G}$. But the reverse is false : take Bernoulli measure $\mu_0 = \mathcal{B}(\frac{3}{4}, \frac{1}{4})$. It corresponds to $g_0 \in \mathcal{G}$ with $g_0(0x) = 3/4$ and $g_0(1x) = 1/4$.*

Consider $g_\eta \in \mathcal{G}_M$ to be such that for $0 < \eta < 1/4$, $g_\eta \circ \Pi$ and g_0 coincide outside balls of radius η centered at 0^∞, 1^∞, 01^∞ and 10^∞, and otherwise let g_η be ergodic and $\geq 1/4$.

Then though $\| g_0 - g_\eta \circ \Pi \|_\infty \geq 1/4$ for all η, our result shows that as $\eta \to 0$,

$$\mu_{g_\eta} \overset{weak\text{-}\star}{\longrightarrow} \mu_{g_0},$$

even in the $\bar{d}$-distance. Hence weak-$\star$ convergence may hold in absence of $\| \ \|_\infty$ accuracy.

3 : Sketch of proof of Theorem 1

The proof of Theorem 1 is quite simple and develops along the following four steps :

($\bullet^1$) : let $Y = X \times X$, $T = S \times S$, and pick a $\tau \in \mathcal{G}_T$ which is a $\mathcal{G}_T$-function for (Y, T) and satisfies for any $(x, y) \in Y$ and $i, j = 0, 1$,

$$\begin{cases} \sum_i \tau(ix, jy) = \tilde{g}(jy), \\ \sum_j \tau(ix, jy) = g(ix). \end{cases}$$

Then by ergodicity assumption, and these two properties, any ν such that $\mathcal{L}_\tau^\star \nu = \nu$ belongs to $J(\mu_g, \mu_{\tilde{g}})$. Here pay attention to choose τ charging the entry to diagonal as much as possible.

$(\bullet^2)$: pick a τ-measure ν, and define for $q \geq 1$,

$$A_q = \{(x, x') \in Y : i < q \Rightarrow x_i = x'_i\}.$$

Then show that

$$\begin{cases} \bar{d}_q(\mu_g, \mu_{\tilde{g}}) \leq 2\nu(A_q^c), \\ \bar{d}(\mu_g, \mu_{\tilde{g}}) \leq \nu(A_q^c). \end{cases}$$

In the sequel we estimate on $\nu(A_m^c)$, for $m \geq 1$ as in Theorem 1.

$(\bullet^3)$: introduce $n_m(z) = \min\{k \geq 1 : T^k z \in A_m\}$. First show, using attractivity - [4], that $\nu(\{n_m < \infty\}) = 1$, whence deduce by Kac's return time theorem [5] that

$$\sum_{k \geq 1} k\nu(A_m \cap \{n_m = k\}) = 1.$$

Remark 4. *There always exists an ergodic τ-measure ν, for which $\nu(\{n_m < \infty\}) = 1$ as soon as $\nu(\{n_m < \infty\}) > 0$, using invariance, and ergodicity. However we think the proof that this holds true for any τ-measure (for chosen τ) is interesting enough, and relevant for the understanding of the treatment of the tail series in $(\bullet^4)$.*

$(\bullet^4)$: observe that $\nu(A_m \cap \{n_m = 1\}) \leq \nu(A_m)$, and using stationarity of ν that for $k \geq 2$, $\nu(A_m \cap \{n_m = k\}) \geq \nu(A_m^c \cap T^{-1}A_m)$. Conclude using attractivity again that the tail series in $(\bullet^3)$ is small, that its first term $(k = 1)$ is about $\nu(A_m)$, and that the intermediate terms are each about $\nu(A_m^c \cap T^{-1}A_m)$: whence if this one is really small we get estimates on the effect on equilibrium of local perturbations of g.

4 : Proof of Theorem 1.

4.1 : The maximal coupling - $(\bullet^1)$.

We let $\pi_1(x, y) = x$ and $\pi_2(x, y) = y$. Define the $\mathcal{G}_T$-function τ for (Y, T) - continuous but not strictly positive - by (cf. [2])

$$\tau(ix, jy) = \begin{cases} \min\{g(ix), \tilde{g}(jy)\} \text{ if } i = j; \\ \frac{(g(ix)-\tilde{g}(iy))^+(\tilde{g}(jy)-g(jx))^+}{(g(0x)-\tilde{g}(0y))^+ + (g(1x)-\tilde{g}(1y))^+} \text{ if } i \neq j \text{ and } g(1x) \neq \tilde{g}(1y); \\ 0 \text{ otherwise.} \end{cases}$$

A few minutes require to check out the required properties of this τ as stated in $(\bullet^1)$. Notice that

$$\tau(1x, 1y) + \tau(0x, 0y) \geq 2\lambda.$$

By [6] (same argument as the one sketched in the introduction) there is at least one τ-measure. Pick one and call it ν.

By ergodicity assumptions on g and $\tilde{g}$, and the properties $(\bullet^1)$, $\pi_1\nu = \mu_g$ and $\pi_2\nu = \mu_{\tilde{g}}$: whence we have the following diagram of measure-theoretical factors :

$$\begin{array}{ccc} & (Y, T, \nu) & \\ & \pi_1 \swarrow \quad \searrow \pi_2 & \\ (X, S, \mu_g) & & (X, S, \mu_{\tilde{g}}) \end{array}$$

4.2 : Variation inequalities - ($\bullet^2$).

We let $A_q = \{(x,y) \in Y : i < q \Rightarrow x_i = y_i\}$, as in ($\bullet^2$), and relate the distance between μ_g and $\mu_{\bar{g}}$ to the quantity $\nu(A_q)$:

Lemma 5. *Let $f \in \mathcal{C}(X)$ and $q \geq 1$. Then*

$$|\mu_g(f) - \mu_{\bar{g}}(f)| \leq 2 \parallel f \parallel_\infty (1 - \nu(A_q)) + var_q(f)\nu(A_q).$$

Hence it follows that

$$\bar{d}_q(\mu_g, \mu_{\bar{g}}) \leq 2(1 - \nu(A_q)).$$

Notice also that since $\nu(A_q) \leq \nu(A_1)$, it follows by ($\bullet^1$) that

$$\bar{d}(\mu_g, \mu_{\bar{g}}) \leq 1 - \nu(A_q).$$

Finally, $\bar{d}_q \leq 2q\bar{d}$.

Proof. We decompose along cylinders and use ($\bullet^1$) to go from integrating on X to integrating on Y $(= X \times X)$:

$$\begin{aligned} |\mu_g(f) - \mu_{\bar{g}}(f)| &\leq \textstyle\sum_{|w|=q} |\int_{[w]} f d\mu_g - \int_{[w]} f d\mu_{\bar{g}}| \\ &= \textstyle\sum_{|w|=q} |\int_{[w]\times X} f \circ \pi_1 d\nu - \int_{X\times[w]} f \circ \pi_2 d\nu| \\ &\leq \textstyle\sum_{\substack{|v|=|w|=q, \\ v\neq w}} \int_{[w]\times[v]} |f \circ \pi_1 - f \circ \pi_2| d\nu \\ &\qquad + \textstyle\sum_{|w|=q} \int_{[w]\times[w]} |f \circ \pi_1 - f \circ \pi_2| d\nu \\ &\leq 2 \parallel f \parallel_\infty (1 - \nu(A_q)) + var_q(f)\nu(A_q). \end{aligned}$$

For the second statement we take f constant on cylinders of length q, and on each such $[w]$ equal to $sign(\mu_g([w]) - \mu_{\bar{g}}([w]))$.

For the third observation of the lemma we notice additionally that $\bar{d}(\mu_g, \mu_{\bar{g}}) \leq 1 - \nu(A_1)$.

Next $\bar{d}_q \leq 2\nu(A_q^c) \leq 2\sum_{i=0}^{q-1} \nu(T^{-i}A_1^c) = 2q\nu(A_1^c)$ by stationarity of ν. Passing to the infimum over $J(\mu_g, \mu_{\bar{g}})$ in this last inequality, we deduce the last statement of the Lemma. ■

4.3 : Attractivity and Kac's return time theorem - ($\bullet^3$).

Define $\delta : Y \to X$ by

$$\delta(x,y) = (|x_i - y_i|)_{i\geq 0}.$$

This is a shift commuting topological factor map.

On X, define the partial order $x \preceq y \Leftrightarrow \forall i,\ x_i \leq y_i$. Say an $f : X \to \mathrm{R}$ is **increasing** if

$$x \preceq y \Rightarrow f(x) \leq f(y).$$

We define μ_λ as in Theorem 1. Then $g_\lambda(1.)$ is increasing (constant) and satisfies as already noticed

$$\sum_{i\neq j} \tau(ix, jy) \leq 1 - 2\lambda = g_\lambda(1\delta(x,y)).$$

Using δ and $\preceq$, heavily inspired by [4, Lemma 4.1], we have the following :

Lemma 6. *Let $\Phi \in \mathcal{C}(Y)$, $f \in \mathcal{C}(X)$ be increasing and together satisfy*

$$\Phi(x,y) \leq f(\delta(x,y)).$$

Then for all $n \geq 0$, $\mathcal{L}_\tau^n \Phi(x,y) \leq \mathcal{L}_{g_\lambda}^n f(\delta(x,y))$, hence

$$\nu(\Phi) \leq \mu_\lambda(f).$$

Remark 7. *The same conclusion holds for g_λ replaced by an ergodic $\bar{g} \in \mathcal{G}$ such that $\bar{g}(1.)$ is increasing - use a remark on the characterization of ergodicity in [4].*

Proof. Set $\tau_{x,y}(i \neq j) = \sum_{i \neq j} \tau(ix, jy)$, and let $\tau_{x,y}(i = j) = 1 - \tau_{x,y}(i \neq j)$. Then compute using the hypothesis of the Lemma and that $g_\lambda(1.)$ increases :

$$\begin{aligned}
\mathcal{L}_\tau \Phi(x,y) &= \left\{\textstyle\sum_{i \neq j} + \sum_{i=j}\right\} \tau(ix, jy)\Phi(ix, jy) \\
&\leq \tau_{x,y}(i \neq j) f(1\delta(x,y)) + \tau_{x,y}(i = j) f(0\delta(x,y)) \\
&= \tau_{x,y}(i \neq j) f(1\delta(x,y)) + (g_\lambda(1\delta(x,y)) - \tau_{x,y}(i \neq j)) f(0\delta(x,y)) \\
&\qquad + g_\lambda(0\delta(x,y)) f(0\delta(x,y)) \\
&\leq \mathcal{L}_{g_\lambda} f(\delta(x,y)).
\end{aligned}$$

By [4, Lemma 2.1], $\mathcal{L}_{g_\lambda} f^n$ is increasing for any $n \geq 0$, whence repeated application of the preceding computations yield that for any such n,

$$\mathcal{L}_\tau^n \Phi(x,y) \leq \mathcal{L}_{g_\lambda}^n f(\delta(x,y)).$$

By [6], since g_λ is Lipschitz, we have that $\mathcal{L}_{g_\lambda}^n f(z) \to \mu_\lambda(f)$ uniformly in z as n goes to infinity. Whence

$$\nu(\Phi) = \nu(\mathcal{L}_\tau^n \Phi) \leq \nu(\mathcal{L}_{g_\lambda}^n f) \to \nu(\mu_\lambda(f)) = \mu_\lambda(f).$$

■

Now we pass to Kac's theorem : we take notations from ($\bullet^3$). Then

$$\nu(A_m) \geq \nu([0^m] \times [0^m]) \geq (2\lambda)^m > 0.$$

Hence Kac's theorem applies :

Theorem 8 (Kac). *[5,10]. $\sum_{k \geq 1} k\nu(A_m \cap \{n_m = k\}) = \nu(n_m < \infty)$.*

We may use attractivity (Lemma 6) to prove that

Lemma 9. $\nu(n_m < \infty) = 1$.

Proof. Take $q \geq 1$ and put

$$\begin{cases} Y(q) = \{n_m \geq qm+1\}; \\ X(q) = \{x \in X : 1 \leq u \leq m \Rightarrow x[um,(u+1)m[\neq 0^m\}. \end{cases}$$

Then define $\Phi_q(x,y) = \mathbf{1}_{Y(q)}(x,y)$ and $f_q(z) = \mathbf{1}_{X(q)}(z)$. It follows that Φ_q and f_q satisfy the conditions for Lemma 6 to apply : whence $\nu(Y(q)) \leq \mu_\lambda(X(q))$, but since μ_λ is Bernoulli with parameter $1-2\lambda$, we get

$$\nu(Y(q)) \leq (1-(2\lambda)^m)^q.$$

Now $\{n_m = \infty\} \subset Y(q)$ whence $0 \leq \nu(\{n_m = \infty\}) \leq \liminf \nu(Y(q)) = 0$. ∎

4.4 : End of Proof of Theorem 1 - ($\bullet^4$).

By stationarity, $\nu(A_m) = \nu(T^{-1}A_m) = \nu(A_m \cap \{n_m = 1\}) + \nu(A_m^c \cap T^{-1}A_m)$. Hence we deduce from ($\bullet^3$) that

$$(\star^1) \qquad \nu(A_m^c) = \sum_{k\geq 2} k\nu(A_m \cap \{n_m = k\}) \; - \nu(A_m^c \cap T^{-1}A_m).$$

Using stationarity of ν and Lemma 9 it follows that for $k \geq 2$,

$$(\star^2) \qquad \nu(A_m^c \cap T^{-1}A_m) = \sum_{k\geq 2} \nu(A_m \cap \{n_m = k\}).$$

Moreover, $A_m \cap \{n_m = k\} = \emptyset$ for $k < m+1$. From ($\star^{1,2}$) we get

$$\nu(A_m^c) = m\nu(A_m^c \cap T^{-1}A_m) + \sum_{k\geq m+2} (k-m-1)\nu(A_m \cap \{n_m = k\}).$$

Now for $K \geq m+1$, the same argument with "$\leq$" instead of "$=$" yields to

$$(\star^3) \qquad \nu(A_m^c) \leq (K-1)\nu(A_m^c \cap T^{-1}A_m) + \sum_{k>K} (k-K)\nu(A_m \cap \{n_m = k\}).$$

Estimating $\nu(A_m^c \cap T^{-1}A_m)$.

Remember $\Delta = \{g \neq \tilde{g}\}$ ($= \cup_i \cup_{v\in F_i} [v]$ in the second statement of Theorem 1). Then decompose

$$\begin{aligned} A_m &= \cup_{|w|=m}(([w] \times ([w] \cap \Delta)) \cup ([w] \times ([w] \setminus \Delta))) \\ &= (A_m \cap (X \times \Delta)) \cup (A_m \cap (X \times (X \setminus \Delta))). \end{aligned}$$

Let us first consider $(x,y)\in A_m\cap(X\times(X\setminus\Delta))$: then $|g(ix)-\tilde{g}(iy)|\le var_{m+1}(g)$, whence $\sum_{i=j}\tau(ix,jy)\ge 1-var_{m+1}(g)$ and $\sum_{i\ne j}\tau(ix,jy)\le var_{m+1}(g)$. Hence

$$(\star^4)\qquad \begin{array}{rcl}\nu(A_m^c\cap T^{-1}A_m) &=& \int_{A_m}\sum_{i\ne j}\tau(ix,jy)d\nu(x,y)\\ &=& \int_{A_m\cap(X\times\Delta)}\sum_{i\ne j}\tau(ix,jy)d\nu(x,y)\\ && \quad +\int_{A_m\cap(X\times(X\setminus\Delta))}\sum_{i\ne j}\tau(ix,jy)d\nu(x,y)\\ &\le& var_{m+1}(g)\nu(A_m)+(1-2\lambda)\mu_{\tilde{g}}(\Delta)\\ &\le& var_{m+1}(g)+(1-2\lambda)\mu_{\tilde{g}}(\Delta)\\ (&\le& var_{m+1}(g)+(1-2\lambda)\big(\sum_i \#F_i(1-\lambda)^{m+p_i}\big)).\end{array}$$

Estimating the tail series in $(\star^3)$.

Define $B_k=T^{-1}A_m^c\cap\ldots\cap T^{-k+1}A_m^c$. Then $A_m\cap\{n_m=k\}\subset B_k$.

Next set $Z_k=\{z\in X: 1\le p\le k-1\Rightarrow z[p,p+m[\ne 0^m\}$. Denote $\Psi_k=\mathbf{1}_{B_k}$ and $f_k=\mathbf{1}_{X_k}$. Then they satisfy conditions for Lemma 6 and therefore we obtain

$$\nu(A_m\cap\{n_m=k\})\le\nu(B_k)\le\mu_\lambda(X_k).$$

Combining this with $(\star^{3,4})$, and Lemma 5, we deduce the first statement of Theorem 1.

To get the second one we first choose $K=um+1$ for some $u\ge 1$. By stationarity we get that for each $v\ge u$,

$$\sum_{t=1}^{m}\mu_\lambda(X_{vm+t+1})\le m\mu_\lambda(T^{-m}X_{vm+1})=m\mu_\lambda(X_{vm+1}).$$

We input this in the tail series of $(\star^3)$ to get with $(\star^4)$ that

$$(\star^5)\qquad \begin{array}{rcl}\nu(A_m^c) &\le& um\big(var_{m+1}(g)+(1-2\lambda)(\sum_i\#F_i(1-\lambda)^{m+p_i})\big)\\ && \quad +\sum_{v\ge u}m^2(v-u+1)\mu_\lambda(X_{vm+1}).\end{array}$$

To end with we compute that by definition of μ_λ,

$$\mu_\lambda(X_{vm+1})\le(1-(2\lambda)^m)^v,\quad\text{and}\quad \sum_{v\ge u}(v+1-u)m^2(1-(2\lambda)^m)^v=m^2\frac{q^u}{q(1-q)^2}$$

with $q=1-(2\lambda)^m$. ∎

References

[1] H. Berbee, *Chains with infinite connections: Uniqueness and Markov representation.*, Probab. Theory Relat. Fields **76** (1987), 243–253.

[2] X. Bressaud, R. Fernàndez & A. Galves, *Speed of $\bar{d}$-convergence for Markov approximations of chains with complete connections. A coupling approach,* To appear, Stochastic Processes & Appl. (1999).

[3] X. Bressaud, R. Fernàndez & A. Galves, *Decay of correlations for non Hölderian dynamics. A coupling approach.*, Electronic Journal of Probability **4** (1999), 1–19.
[4] P. Hulse, *A class of unique g-measures*, Erg. Th. & Dyn. Syst. **17** (1997), 1383–1392.
[5] M. Kac, *On the notion of recurrence in discrete stochastic processes*, Bull. A.M.S. **53** (1947), 1002–1010.
[6] M. Keane, *Strongly mixing g-measures*, Invent. Math. (1972), 309–324.
[7] Y. Lacroix, *Une remarque sur la perturbation ⋆-faible des g-mesures*, Submitted (1999).
[8] D.S. Ornstein, *Ergodic theory, randomness, and dynamical systems*, Yale Mathematical Monographs, vol. 5, Yale University Press, 1974.
[9] D. Ruelle, *Thermodynamic formalism*, Encyclopedia of Mathematics and its Applications, vol. 5, 1978.
[10] B. Saussol, *Etude statistique de systèmes dynamiques dilatants*, PHD Thesis, Université de Toulon et du Var, 1998.
[11] P. Walters, *A variational principle for the pressure of continuous transformations*, Amer. J. Math. **97** (1976), 937–971.
[12] L.-S. Young, *Recurrence times and rates of mixing*, Israel J. Math. **110** (1999), 153–188.

U.B.O., Fac. des Sciences et Techniques, Departement de Maths, 6 Av. V. Le Gorgeu, B.P. 809, 29285 Brest Cedex, France.
E-mail address: lacroix@univ-brest.fr

M P E J

MATHEMATICAL PHYSICS ELECTRONIC JOURNAL

ISSN 1086-6655
Volume 5, 1999

Paper 6
Received: Sep 21, 1999, Accepted: Dec 7, 1999
Editor: R. de la Llave

STABILITY OF THE BROWN-RAVENHALL OPERATOR

GEORG HOEVER AND HEINZ SIEDENTOP

ABSTRACT. The Brown-Ravenhall Hamiltonian is a model for the behavior of N electrons in a field of K fixed nuclei having the atomic numbers $\mathbf{Z} = (Z_1, \dots, Z_K)$, which is written, in appropriate units, as

$$B = \Lambda_{+,N} \left(\sum_{n=1}^{N} D_0^{(n)} + \alpha V_c \right) \Lambda_{+,N}$$

acting on the N-fold antisymmetric tensor product $\mathfrak{H}_N$ of $\Lambda_+(L^2(\mathbb{R}^3) \otimes \mathbb{C}^4)$, where $D_0^{(n)}$ denotes the free Dirac operator D_0 acting on the n-th particle, Λ_+ denotes the projection onto the positive spectral subspace of D_0, $\Lambda_{+,N}$ the projection onto $\mathfrak{H}_N$ and the potential V_c is the usual Coulomb interaction of the particles, coupled by the constant α. It is proved in the massless case that for any $\gamma < 2/(2/\pi + \pi/2)$ there exists an α_0 such that for all $\alpha < \alpha_0$ and $\alpha Z_k \leq \gamma$ $(k = 1, \dots K)$ we have stability, i.e., $B \geq 0$. Using numerical calculations we get stability for the physical value $\alpha \approx 1/137$ up to $Z_k \leq 88$ $(k = 1, \dots K)$.

1. INTRODUCTION

A basic requirement of thermodynamics is the extensivity of the energy. To be able to show this property, it is essential that on a microscopic level the energy per particle is bounded from below independently of the size of the considered system. This property, also referred to as stability of matter, as been proven in the literature for a wide class of models starting from the pioneering work of Dyson and Lenard [6, 7] in the non-relativistic case and of Conlon [3] and Fefferman and de la Llave [9] in the relativistic case. (See [10, 12, 14, 16] for an overview and more references.) A particular interesting work for our purposes is the work of Lieb and Yau [15] who consider the stability of matter of a relativistic system of N (spinless) electrons in a field of K fixed nuclei having atomic numbers $\mathbf{Z} = (Z_1, \dots, Z_K)$ and positions

Key words and phrases. Dirac operator, stability of relativistic matter.

$\mathbf{R} = (R_1, \ldots, R_K)$. Hence the relevant Coulomb potential is

(1)
$$V_{(R_1,\ldots,R_K;Z_1,\ldots,Z_K)}(x_1,\ldots,x_N) := \sum_{\substack{n,m=1\\ n<m}}^{N} \frac{1}{|x_n - x_m|} - \sum_{n=1}^{N}\sum_{k=1}^{K} \frac{Z_k}{|x_n - R_k|} + \sum_{\substack{k,l=1\\ k<l}}^{K} \frac{Z_k Z_l}{|R_k - R_l|}.$$

As an expression for the kinetic energy they use $\sum_{n=1}^{N} |p|^{(n)}$, where $|p| = \sqrt{-\Delta}$ and $|p|^{(n)}$ acts on the n-th particle, i.e., they examine the Hamiltonian

$$H_{N,K,\mathbf{R},\mathbf{Z}} = \sum_{n=1}^{N} |p|^{(n)} + \alpha V_{\mathbf{R};\mathbf{Z}},$$

where α is a coupling constant. The physical value of α, the Sommerfeld fine structure constant, is approximately $1/137.037$.

The goal is to prove stability, i.e., the existence of a constant c such that for all K and N

$$H_{N,K,\mathbf{R},\mathbf{Z}} \geq -c(N+K).$$

Due to scaling properties this is equivalent to $H_{N,K,\mathbf{R},\mathbf{Z}} \geq 0$. In [15] stability is proved if $Z\alpha \leq 2/\pi$ and $\alpha < 1/47q$, where $Z := \max\{Z_1, \ldots, Z_K\}$ and q is the number of spin states. The bound $2/\pi$ is sharp for this kind of kinetic energy.

The above model has the problem that it does not really account for the spin of the electrons and becomes instable if one of the atomic numbers exceeds $2\alpha/\pi$ (which is about 87.2 for the physical value of α). A model that does not have the problem on the one-particle level for any physical atomic number has been proposed by Brown and Ravenhall [2]. This has been shown on the one-particle level by Evans et al [8] and improved by Tix [17, 18]. Instead of expectations of $|p|$ expectations of the Dirac operator are considered. A collapse to the negative spectral subspace is prevented by restricting to positive energy states only which is a particular way of implementing Dirac's idea of filling the sea of negative states. The precise definition of the Brown-Ravenhall operator is as follows:

Let D_0 denote the free (massless) Dirac operator acting on $L^2(\mathbb{R}^3) \otimes \mathbb{C}^4$, Λ_+ the orthogonal projection onto the positive spectral subspace of D_0 and $\mathfrak{H} := \Lambda_+(L^2(\mathbb{R}^3) \otimes \mathbb{C}^4)$. Let $\mathfrak{H}_N := \bigwedge_{n=1}^{N} \mathfrak{H}$ be the N-fold antisymmetric tensor product of $\mathfrak{H}$ and $\Lambda_{+,N}$ the orthogonal projection from $\bigotimes_{n=1}^{N}(L^2(\mathbb{R}^3) \otimes \mathbb{C}^4)$ onto $\mathfrak{H}_N$. We consider the operator

(2)
$$B_{N,K,\mathbf{R},\mathbf{Z}} = \Lambda_{+,N} \left(\sum_{n=1}^{N} D_0^{(n)} + \alpha V_{\mathbf{R};\mathbf{Z}} \right) \Lambda_{+,N}$$

on $\mathfrak{H}_N$, where $D_0^{(n)}$ denotes the free Dirac operator D_0 acting on the n-th particle. (The $\Lambda_{+,N}$ on the very right in (2) is superfluous but we write it here and in the following to stress the reduction to the positive subspace.)

With the results of [15] one gets stability also for this model if $Z\alpha \leq 2/\pi$. But it is guessed that $B_{N,K,\mathbf{R},\mathbf{Z}} \geq 0$ if $\alpha Z \leq \gamma_c := \frac{2}{2/\pi + \pi/2}$ perhaps under some restrictions on α. Indeed in [8] this is proved for the special case $N = K = 1$ for the massless case and in [17, 18] in the massive case. In [8] it is also shown that γ_c is optimal: If $\alpha Z > \gamma_c$ then we have instability, i.e., $B_{1,1,R,Z}$ is unbounded from below. The one-electron molecule ($N = 1$, K arbitrary) was treated by Balinsky and Evans [1]; they prove stability for $\alpha Z \leq \gamma_c$ under a constraint on α, which is satisfied by the physical value. We consider the general case (N, K arbitrary):

Theorem 1. *Let $\gamma < \frac{2}{2/\pi+\pi/2}$. Then there exists an α_0 such that for all $\alpha < \alpha_0$ and $Z_1, \ldots, Z_K$ with $\alpha Z_k \leq \gamma$ ($k = 1, \ldots K$) we have $B_{N,K,\mathbf{R},\mathbf{Z}} \geq 0$.*

Remark. For the physical value 1/137.037 of α we get stability up to $Z_k \le 88$ $(k = 1, \dots, K)$. This exceeds the maximal atomic number one would get with the mentioned methods of [15]

It remains as a challenge to improve this to get stability in our model for the physical value of α up to the largest possible atomic number $Z = \gamma_c \alpha^{-1} \approx 124$.

The paper is organized as follows: In Section 2 we introduce some notations and summarize known results, that we will use, so that the rest of the paper should be self-contained. The third section treats a commutator estimate, that will be a crucial ingredient in the proof of our Theorem. This main proof is done in Section 4. The last section contains the numerical results that give the mentioned stability up to $Z = 88$ for the physical value of α.

2. Preliminaries

Here we introduce the notations we will use below and summarize some known results.

Fix $K \ge 2$ nuclei located at distinct points $\mathbf{R} = (R_1, \dots, R_K) \in \mathbb{R}^{3K}$ all with the same atomic number Z. Decompose $\mathbb{R}^3$ in K Voronoi cells Γ_k, which are the nearest neighborhoods to the nuclei:

$$\Gamma_k := \{x : |x - R_k| \le |x - R_l|, l = 1, \dots, K\}.$$

Let B_k denote the biggest open ball with center R_k in Γ_k and let D_k be its radius:

$$D_k := \mathrm{dist}(R_k, \partial\Gamma_k) = \frac{1}{2}\min\{|R_k - R_l| : l \ne k\}, \quad B_k := \{x : |x - R_k| < D_k\}.$$

Further, for $\sigma \in (0,1)$ let

$$B_k^{(\sigma)} := \{x : |x - R_k| \le (1-\sigma)D_k\}.$$

We identify sets with their characteristic functions.

Now consider N electrons and let $V_Z := V_{(R_1,\dots,R_K;Z,\dots,Z)}(x_1, \dots, x_N)$ be the Coulomb potential induced by the nuclei with all the same atomic number Z and the electrons as defined in (1).

Proposition 1 ([15], Theorem 6). *For any* $0 < \lambda < 1$

$$V_Z \ge -\sum_{n=1}^{N} W_\lambda(x_n) + \frac{1}{8}Z^2\sum_{k=1}^{K}\frac{1}{D_k},$$

where, for $x \in \Gamma_k$, $W_\lambda(x) := W_{\lambda,k}(x) := \frac{Z}{|x-R_k|} + F_{\lambda,k}(x)$ *with*

$$F_{\lambda,k} := \begin{cases} \frac{1}{2D_k(1-\frac{1}{D_k{}^2}|x-R_k|^2)} & \textit{for} \quad |x-R_k| \le \lambda D_k, \\ (\sqrt{2Z}+\frac{1}{2})\frac{1}{|x-R_k|} & \textit{for} \quad |x-R_k| > \lambda D_k. \end{cases}$$

We will localize the kinetic energy to control the Coulomb singularities. More exactly we shall consider $\mathrm{tr}\gamma|p|$, where γ is a density matrix, i.e., a positive definite trace class operator, on $L^2(\mathbb{R}^3) \otimes \mathbb{C}^4$ or on $L^2(\mathbb{R}^3)$.

Proposition 2 ([15],Theorem 10). *Let* $0 < \sigma < 1$ *and* χ_0, χ_1 *be Lipschitz continuous nonnegative functions with* ${\chi_0}^2 + {\chi_1}^2 = 1$ *and* χ_1 *supported in* $B_1^{(\sigma)}$. *Define*

$$L(x,y) := \frac{1}{\pi^2}\frac{1 - \chi_0(x)\chi_0(y) - \chi_1(x)\chi_1(y)}{|x-y|^4},$$

$$L_1(x,y) := \begin{cases} L(x,y) & \textit{for } \big||x| - |y|\big| \le \sigma D_1,\ x,y \in B_1, \\ 0 & \textit{otherwise}, \end{cases}$$

and $L_0 := L - L_1$. *For any positive function* h *on* B_1 *and for arbitrary* $\varepsilon > 0$ *let*

$$U_1(x) := \begin{cases} \frac{\varepsilon}{D_1} B_1^{(\sigma)}(x) + \frac{1}{h(x)} \int_{B_1} dy\, L_1(x,y)h(y) & \text{for } x \in B_1, \\ 0 & \text{otherwise.} \end{cases}$$

Then, for any density matrix γ *acting on* $L^2(\mathbb{R}^3)$,

$$\mathrm{tr}\gamma|p| \geq \mathrm{tr}\chi_1\gamma\chi_1(|p| - U_1(x)) + \mathrm{tr}\chi_0\gamma\chi_0(|p| - U_1(x)) - \frac{\|\gamma\|}{\varepsilon D_1}\Omega,$$

where $\Omega = \frac{1}{2}{D_1}^2 \int\int dx dy\, L_0(x,y)^2$.

(Note the typing error in [15, (3.12)].)

To control the remaining potentials we will use Daubechies' inequality:

Proposition 3 (Daubechies [4], see also [15], Theorem 8). *For any density matrix* γ *on* $L^2(\mathbb{R}^3)$, *any positive function* $U \in L^4(\mathbb{R}^3)$ *and any* $\mu > 0$ *we have*

$$\mathrm{tr}\gamma(\mu|p| - U) \leq -0.0258 \cdot \|\gamma\|\mu^{-3} \int dx\, U(x)^4.$$

We want to apply the above two propositions to the reduced density matrix $\hat{\gamma}$ (without spin) of a density matrix $\Lambda_+\gamma\Lambda_+$ acting on $L^2(\mathbb{R}^3) \otimes \mathbb{C}^4$ (using the kernels and indicating the space-spin variables by $(x,s) \in \mathbb{R}^3 \times \{1,\ldots,4\}$ it is $\hat{\gamma}(x,y) := \sum_{s=1}^4 (\Lambda_+\gamma\Lambda_+)((x,s),(y,s)))$. If one had no restriction (i.e., without the Λ_+) one could only conclude that $\|\hat{\gamma}\| \leq 4 \cdot \|\gamma\|$. But due to the projection onto the positive spectral subspace we have:

Proposition 4 ([14], Appendix B). *Let* $\hat{\gamma}$ *be the reduced density matrix of* $\Lambda_+\gamma\Lambda_+$ *as above. Then* $\|\hat{\gamma}\| \leq 2 \cdot \|\gamma\|$.

The key estimate we will use is the positivity of the one particle operator ($N = K = 1$ in (2)). Using the notations of the introduction we have:

Proposition 5 ([8], Theorem 1).

$$B_{1,1,R,Z} \geq 0 \quad \text{if} \quad \alpha Z \leq \gamma_c = \frac{2}{\frac{2}{\pi} + \frac{\pi}{2}}.$$

3. A Useful Commutator Estimate

Here we prove a proposition which will be essential in the proof of our main Theorem in Section 4.

Proposition 6. *Let* $0 < \sigma < 1$ *and* χ *be a Lipschitz continuous radial real function with* $\chi(x) = 0$ *if* $|x| \geq 1 - \sigma$, *so that* $\chi_k(x) := \chi(\frac{x - R_k}{D_k})$ *is supported in* $B_k^{(\sigma)}$. *Let* $\Lambda_- := 1 - \Lambda_+$ *and*

$$K := [\chi_k, \Lambda_-]\frac{1}{|x - R_k|}\chi_k - \chi_k\frac{1}{|x - R_k|}[\chi_k, \Lambda_-].$$

Then, for all density matrices γ *acting on* $L^2(\mathbb{R}^3) \otimes \mathbb{C}^4$ *and arbitrary* $\delta > 0$,

$$\mathrm{tr}\Lambda_+\gamma\Lambda_+ K \leq \mathrm{tr}\Lambda_+\gamma\Lambda_+\frac{\delta}{D_k}B_k^{(\sigma)} + \frac{\|\gamma\|}{\delta D_k}C,$$

where (writing $\chi(x) = \chi(|x|)$*)*

$$C = \frac{8}{\pi^2}\int_0^{1-\sigma} dr \int_r^\infty ds\, \frac{|\chi(r) - \chi(s)|^2 |r\chi(s) - s\chi(r)|^2 (r^2 + s^2)}{(r^2 - s^2)^4}$$

Proof. Without loss of generality we can take $R_k = 0$. Because of $\Lambda_- = \frac{1}{2}(1 - \frac{\alpha\cdot p}{|p|})$ with the Dirac matrices α (see [8]) we have

$$[\chi_k, \Lambda_-] = \chi_k\Lambda_- - \Lambda_-\chi_k = \frac{1}{2}\left((\alpha\cdot p)\frac{1}{|p|}\chi_k - \chi_k(\alpha\cdot p)\frac{1}{|p|}\right).$$

Since $\frac{1}{|p|}$ has the integral kernel $\frac{1}{2\pi^2|x-y|^2}$ we could get as an integral kernel of $[\chi_k, \Lambda_-]$ by differentiating

$$\begin{aligned} &\frac{1}{2}\left(\frac{-2\alpha\cdot(x-y)}{2\pi^2 i|x-y|^4}\chi_k(y) - \chi_k(x)\frac{-2\alpha\cdot(x-y)}{2\pi^2 i|x-y|^4}\right)\\ (3)\qquad &= \frac{1}{2\pi^2 i}\frac{\alpha\cdot(x-y)}{|x-y|^4}(\chi_k(x)-\chi_k(y)). \end{aligned}$$

To justify this in spite of the non-integrable singularity in the first line (notice that due to the Lipschitz continuity of χ the appearing singularity in the second line is integrable) one can argue as follows:

First we regularize $\frac{1}{|p|}$ by considering $\frac{e^{-\varepsilon|p|}}{|p|}$ for $\varepsilon > 0$. By an easy calculation one gets $\frac{1}{2\pi^2|x-y|^2+\varepsilon^2}$ as the integral kernel of $\frac{e^{-\varepsilon|p|}}{|p|}$. Differentiation yields $-\frac{1}{\pi^2 i}\frac{\alpha\cdot(x-y)}{(|x-y|^2+\varepsilon^2)^2}$ as the integral kernel of $\frac{\alpha\cdot p}{|p|}e^{-\varepsilon|p|}$ and for $f \in L^2(\mathbb{R}^2)\otimes\mathbb{C}^4$ we have

$$\frac{\alpha\cdot p}{|p|}e^{-\varepsilon|p|}\chi_k f - \chi_k\frac{\alpha\cdot p}{|p|}e^{-\varepsilon|p|}f = \int dy\frac{1}{\pi^2 i}\frac{\alpha\cdot(x-y)}{(|x-y|^2+\varepsilon^2)^2}(\chi_k(x)-\chi_k(y))f(y).$$

Tending ε to zero we get $e^{-\varepsilon|p|}f \to f$ and $e^{-\varepsilon|p|}\chi_k f \to \chi_k f$ in L^2 and hence – by dominated convergence – (3) as the integral kernel of $[\chi_k, \Lambda_-]$.

So, K has the kernel

$$(4)\qquad K(x,y) := \frac{1}{2\pi^2 i}\frac{\alpha\cdot(x-y)}{|x-y|^4}(\chi_k(x)-\chi_k(y))\left(\frac{\chi_k(y)}{|y|} - \frac{\chi_k(x)}{|x|}\right).$$

We now proceed as in Section VI of [15], but in contrast to [15] we have to treat variables with spin. Let $\underline{x} := (x,s) \in \mathbb{R}^3\times\{1,\dots,4\}$ be a space-spin variable and let $\int d\underline{x}$ indicate the integration over $x\in\mathbb{R}^3$ (if not stated otherwise) and summation over $s\in\{1,\dots,4\}$. Let $\gamma_1 := \Lambda_+\gamma\Lambda_+$ and $\gamma_1^{\frac{1}{2}}$ denote the operator square root of γ_1. We identify the operators with their kernels. Since $\gamma_1(\underline{x},\underline{y}) = \overline{\gamma_1(\underline{y},\underline{x})}$ and $K(\underline{x},\underline{y}) = \overline{K(\underline{y},\underline{x})}$ we get

$$\begin{aligned} \mathrm{tr}\gamma_1 K &= \int d\underline{x}\int d\underline{y}\,\gamma_1(\underline{x},\underline{y})K(\underline{y},\underline{x})\\ &= 2\Re\int d\underline{x}\int_{y:|y|>|x|} d\underline{y}\,\gamma_1(\underline{x},\underline{y})K(\underline{y},\underline{x})\\ &= 2\Re\int d\underline{x}\int_{y:|y|>|x|} d\underline{y}\int d\underline{z}\,\gamma_1^{\frac{1}{2}}(\underline{x},\underline{z})\gamma_1^{\frac{1}{2}}(\underline{z},\underline{y})K(\underline{y},\underline{x}). \end{aligned}$$

If $(1-\sigma)D_k < |x|$ and $|x| < |y|$ then $\chi_k(x) = \chi_k(y) = 0$, hence $K(\underline{x},\underline{y}) = 0$. Therefore we can add a factor $B_k^{(\sigma)}(x)$ in the integrand. By applying Minkowski's inequality we get

$$
\begin{aligned}
&\mathrm{tr}\gamma_1 K\\
= &\ 2\Re \int d\underline{x} \int d\underline{z} \left(\gamma_1^{\frac12}(\underline{x},\underline{z}) B_k^{(\sigma)}(x)\right) \cdot \left(B_k^{(\sigma)}(x) \int_{|y|>|x|} d\underline{y}\, \gamma_1^{\frac12}(\underline{z},\underline{y}) K(\underline{y},\underline{x})\right)\\
\le &\ \varepsilon \int d\underline{x} \int d\underline{z}\, |\gamma_1^{\frac12}(\underline{x},\underline{z}) B_k^{(\sigma)}(x)|^2\\
&+ \frac{1}{\varepsilon} \int d\underline{x} \int d\underline{z} \left| B_k^{(\sigma)}(x) \int_{|y|>|x|} d\underline{y}\, \gamma_1^{\frac12}(\underline{z},\underline{y}) K(\underline{y},\underline{x})\right|^2\\
= &\ \varepsilon \int d\underline{x} \int d\underline{z}\, \gamma_1^{\frac12}(\underline{x},\underline{z}) \gamma_1^{\frac12}(\underline{z},\underline{x}) B_k^{(\sigma)}(x)\\
&+ \frac{1}{\varepsilon} \int d\underline{y} \int d\underline{y}' \int_{\substack{|x|< \\ \min\{|y|,|y'|\}}} d\underline{x} \int d\underline{z}\, B_k^{(\sigma)}(x) \gamma_1^{\frac12}(\underline{z},\underline{y}) K(\underline{y},\underline{x}) \overline{\gamma_1^{\frac12}(\underline{z},\underline{y}') K(\underline{y}',\underline{x})}\\
= &\ \varepsilon \int d\underline{x}\, \gamma_1(\underline{x},\underline{x}) B_k^{(\sigma)}(x)\\
&+ \frac{1}{\varepsilon} \int d\underline{y} \int d\underline{y}' \left(\int d\underline{z}\, \gamma_1^{\frac12}(\underline{y}',\underline{z}) \gamma_1^{\frac12}(\underline{z},\underline{y}) \right) \cdot \left(\int_M d\underline{x}\, K(\underline{y},\underline{x}) K(\underline{x},\underline{y}') \right)\\
= &\ \varepsilon \mathrm{tr}\gamma_1 B_k^{(\sigma)} + \frac{1}{\varepsilon} \int d\underline{y} \int d\underline{y}' \gamma_1(\underline{y}',\underline{y}) \left(\int_M d\underline{x}\, K(\underline{y},\underline{x}) K(\underline{x},\underline{y}') \right),
\end{aligned}
$$

with $M := \{\underline{x} : x \in B_k^{(\sigma)}, |x| \le \min\{|y|,|y'|\}\}$. The last summand is equal to $\frac{1}{\varepsilon}\mathrm{tr}\gamma_1 \tilde{K}^* \tilde{K}$, where $\tilde{K}$ has the kernel

$$
\tilde{K}(x,y) := K(x,y) \cdot 1_{\{(x,y): x \in B_k^{(\sigma)}, |x| \le |y|\}}(x,y) \tag{5}
$$

with the characteristic function 1_A of the set A. Since $\mathrm{tr}\gamma_1 \tilde{K}^*\tilde{K} = \mathrm{tr}\gamma\Lambda_+ \tilde{K}^*\tilde{K}\Lambda_+$ we get, using $|\mathrm{tr}AB| \le \|A\| \mathrm{tr}B$ if B is positive and taking $\varepsilon = \frac{\delta}{D_k}$ above,

$$
\mathrm{tr}\gamma_1 K \le \frac{\delta}{D_k} \mathrm{tr}\gamma_1 B_k^{(\sigma)} + \frac{D_k}{\delta} \|\gamma\| \mathrm{tr}\Lambda_+ \tilde{K}^*\tilde{K}\Lambda_+. \tag{6}
$$

Similar to [14], Appendix B, we claim

$$
\mathrm{tr}\Lambda_+ \tilde{K}^*\tilde{K}\Lambda_+ = \frac{1}{2}\mathrm{tr}\tilde{K}^*\tilde{K}. \tag{7}
$$

Because of

$$
\mathrm{tr}\tilde{K}^*\tilde{K} = \mathrm{tr}(\Lambda_+^2 + \Lambda_-^2)\tilde{K}^*\tilde{K} = \mathrm{tr}\Lambda_+\tilde{K}^*\tilde{K}\Lambda_+ + \mathrm{tr}\Lambda_-\tilde{K}^*\tilde{K}\Lambda_-
$$

it is enough to prove $\mathrm{tr}\Lambda_+\tilde{K}^*\tilde{K}\Lambda_+ = \mathrm{tr}\Lambda_-\tilde{K}^*\tilde{K}\Lambda_-$.

Let $U := \begin{pmatrix} 0 & 1 \\ -1 & 0 \end{pmatrix}$ where 1 indicates the 2×2 unit matrix. Then it is easy to verify that $\Lambda_- = U^*\Lambda_+ U$, hence

$$
\mathrm{tr}\Lambda_-\tilde{K}^*\tilde{K}\Lambda_- = \mathrm{tr}U^*\Lambda_+ U\tilde{K}^*\tilde{K}U^*\Lambda_+ U = \mathrm{tr}\Lambda_+ U\tilde{K}^*U^*U\tilde{K}U^*\Lambda_+.
$$

Now, U commutes with scalar multiplications. Hence, having (5) and (4) in mind, the fact

$$
U^*(\boldsymbol{\alpha}\cdot(x-y))U = U(\boldsymbol{\alpha}\cdot(x-y))U^* = -\boldsymbol{\alpha}\cdot(x-y)
$$

yields $U\tilde{K}U^* = -\tilde{K}$ and $U\tilde{K}^*U^* = -\tilde{K}^*$. Thus

$$
\mathrm{tr}\Lambda_-\tilde{K}^*\tilde{K}\Lambda_- = \mathrm{tr}\Lambda_+\tilde{K}^*\tilde{K}\Lambda_+
$$

and we get (7).

Now, continuing (6), we have

$$
\mathrm{tr}\gamma_1 K \le \frac{\delta}{D_k}\mathrm{tr}\gamma_1 B_k^{(\sigma)} + \frac{D_k}{\delta}\|\gamma\| \cdot \frac{1}{2}\mathrm{tr}\tilde{K}^*\tilde{K}
$$

and it remains to calculate the last trace. Using (4) and the fact that $(\boldsymbol{\alpha}\cdot(x-y))^2 = (x-y)^2$, elementary calculations give

$$\begin{aligned}\frac{1}{2}\mathrm{tr}\tilde{K}^*\tilde{K} &= \frac{1}{2}\int d\underline{y}\int_{\substack{x\in B_k^{(\sigma)}:\\ |x|\le|y|}} d\underline{x}\,|K(x,y)|^2\\ &= \frac{1}{{D_k}^2}\frac{8}{\pi^2}\int_0^{1-\sigma}dr\int_r^\infty ds\,\frac{|\chi(r)-\chi(s)|^2|r\chi(s)-s\chi(r)|^2(r^2+s^2)}{(r^2-s^2)^4},\end{aligned}$$

and we are done. □

4. Proof of the Theorem

We want to prove

$$B_{N,K,\mathbf{R},\mathbf{Z}} = \Lambda_{+,N}\left(\sum_{n=1}^{N} D_0^{(n)} + \alpha V_{\mathbf{R};\mathbf{Z}}\right)\Lambda_{+,N} \geq 0$$

for $\mathbf{Z} = (Z_1,\dots,Z_K)$. Using the same convexity arguments as in [5, p. 507] it is enough to prove $B_{N,K,\mathbf{R},\mathbf{Z_c}} \geq 0$ with $\mathbf{Z_c} = (Z,\dots,Z)$, $Z \geq \max\{Z_1,\dots,Z_K\}$. Applying Proposition 1 and the fact that $\Lambda_+ D_0 \Lambda_+ = \Lambda_+|p|\Lambda_+$ we get

$$\begin{aligned}B_{N,K,\mathbf{R},\mathbf{Z_c}} &\geq \Lambda_{+,N}\left(\sum_{n=1}^{N} D_0^{(n)} - \alpha\sum_{n=1}^{N} W_\lambda(x_n) + \frac{1}{8}\alpha Z^2\sum_{k=1}^{K}\frac{1}{D_k}\right)\Lambda_{+,N}\\ &= \Lambda_{+,N}\left(\sum_{n=1}^{N} h_n\right)\Lambda_{+,N} + D\end{aligned}$$

with $h_n := |p|^{(n)} - \alpha W_\lambda(x_n)$ and $D := \frac{1}{8}\alpha Z^2\sum_{k=1}^{K}\frac{1}{D_k}$ (note that we are acting on $\mathfrak{H}_N$ and $\Lambda_{+,N}|_{\mathfrak{H}_N} = \mathrm{Id}$). Hence the positivity of $B_{N,K,\mathbf{R},\mathbf{Z}}$ is implied by $\Lambda_{+,N}\left(\sum_{n=1}^{N} h_n\right)\Lambda_{+,N} \geq -D$. Proving this for all N is equivalent to showing

$$\mathrm{tr}\gamma\Lambda_+ h\Lambda_+ \geq -D = -\frac{1}{8}\alpha Z^2\sum_{k=1}^{K}\frac{1}{D_k} \tag{8}$$

with $h := |p| - \alpha W_\lambda$ (acting componentwise on $L^2(\mathbb{R}^3)\otimes\mathbb{C}^4$) for all density matrices γ on $L^2(\mathbb{R}^3)\otimes\mathbb{C}^4$ with $0 \leq \gamma \leq 1$ (cf. [11] and [15, (2.22)]). This is, what we want to do now.

Fix some Lipschitz continuous nonnegative radial function $0 \leq \chi \leq 1$ with $\chi(x) = 1$ if $|x| \leq 1 - 3\sigma$ and $\chi(x) = 0$ if $|x| \geq 1 - \sigma$ and so that $\sqrt{1-\chi^2}$ is also Lipschitz continuous. Define $\chi_k(x) := \chi(\frac{x-R_k}{D_k})$ $(k = 1,\dots,K)$ as in Proposition 6 and let $\chi_{0,k}$ be nonnegative functions with $\chi_k^2 + \chi_{0,k}^2 = 1$. (The Lipschitz continuity of $\sqrt{1-\chi^2}$ implies the same for $\chi_{0,k}$).

We borrow a part $\mu|p|$ $(\mu \in (0,1))$ of the kinetic energy to control the remaining potentials in the end and set $\nu := 1 - \mu$. Now we fix a density matrix γ acting on $L^2(\mathbb{R}^3)\otimes\mathbb{C}^4$ with $0 \leq \gamma \leq 1$. First we observe that – for estimating $\mathrm{tr}\Lambda_+\gamma\Lambda_+(\nu|p| - \alpha W_\lambda)$ – it is enough to consider the reduced density matrix $\hat{\gamma}$ without spin (see the definition before Proposition 4) instead of $\Lambda_+\gamma\Lambda_+$. We then can apply Proposition 2 to $\hat{\gamma}$. Using the fact that $\|\hat{\gamma}\| \leq 2$ (Proposition 4) and going back to the full density matrix we get with $\tilde{\alpha} := \nu^{-1}\alpha$

$$
\begin{aligned}
& \mathrm{tr}\Lambda_+\gamma\Lambda_+\Big(\nu|p| - \alpha W_\lambda\Big) \\
\geq\ & \mathrm{tr}\chi_1\Lambda_+\gamma\Lambda_+\chi_1\Big(\nu|p| - \frac{\alpha Z}{|x-R_1|} - \alpha F_{\lambda,1}(x) - \nu U_1(x)\Big) \\
& +\mathrm{tr}\chi_{0,1}\Lambda_+\gamma\Lambda_+\chi_{0,1}\Big(\nu|p| - \alpha W_\lambda(x) - \nu U_1(x)\Big) - \frac{2\nu\Omega}{\varepsilon D_1}
\end{aligned}
$$

$$
= \nu\mathrm{tr}\gamma\Lambda_+\chi_1\Big(|p| - \frac{\tilde{\alpha}Z}{|x-R_1|}\Big)\chi_1\Lambda_+ \tag{9}
$$

$$
+\mathrm{tr}\chi_{0,1}\Lambda_+\gamma\Lambda_+\chi_{0,1}\Big(\nu|p| - \alpha W_\lambda(x)\Big) \tag{10}
$$

$$
-\mathrm{tr}\Lambda_+\gamma\Lambda_+\Big(\alpha{\chi_1}^2F_{\lambda,1} + \nu U_1(x)\Big) - \frac{2\nu\Omega}{\varepsilon D_1}.
$$

To handle the first summand (9) we want to apply Proposition 5. Using the definition of the operator K in Proposition 6 (with $k = 1$) and $\Lambda_+ + \Lambda_- = 1$, $\Lambda_+\Lambda_- = 0$ it is easy to see that

$$
\begin{aligned}
-\Lambda_+\chi_1\frac{1}{|x-R_1|}\chi_1\Lambda_+ &= -\Lambda_+\chi_1\Lambda_+\frac{1}{|x-R_1|}\Lambda_+\chi_1\Lambda_+ - \Lambda_+K\Lambda_+ \\
&\quad +\Lambda_+\chi_1\Lambda_-\frac{1}{|x-R_1|}\Lambda_-\chi_1\Lambda_+.
\end{aligned}
$$

Neglecting the last term, which is positive, and using $|p| \geq \Lambda_+|p|\Lambda_+$ and Proposition 5 we get, as soon as $\tilde{\alpha}Z \leq \gamma_c$,

$$
\begin{aligned}
& \Lambda_+\chi_1\Big(|p| - \frac{\tilde{\alpha}Z}{|x-R_1|}\Big)\chi_1\Lambda_+ \\
\geq\ & \Lambda_+\chi_1\Big(\Lambda_+|p|\Lambda_+ - \Lambda_+\frac{\tilde{\alpha}Z}{|x-R_1|}\Lambda_+\Big)\chi_1\Lambda_+ - \tilde{\alpha}Z\Lambda_+K\Lambda_+ \\
\geq\ & -\tilde{\alpha}Z\Lambda_+K\Lambda_+.
\end{aligned}
$$

Thus, Proposition 6 yields

$$
(9) \geq -\nu\tilde{\alpha}Z\mathrm{tr}\Lambda_+\gamma\Lambda_+K \geq -\alpha Z\mathrm{tr}\Lambda_+\gamma\Lambda_+\frac{\delta}{D_1}B_1^{(\sigma)} - \frac{\alpha Z}{\delta D_1}C. \tag{11}
$$

We now repeat the whole strategy treating the second summand (10) and the second nucleus: Applying Proposition 2 to the reduced density matrix (without spin) of $\tilde{\gamma} := \chi_{0,1}\Lambda_+\gamma\Lambda_+\chi_{0,1}$ we get with the corresponding definition of U_2 (notice that by scaling Ω does only depend on χ and not on D_k)

$$
\begin{aligned}
(10)\ \geq\ & \mathrm{tr}\chi_2\tilde{\gamma}\chi_2\Big(\nu|p| - \frac{\alpha Z}{|x-R_2|} - \alpha F_{\lambda,2}(x) - \nu U_2(x)\Big) \\
& +\mathrm{tr}\chi_{0,2}\tilde{\gamma}\chi_{0,2}\Big(\nu|p| - \alpha W_\lambda(x) - \nu U_2(x)\Big) - \frac{2\nu\Omega}{\varepsilon D_2} \\
=\ & \nu\mathrm{tr}\gamma\Lambda_+\chi_2\Big(|p| - \frac{\tilde{\alpha}Z}{|x-R_2|}\Big)\chi_2\Lambda_+ \\
& +\mathrm{tr}\chi_{0,2}\chi_{0,1}\Lambda_+\gamma\Lambda_+\chi_{0,1}\chi_{0,2}\Big(\nu|p| - \alpha W_\lambda(x)\Big) \\
& -\mathrm{tr}\Lambda_+\gamma\Lambda_+\Big(\alpha{\chi_2}^2F_{\lambda,2} + \nu U_2(x)\Big) - \frac{2\nu\Omega}{\varepsilon D_2}.
\end{aligned}
$$

(For the last equality notice $\chi_2\chi_{0,1} = \chi_2$ and $U_2\chi_{0,1} = U_2$.) We can estimate the first summand similarly to above (cf. (11)). Repeating this procedure for all other nuclei we get

$$\begin{aligned}
&\mathrm{tr}\Lambda_+\gamma\Lambda_+\Big(\nu|p| - \alpha W_\lambda\Big) \\
\geq\ & -\alpha Z\ \mathrm{tr}\Lambda_+\gamma\Lambda_+\left(\sum_{k=1}^K \frac{\delta}{D_k} B_k^{(\sigma)}\right) - \alpha Z\frac{C}{\delta}\sum_{k=1}^K \frac{1}{D_k} \\
& -\mathrm{tr}\Lambda_+\gamma\Lambda_+\left(\sum_{k=1}^K (\alpha{\chi_k}^2 F_{\lambda,k} + \nu U_k)\right) - \frac{2\nu\Omega}{\varepsilon}\sum_{k=1}^K \frac{1}{D_k} \\
& +\mathrm{tr}\chi_{0,K}\cdots\chi_{0,1}\Lambda_+\gamma\Lambda_+\chi_{0,1}\cdots\chi_{0,K}\Big(\nu|p| - \alpha W_\lambda\Big).
\end{aligned}$$

Introducing a new parameter $\beta \in (0,1)$ we split the potential W_λ in the last summand into two parts. One part is joined with the other potentials. To the other part we can add a factor R, where R is the characteristic function of the complement of $\bigcup_{k=1}^K B_k^{(3\sigma)}$. Together with the remaining $\mu|p|$ and using the abbreviations

$$G_k := \alpha{\chi_k}^2 F_{\lambda,k} + (1-\mu)U_k + \alpha Z\frac{\delta}{D_k}B_k^{(\sigma)}, \quad A := \frac{2(1-\mu)\Omega}{\varepsilon} + \alpha Z\frac{C}{\delta}$$

we get

$$\begin{aligned}
&\mathrm{tr}\Lambda_+\gamma\Lambda_+\Big(|p| - \alpha W_\lambda\Big) \\
\geq\ & \mathrm{tr}\Lambda_+\gamma\Lambda_+\Big(\mu|p| - (1-\beta)\alpha\chi_{0,1}^2\cdots\chi_{0,K}^2 W_\lambda - \sum_{k=1}^K G_k\Big) - A\sum_{k=1}^K \frac{1}{D_k} \\
& +\mathrm{tr}\chi_{0,K}\cdots\chi_{0,1}\Lambda_+\gamma\Lambda_+\chi_{0,1}\cdots\chi_{0,K}\Big((1-\mu)|p| - \beta\alpha W_\lambda R\Big)
\end{aligned}$$

Finally we estimate the first and last summand applying Daubechies' inequality (Proposition 3). As above, we make again use of the fact that only the reduced density matrix without spin is relevant and that its norm does not exceed two (cf. Proposition 4):

$$\begin{aligned}
&\mathrm{tr}\Lambda_+\gamma\Lambda_+ h \\
\geq\ & -0.0258\cdot 2\mu^{-3}\int\Big((1-\beta)\alpha\chi_{0,1}(x)^2\cdots\chi_{0,K}(x)^2 W_\lambda(x) + \sum_{k=1}^K G_k(x)\Big)^4 dx \\
& -0.0258\cdot 2(1-\mu)^{-3}\int \beta^4\alpha^4 W_\lambda(x)^4 R(x)\,dx - A\sum_{k=1}^K \frac{1}{D_k}.
\end{aligned}$$

To estimate the integrals we consider each Voronoi cell Γ_k separately. Since the support of G_k lies in B_k, outside of B_k we have only to consider W_λ. There

$$W_\lambda(x) = \frac{Z_0}{|x - R_k|} \quad \text{with} \quad Z_0 := Z + \sqrt{2Z} + \frac{1}{2}.$$

If we estimate the integral over $\Gamma_k \setminus B_k$ by integrating over one side of the mid-plane defined by the nearest nucleus (cf. [13, p. 982]) minus B_k we get

$$\int_{\Gamma_k\setminus B_k} dx\, W_\lambda(x)^4 \leq {Z_0}^4\frac{3\pi}{D_k}.$$

Since all terms scale in the right way, the integral over B_k gives a factor $\frac{1}{D_k}$ times an integral over normalized functions (with 1 instead of D_k and 0 instead of R_k).

Hence we get, indicating the normalized functions by the symbols without indices

$$
\begin{aligned}
&\mathrm{tr}\Lambda_+\gamma\Lambda_+ h \\
(12)\quad &\geq -\Big\{0.0516\mu^{-3}\Big[3\pi(1-\beta)^4\alpha^4 {Z_0}^4 \\
(13)\quad &\qquad + \int_B dx\Big((1-\beta)\alpha(1-\chi(x)^2)\big(F_\lambda(x)+\frac{Z}{|x|}\big)+G(x)\Big)^4\Big] \\
&+0.0516(1-\mu)^{-3}\beta^4\alpha^4\Big[3\pi {Z_0}^4 + \int_{B\setminus B(3\sigma)} dx\big(F_\lambda(x)+\frac{Z}{|x|}\big)^4\Big] + A\Big\} \\
&\cdot\sum_{k=1}^{K}\frac{1}{D_k}.
\end{aligned}
$$

In view of (8) we get stability, if the above expression in braces is smaller than or equal to $\frac{1}{8}\alpha Z^2$ and $\tilde{\alpha}Z = \alpha Z(1-\mu)^{-1} \leq \gamma_c$.

To get the statement of the Theorem we choose $\gamma < \gamma_c$. We want to consider $\alpha Z_k \leq \gamma$ ($k = 1, \dots, K$). Due to the convexity argument mentioned in the beginning it is enough to consider $Z := \gamma\alpha^{-1}$. We choose $(1-\mu) = \frac{\gamma}{\gamma_c}$ (thus $\alpha Z \leq (1-\mu)\gamma_c$ is satisfied) and fix all other parameters arbitrarily. With these choices and using $Z_0 \leq 4Z$ for $Z \geq 1$, which is valid for small values of α, all expressions inside the braces above can be estimated by a constant (independent of α). Now, we only have to guarantee that this constant is less than or equal to $\frac{1}{8}\alpha Z^2 = \frac{\gamma^2}{8}\alpha^{-1}$, which is fulfilled for small values of α.

5. Numerical Calculations.

To show the mentioned stability for the physical value $\alpha = 1/137.037$ up to $Z = 88$ we choose as in [15] $\sigma = 0.3$,

$$
\chi(r) = \begin{cases} 1 & \text{for } r \leq 0.1, \\ \cos\left((x-0.1)\frac{\pi}{1.2}\right) & \text{for } 0.1 < r \leq 0.7, \\ 0 & \text{for } r > 0.7, \end{cases}
$$

and

$$
rh(r) = \begin{cases} 1 & \text{for } r \leq 0.1 \text{ and } 0.7 \leq r \leq 1, \\ 2 - \frac{|r-0.4|}{0.3} & \text{for } 0.1 < r < 0.7, \end{cases}
$$

so that we can also use the estimate (see [15, Section VIII(B)])

$$
U(r) \leq \varepsilon + \begin{cases} 0.5751 & \text{for } r \leq 0.7, \\ \frac{\pi}{64\cdot 0.3^5}(1.6-r)(1-r)^3 & \text{for } 0.7 < r \leq 1. \end{cases}
$$

For $Z = 88$ we choose $\mu = 0.291$ (then $\alpha Z \leq (1-\mu)\gamma_c$) and $\lambda = 0.98$, $\varepsilon = 0.159$, $\delta = 0.374$ and $\beta = 0.874$. We do the angular integrations of the integrals analytically (cf. [15, Section VIII(A)]) and the remaining integrations on a computer. (The numerical reliability of our results is enhanced by the fact that all occurring integrands are regular.)

So we get $\Omega < 0.116$, $C < 1.289$, hence $A < 3.248$. For the integrals the result is

$$
\int_B dx\Big((1-\beta)\alpha(1-\chi(x)^2)\big(F_\lambda(x)+\frac{Z}{|x|}\big)+G(x)\Big)^4 < 0.861
$$

and

$$
\int_{B\setminus B(3\sigma)} dx\big(F_\lambda(x)+\frac{Z}{|x|}\big)^4 < 6.864\cdot 10^9.
$$

Thus the first summand in the braces of (12) and (13) is smaller than 1.805, the second one smaller than 1.887 and the whole expression is bounded by 6.94 whereas $\frac{1}{8}\alpha Z^2 > 7.06$ yielding the desired estimate.

References

[1] A. A. Balinsky and W. D. Evans. Stability of one-electron molecules in the Brown-Ravenhall model. *Commun. Math. Phys.*, 202(2):481–500, April 1999.

[2] G. E. Brown and D. G. Ravenhall. On the interaction of two electrons. *Proc. Roy. Soc. London A*, 208(A 1095):552–559, September 1951.

[3] Joseph G. Conlon. The ground state energy of a classical gas. *Commun. Math. Phys.*, 94(4):439–458, August 1984.

[4] Ingrid Daubechies. An uncertainty principle for Fermions with generalized kinetic energy. *Commun. Math. Phys.*, 90:511–520, September 1983.

[5] Ingrid Daubechies and Elliott H. Lieb. One-electron relativistic molecules with Coulomb interaction. *Comm. Math. Phys.*, 90(4):497–510, 1983.

[6] Freeman J. Dyson and Andrew Lenard. Stability of matter I. *J. Math. Phys.*, 8:423–434, 1967.

[7] Freeman J. Dyson and Andrew Lenard. Stability of matter II. *J. Math. Phys.*, 9:698–711, 1967.

[8] William Desmond Evans, Peter Perry, and Heinz Siedentop. The spectrum of relativistic one-electron atoms according to Bethe and Salpeter. *Commun. Math. Phys.*, 178(3):733–746, July 1996.

[9] C. L. Fefferman and R. de la Llave. Relativistic stability of matter – I. *Revista Matematica Iberoamericana*, 2(1, 2):119–161, 1986.

[10] Elliott H. Lieb. The stability of matter. *Rev. Mod. Phys.*, 48:553–569, 1976.

[11] Elliott H. Lieb. Sharp constants in the Hardy-Littlewood-Sobolev and related inequalities. *Annals of Mathematics*, 118:349–374, 1983.

[12] Elliott H. Lieb. The stability of matter: from atoms to stars. *Bull. Amer. Math. Soc. (N.S.)*, 22(1):1–49, 1990.

[13] Elliott H. Lieb, Michael Loss, and Heinz Siedentop. Stability of relativistic matter via Thomas-Fermi theory. *Helv. Phys. Acta*, 69(5/6):974–984, December 1996.

[14] Elliott H. Lieb, Heinz Siedentop, and Jan Philip Solovej. Stability and instability of relativistic electrons in classical electromagnetic fields. *J. Statist. Phys.*, 89(1-2):37–59, 1997. Dedicated to Bernard Jancovici.

[15] Elliott H. Lieb and Horng-Tzer Yau. The stability and instability of relativistic matter. *Commun. Math. Phys.*, 118:177–213, 1988.

[16] Michael Loss. The stability of matter interacting with fields. *Notices Amer. Math. Soc.*, 44(10):1288–1293, 1997.

[17] C. Tix. Lower bound for the ground state energy of the no-pair Hamiltonian. *Phys. Lett. B*, 405(3-4):293–296, 1997.

[18] C. Tix. Strict positivity of a relativistic Hamiltonian due to Brown and Ravenhall. *Bull. London Math. Soc.*, 30(3):283–290, 1998.

Mathematik I, Universität Regensburg, D-93040 Regensburg, Germany
E-mail address: `georg.hoever@mathematik.uni-regensburg.de`

Mathematik I, Universität Regensburg, D-93040 Regensburg, Germany
E-mail address: `heinz.siedentop@mathematik.uni-regensburg.de`

MPEJ

MATHEMATICAL PHYSICS ELECTRONIC JOURNAL*

ISSN 1086-6655

Volume 6, 2000

Chief Editors

P. Collet
Ecole Polytechnique
Palaiseau

H. Koch
The University of Texas
Austin

C.E. Wayne
Boston University
Boston

Editorial Board

J.E. Avron
Israel Institute of Technology
Haifa

G. Benettin
Università di Padova
Padua

P. Constantin
The University of Chicago
Chicago

J.-P. Eckmann
Université de Genève
Geneva

J. Feldman
University of British Columbia
Vancouver

G. Gallavotti
Università di Roma La Sapienza
Rome

R. Kotecký
Charles University
Prague

A. Kupiainen
Helsinki University
Helsinki

R. de la Llave
The University of Texas
Austin

H. Spohn
Universität München
Munich

H. Tasaki
Gakushin University
Tokyo

* For an electronic version of this volume, and for information on subscriptions and other matters, see `http://www.ma.utexas.edu/mpej/` or `http://mpej.unige.ch/mpej/` or `http://www.maia.ub.es/mpej/` or send an empty e-mail message to `mpej@maia.ub.es` for instructions.

Contents
Volume 6 (2000)

M P E J

MATHEMATICAL PHYSICS ELECTRONIC JOURNAL

ISSN 1086-6655
Volume 6, 2000

Paper 1
Received: June 21, 1999, Revised: Jan 9, 2000, Accepted: Jan 9, 2000.
Editor: A. Kupiainen

Construction of the renormalized $GN_{2-\epsilon}$ trajectory

M. Salmhofer[1] and Chr. Wieczerkowski[2]

[1] Mathematik, ETH Zürich
CH-8092 Zürich, Switzerland
manfred@math.ethz.ch

[2] Institut für Theoretische Physik I, Universität Münster
Wilhelm-Klemm-Straße 9, D-48149 Münster
wieczer@uni-muenster.de

Abstract

We construct the renormalized Gross-Neveu trajectory in $2-\epsilon$ dimensions. Our construction uses a contraction mapping for an extended renormalization group. The extension is a running coupling with linear step β-function. The contraction mapping relies on norm estimates for a fermionic momentum space renormalization group.

1 Introduction

In this paper, we construct the renormalized trajectory of the (chiral) Gross–Neveu model in $2-\epsilon$ dimensions. The two dimensional model was introduced by Gross and Neveu [1] and by Mitter and Weisz [2], both as a model for asymptotic freedom and for dynamical mass generation. In this paper, we consider a super-renormalizable deformation of its renormalization flow. The deformation mimicks a dimensional continuation, without being a regularization. We use it to illustrate how rigorous control of a fermionic ultraviolet limit can be gained by general norm bounds on fermionic renormalization groups. Our construction relies on a cumulant bound, which was proved first by Gawedzki and Kupiainen in [3]. A simplified proof by Lesniewski appeared later in [4]. In an accompanying paper [5], we will give a fresh proof of the cumulant bound and its implications on norm estimates for fermionic renormalization groups.

Our construction is furthermore based on a non-perturbative implementation of the beta function method of [6, 7]. In this approach, one computes renormalized field theories directly as invariant curves emerging from a renormalization fixed point. The fundamental dynamical equations are the condition of renormalization invariance and a tangent (or first order) condition, which selects a particular curve.

The two dimensional Gross-Neveu model has attracted a lot of attention by rigorous renormalization theorists, both because of its simplicity and its interesting non-perturbative features. We mention the work of Gawedzki and Kupiainen [3]; Feldman, Magnen, Rivasseau, and Seneor [8]; Iagolnitzer and Magnen [9, 10]; Kopper, Magnen and Rivasseau [11]. Recent work of Disertori and Rivasseau [12] simplifies earlier constructions by avoiding the use of phase space expansion technology. Our work has the same intention although it proceeds along a different route. Another simple, and conceptually rather different, approach, which organizes perturbation theory in a *ring expansion*, was developed in [13]. Although it has not been applied to a construction of the Gross-Neveu model, it is an alternative to our bounds.

In the following, we briefly describe the model and our main result, leaving detailed definitions for later sections. The ϵ comes from a modification of the massless free propagator, which reads

$$\widehat{C}(p) = \frac{\zeta \not{p}}{|p|^{2+\epsilon}} \tag{1}$$

in momentum space. Here $\zeta \in \mathbb{C}$; $\not{p} = p_1\gamma_1 + p_2\gamma_2$, where γ_1 and γ_2 are two–dimensional hermitean Dirac matrices, with $\{\gamma_\mu, \gamma_\nu\} = 2\delta_{\mu,\nu}$; and finally $0 < \epsilon < \frac{2}{3}$. As an interaction, we can take any chirally invariant four-fermion interaction. For instance, the two choices

$$\mathcal{O}_{GN}(\psi) = \int \mathrm{d}x \Big(\bar{\psi}\psi(x)\Big)^2 \tag{2}$$

and

$$\mathcal{O}_{GN}(\psi) = \int \mathrm{d}x \left\{ \Big(\bar{\psi}\psi(x)\Big)^2 + \Big(\bar{\psi}\mathrm{i}\gamma_5\psi(x)\Big)^2 \right\}, \tag{3}$$

which correspond to the Gross–Neveu model with discrete or continuous chiral symmetry, are allowed. Here $\bar{\psi}\psi(x) = \sum_{a,\sigma} \bar{\psi}_{\sigma,a}(x)\psi_{\sigma,a}(x)$, where $\sigma \in \{1, 2\}$ is the spin and $a \in \{1, \ldots, N\}$ is the colour index. A function of $F(\psi, \bar{\psi})$ has a continuous chiral invariance if

$$F(\psi, \bar{\psi}) = F\big(\mathrm{e}^{\mathrm{i}\alpha\gamma_5}\psi, \bar{\psi}\mathrm{e}^{\mathrm{i}\alpha\gamma_5}\big) \tag{4}$$

for all $\alpha \in \mathbb{R}$, regarding ψ as a column vector and $\bar{\psi}$ as a row vector with respect to the spin indices. γ_5 is a hermitian matrix that anticommutes with the matrices γ_1 and γ_2 and whose square is $(\gamma_5)^2 = 1$. The interaction (2) is only invariant under the above transformation if $\alpha = \pi$. But even this discrete symmetry forbids a mass term $m \int \mathrm{d}x\ \bar{\psi}(x)\psi(x)$.

A construction of the model can be obtained by iteration of a renormalization group transformation R_L, which combines an integration over fluctuations with a rescaling step. We use the scaled momentum space renormalization group as in [3], given by

$$R_L(V)(\Psi) = \log \int \mathrm{d}\mu_{C_L}(\Phi)\ \exp\Big(V(S_L\Psi + \Phi)\Big) - \text{const.}, \tag{5}$$

Here C_L is a two sided regularization of (1), with unit ultraviolet cutoff and infrared cutoff L^{-1} in units of mass, $\mathrm{d}\mu_{C_L}(\Phi)$ is the corresponding fermionic Gaussian measure (10), S_L is a dilatation by a scale factor of L, and $V(\Psi)$ is a fermionic potential, which perturbs the free model (details follow below). The constant is subtracted to make $R_L(V)(0) = 0$. Eq. (5) satisfies the semi-group law $R_L R_{L'} = R_{L\,L'}$. Consequently, an n-fold iteration of (5) is equivalent to a single step with scale L^n. For technical reasons, we prefer a discrete renormalization group with a rather large scale L. The coupling constant g in front of the interaction will have to be small, its maximal value γ depending on L. It may be complex, but since our bounds involve only $|g|$ and the beta function will only amount to multiplication by a real scale factor, we may take $g > 0$ without loss of generality.

Theorem. *There are $L > 1$ and $\gamma \leq 1$ such that, for all $0 \leq g \leq \gamma$, the following holds.*

1. *Let $g_N = L^{-2\epsilon N} g$, and $V_0^{(N)} = g_N\, \mathcal{O}_{GN}(\psi)$, with $\mathcal{O}_{GN}(\psi)$ given by (2) or (3). Then the limit*

$$V(\psi, g) = \lim_{N\to\infty} (R_L)^N (V_0^{(N)})(\psi) \tag{6}$$

exists.

2. *Let $\beta_L(g) = L^{-2\epsilon} g$. Then the composition of R_L with the application of the step beta function β_L has a fixed point. For all $g < L^{-2\epsilon}\gamma$,*

$$R_L\big(V(\psi, g)\big) = V(\psi, \beta_{L^{-1}}(g)). \tag{7}$$

3.

$$V(\psi, g) = g\, \mathcal{O}_{GN}(\psi) + g^{\frac{7}{4}} \mathcal{V}(\psi, g), \tag{8}$$

where $\mathcal{V}$ is small in a norm that depends on L (the details will be given in Section 4).

We prove the Theorem by showing that, in an appropriate Banach space of coupling constants g and interactions, the extended renormalization group T_L, defined by

$$V(\psi, g) \mapsto R_L(V)(\psi, \beta_L(g)), \tag{9}$$

is a contraction mapping on a cone emerging from the free field fixed point, which corresponds to a ball of second order perturbations $\mathcal{V}$. The interactions in this Banach space are analytic in the fields, chirally invariant, and have exponential spatial decay. Their decay length is determined by that of the fluctuation covariance C_L, and is of the order $O(L)$. [1]

1.1 Setup

We consider continuum functional integrals with ultraviolet and infrared cutoff. [2] Our cutoffs will be built into the propagator. The model is then defined by a regularized propagator together with an effective (inter-) action. The details of this standard setup are, for example, given in [14]. It is also possible to regard the Grassmann variables merely as a convenient way of organizing infinite systems of equations for antisymmetric functions.

Our fermionic fields Ψ are indexed by $\mathbb{X} = \mathbb{R}^2 \times \Lambda$, where Λ is a discrete set, in our case $\Lambda = \{1, -1\} \times \{1, 2, \ldots, N\} \times \{1, -1\}$, where the first index is the spin index, the second a colour index, and the third distinguishes between ψ and $\bar{\psi}$ according to $\Psi(x, \sigma, a, 1) = \bar{\psi}_{\sigma,a}(x)$ and $\Psi(x, \sigma, a, -1) = \psi_{\sigma,a}(x)$. The Grassmann Gaussian integral corresponding to a free theory with a propagator C is determined by

$$\int \mathrm{d}\mu_C(\Psi)\, \mathrm{e}^{(\eta, \Psi)} = \mathrm{e}^{\frac{1}{2}(\eta, C\eta)}. \tag{10}$$

Here the $\eta(X)$ are Grassmann source fields labelled by $X \in \mathbb{X}$, and (η, Ψ) is an abbreviation for $\int_{\mathbb{X}} \mathrm{d}X\, \eta(X)\, \Psi(X)$, the integral over $\mathbb{X}$ meaning $\int \mathrm{d}X F(X) = \int \mathrm{d}^2 x \sum_\lambda F(x, \lambda)$.

The fluctuation integral in (5) is well–defined if the covariance $C(X, X')$ is a bounded function of X and X'. The inverse Fourier transform of (1) is not bounded; the construction proceeds by first replacing it by an ultraviolet cutoff covariance which is a finite sum of bounded covariances. The ultraviolet cutoff is removed by taking the number of terms in the sum to infinity, and at the same time letting the coupling constant flow in the way described in the Theorem. The terms in the sum are given by the single–scale covariance of our model,

$$C_L((x, \sigma, a, -1), (x', \sigma', a', 1)) = \mathbf{C}_{1,L}(x, \sigma, a; x', \sigma', a') = -C_L((x', \sigma', a', 1), (x, \sigma, a, -1)), \tag{11}$$

[1] In the scaled renormalization group, one iterates the same transformation, and localization properties depend on this iterated transformation rather than on a flowing scale. Translated to a non-scaled renormalization group, where fluctuation propagators come on different scales, the localization scale becomes proportional to the ultraviolet cutoff.

[2] The continuum regularized functional integral again can be defined by discretizing the regularized field theory to a finite lattice. One then performs both its infinite-volume and zero lattice spacing limit in the presence of continuum cutoffs. We remark that the bounds given in Section 2 imply that the effective action converges as the lattice cutoff is removed.

and zero when the charge indices coincide: $C_L((\cdot,j),(\cdot,j)) = 0$. It is given by the following Dirac propagator

$$\mathbf{C}_{L,L'}(x,\sigma,a;x',\sigma',a') = \delta_{aa'} \int \text{đ}p \; e^{ip(x-x')} \frac{\not{p}_{\sigma,\sigma'}}{|p|^{2+\epsilon}} \left(\hat{\chi}(Lp) - \hat{\chi}(L'p)\right), \tag{12}$$

which is two-sided regularized in momentum space with the help of the cutoff function

$$\hat{\chi}(p) = \frac{1}{\Gamma(1+\frac{\epsilon}{2})} \int_{p^2}^{\infty} \mathrm{d}t \; e^{-t} t^{\frac{\epsilon}{2}}. \tag{13}$$

(This particular regulator has the advantage that the cutoff propagator (12) becomes analytic in momentum space.)

The covariance with unit infrared cutoff and ultraviolet cutoff L^N can be written as a telescope sum

$$\mathbf{C}_{L^{-N},1} = \sum_{m=1}^{N} \mathbf{C}_{L^{-m},L^{-m+1}}; \tag{14}$$

in terms of self-similar **C**s, which are supported on narrow momentum slices,

$$\mathbf{C}_{L^{-m},L^{-m+1}}(x,\sigma,a;x',\sigma',a') = L^{2m\epsilon} \mathbf{C}_{1,L}(L^m x,\sigma,a;L^m x',\sigma',a'). \tag{15}$$

1.2 The RG transformation

The exponent σ denotes the scaling dimension of the massless free fermionic field. In our model, $\sigma = \frac{1}{2}(1-\epsilon)$. The associated scale transformation of fields reads

$$S_L(\Psi)(x,\lambda) = L^{-\sigma}\, \Psi\left(L^{-1}x,\lambda\right). \tag{16}$$

With its help, the self-similarity property of the telescoped covariances (15) becomes

$$\mathbf{C}_{L^{-m},L^{-m+1}} = S_{L^{-m}} \mathbf{C}_{1,L} (S_{L^{-m}})^T \tag{17}$$

in operator notation, where T denotes the transposition. The additive decomposition (14) of the covariance implies that the exponential of the effective action at scale 1 is

$$\int \mathrm{d}\mu_{C_{L^{-N},1}}(\Psi) e^{\mathbf{V}(\Psi+\Phi)} = \int \prod_{m=1}^{N} \mathrm{d}\mu_{C_{1,L}}(\Psi_m) e^{\mathbf{V}\left(S_{L^{-1}}\Psi_1+\ldots+S_{L^{-m}}\Psi_m+\Phi\right)} \tag{18}$$

and is thus equal to $e^{(R_L\circ\cdots\circ R_L)(V)(\Phi)}$, provided that V is a scaled version of the bare potential, namely $V(\Psi) = \mathbf{V}(S_{L^{-m}}\Psi)$. Notice that because of this rescaling, the infrared cutoff of the theory, defined by the left hand side of (18), is one and not L^{-N}, and is not changed by the scaling of the bare potential. More generally, one obtains the scaled renormalization group by a multi-scale transformation, where each multi-scale component is rescaled to a unit scale.

Conversely, one reconstructs the non-scaled renormalization flow by the introduction of a (physical) renormalization scale, often together with a renormalization condition on a coupling parameter. We emphasize that the converse step thus requires an additional datum. [3]

We may decompose R_L into two parts, $R_L = S_L \circ F_L$, with F_L an integration over the fluctuations

$$F_L(V)(\Phi) = \log \int \mathrm{d}\mu_{C_{1,L}}(\Psi) \mathrm{e}^{V(\Psi+\Phi)} - \text{const.} \tag{19}$$

We will derive an estimate on the renormalization group in terms of estimates on these two parts. A field independent constant, which is proportional to the volume, is subtracted in order to preserve the condition $V(0) = 0$ in the RG flow. In statement 1 of the Theorem, the rescaling of the initial coupling constant as a function of the renormalized coupling constant g is given. In the next section, we specify a set of potentials V to which the RG transformation can be applied.

An important property is that the cutoff covariances are of the form $\not{d}$, so that

$$\mathrm{e}^{\mathrm{i}\alpha\gamma_5}\, \mathbf{C}_{L-m,L-m+1}\, \mathrm{e}^{\mathrm{i}\alpha\gamma_5} = \mathbf{C}_{L-m,L-m+1}. \tag{20}$$

Thus, any discrete or continuous chiral invariance of V is preserved under the RG transformation. In other words, if V obeys (4), then the same holds for $F_L(V)$ and $R_L(V)$.

2 Estimate on the renormalization flow

We now give a norm estimate on the renormalization group transformation $R_L = S_L \circ F_L$ built from two separate norm estimates, an estimate on the scale transformation S_L and an estimate on the fluctuation integral F_L. It will serve as a template for the refined estimates presented thereafter.

2.1 Banach space $\mathbb{V}_{h,\kappa}$

We consider potentials of the following general (power series) type. Let $V(\Psi)$ be given by an infinite sum $V(\Psi) = \sum_{f=1}^{\infty} V_f(\Psi)$ of f-point vertices

$$V_f(\Psi) = \int \mathrm{d}X_1 \Psi(X_1) \cdots \int \mathrm{d}X_f \Psi(X_f)\, V_f(X_1, \ldots, X_f), \tag{21}$$

where the vertices are distributional kernels. We will restrict our attention to vertex functions of the general form

$$V_f(X_1, \ldots, X_f) = \sum_{l=0}^{k-1} \overline{V}_{f,l}(X_1, \ldots, X_f) \sum_{\pi \in \mathfrak{S}_f} \prod_{i=1}^{l} \delta(x_{\pi(1)} - x_{\pi(1+i)}), \tag{22}$$

[3]The scaled renormalization group is best thought of as a block spin transformation on lattice theories, which live on an infinite unit lattice, but encode exact continuum information.

where $\overline{V}_{f,l} \in L^1_{loc}(\mathbb{X} \times \cdots \times \mathbb{X}, \mathbb{C})$, and has the usual properties of a fermionic theory (anti-symmetry, Euclidean covariance). We also assume that V is even, that is, $V_f(\Psi) = 0$ for $f \in 2\mathbb{N}+1$. Then $R_L(V)$ is also even. Temporarily, Ψ denotes the fermionic field without derivatives. Later, we will encorporate derivative fields by enlarging Λ to an appropriate multiplet. Let $\|V\|_{h,\kappa} = \sum_{f=1}^{\infty} h^f \, \|V_f\|_\kappa$, where

$$\|V_f\|_\kappa = \sup_{x_0 \in \mathbb{R}^2} \int \mathrm{d}X_1 \cdots \mathrm{d}X_f \, \delta(x_0 - x_1) \, |V_f(X_1, \ldots, X_f)| \, \exp\Big(\kappa \, \mathcal{L}(x_1, \ldots, x_f)\Big). \tag{23}$$

Here $\mathcal{L}(x_1, \ldots, x_f)$ denotes the tree distance of $(x_1, \ldots, x_f)$. The tree distance on an f-tuple is defined as

$$\mathcal{L}(x_1, \ldots, x_n) = \inf_{\tau \in \mathcal{T}_n} \sum_{b \in \tau} \|x_{b_1} - x_{b_2}\|, \tag{24}$$

where $\mathcal{T}_n$ is the set of trees on $\{1, 2, \ldots, n\}$, and where $b = (b_1, b_2) \in \tau$ are the bonds of τ. The potentials with $\|V\|_{h,\kappa} < \infty$ form a Banach space $\mathbb{V}_{h,\kappa}$. It depends on two parameters h and κ, where h can be thought of as an inverse radius of convergence in field space and κ as an inverse exponential rate of decay. We will show that there exists a choice such that the action of R_L is well–defined on a suitable ball around zero in $\mathbb{V}_{h,\kappa}$.

2.2 Estimate on S_L

Let $S_L(V_f)(\Psi) = V_f\Big(S_L(\Psi)\Big)$. Then [4] $\|S_L(V_f)\|_\kappa = L^{2-f\sigma} \, \|V_f\|_{L^{-1}\kappa}$, so S_L performs the following simple scale transformation on our norm

$$\|S_L(V)\|_{h,\kappa} = L^2 \, \|V\|_{L^{-\sigma}h, L^{-1}\kappa}. \tag{25}$$

Derivatives produce additional inverse powers of L. Because $V_f(\Psi) = 0$ for $f \in 2\mathbb{N}+1$,

$$\|S_L(V)\|_{h,\kappa} \leq L^{1+3\epsilon} \, \|V\|_{L^{-\frac{\epsilon}{4}}h, L^{-1}\kappa}, \tag{26}$$

at least under the wasteful condition that $0 < \epsilon < \frac{2}{3}$. Here we saved a small amount of the scale factor to control an anticipated shift of the field, which will come about in the integral over fluctuations.

2.3 Estimate on F_L

As shown in the Appendix, the one–scale propagator C_L satisfies

$$|C_L(X, Y)| \leq O(1) \, L^{-2\sigma} \, \exp\Big(-L^{-1} \, \|x - y\|\Big) \tag{27}$$

[4]The exponent $2 - f\sigma$ is an old friend from perturbative renormalization theory, namely the scaling dimension of V_f.

(here and in the following, $O(1)$ denotes constants which are independent of L), and it has a Gram representation

$$\mathbf{C}_{1,L}(x,\sigma,a;x',\sigma',a') = \langle \varphi_L(x,\sigma,a) | \tilde{\varphi}_L(x',\sigma',a') \rangle \tag{28}$$

where $\|\varphi_L(x,\sigma,a)\|$ and $\|\tilde{\varphi}_L(x',\sigma',a')\|$ are $O(1)$ uniformly in X. The fluctuation transformation is defined as follows. In an expansion in the fields, $F_L(V)(\Psi) = \sum_{f=1}^{\infty} F_L(V)_f(\Psi)$ with

$$\begin{aligned} F_L(V)_f(\Psi) &= \int \mathrm{d}Z_1 \Psi(Z_1) \cdots \int \mathrm{d}Z_f \Psi(Z_f) \\ \sum_{n=1}^{\infty} \frac{1}{n!} \sum_{f_1=1}^{\infty} \sum_{e_1=0}^{f_1-1} \binom{f_1}{e_1} &\cdots \sum_{f_n=1}^{\infty} \sum_{e_n=0}^{f_n-1} \binom{f_n}{e_n} \delta_{f,e_1+\cdots e_n} \Theta_{f,1} (-1)^{\alpha_n(f_1,e_1,\ldots,f_n,e_n)} \\ &\int \mathrm{d}Y_{1,1} \cdots \mathrm{d}Y_{1,i_1} V_{f_1}(X_{1,1},\ldots,X_{1,e_1},Y_{1,1},\ldots,Y_{1,i_1}) \\ &\cdots \int \mathrm{d}Y_{n,1} \cdots \mathrm{d}Y_{n,i_n} V_{f_n}(X_{n,1},\ldots,X_{n,e_n},Y_{n,1},\ldots,Y_{n,i_n}) \\ &\Big\langle \Phi(Y_{1,1}) \cdots \Phi(Y_{1,i_1}); \cdots ; \Phi(Y_{n,1}) \cdots \Phi(Y_{n,i_n}) \Big\rangle_{C_L}^{T}, \end{aligned} \tag{29}$$

where $(Z_1,\ldots,Z_f) = (X_{1,1},\ldots,X_{1,e_1},\ldots,X_{n,1},\ldots,X_{n,e_n})$, $(-1)^{\alpha_n}$ is a sign factor, and where we use the notation $f_l = e_l + i_l$. (f_l is the power of fields of the l'th vertex, e_l is the number of external fields chosen therefrom, and i_l is the number of the remaining internal fields.) The fluctuation integral produces effective vertices $F_L(V)_f(Z_1,\ldots,Z_f)$ as the anti-symmetrized kernels given by the integrand of the expression (29). They are infinite sums of convolutions of the original kernels with propagators. The norm estimates in the remainder of this section imply that these infinite sums converge if $\|V\|_{h,\kappa}$ is small enough.

2.3.1 Estimate on partially truncated correlators

The cumulant expansion (29) involves partially truncated correlators. They obey the following beautiful bound due to Gawedzki and Kupianen [3], equations (109) and (112), and Lesniewski [4]. For a simplified proof consult Theorem 4 in [5].

There are positive constants κ_1, C_1 and C_2, all independent of L, such that

$$\begin{aligned} &\left| \Big\langle \Phi(Y_{1,1}) \cdots \Phi(Y_{1,i_1}); \cdots ; \Phi(Y_{n,1}) \cdots \Phi(Y_{n,i_n}) \Big\rangle_{C_L}^{T} \right| \\ &\leq n!\, C_1^{i_1+\cdots+i_n} \left(L^{-2\sigma} C_2 \right)^{n-1} \exp\Big(-L^{-1} \kappa_1 \mathcal{L}(y_{1,1},\ldots,y_{1,i_l} \mid \cdots \mid y_{n,1},\ldots,y_{n,i_n}) \Big) \end{aligned} \tag{30}$$

In an accompanying paper, we derive these bounds, and the norm bounds that follow from them, in a simplified way [5]. Here $\mathcal{L}(\underline{y_1} \mid \ldots \mid \underline{y_n})$ denotes the inter-tuple tree distance of the tuples $\underline{y_l} = (y_{l,1},\ldots,y_{l,i_l})$ defined as

$$\mathcal{L}(\underline{y_1} \mid \ldots \mid \underline{y_n}) = \inf_{j_l \in \{1,\ldots,i_l\}} \mathcal{L}(y_{1,j_1},\ldots,y_{n,j_n}). \tag{31}$$

One selects a point in each tuple and computes the ordinary tree distance for this selection. The selection with a minimal tree distance defines the inter-tuple tree distance. It will be important that κ_1, C_1, and C_2 do not depend on L because we shall use a large L argument later on. The constant C_1 is proportional to the Gram constant (see the Appendix).

2.3.2 Estimates on tree distances

Since the inter-tuple distance is a tree distance with respect to one particular tree, which may or may not be the minimal one, we have that

$$\sum_{l=1}^{n} \mathcal{L}(\underline{x_l}, \underline{y_l}) + \mathcal{L}(\underline{y_1} \mid \dots \mid \underline{y_n}) \geq \mathcal{L}(\underline{x_1}, \underline{y_1}, \dots, \underline{x_n}, \underline{y_n}). \tag{32}$$

In addition to (32), we need a bound which tells how tree distances behave under the removal of points. [5] There exists a constant α, with $1 < \alpha \leq 2$, such that

$$\alpha \, \mathcal{L}(\underline{x_1}, \underline{y_1}, \dots, \underline{x_n}, \underline{y_n}) \geq \mathcal{L}(\underline{x_1}, \dots, \underline{x_n}). \tag{33}$$

Simple examples show that (33) cannot hold with $\alpha = 1$. For $\alpha = 2$, (33) is readily proved by grouping the removed points into trees and reconnecting the connected components in a way that can be estimated by twice the length of the removed trees.

Let

$$\kappa = L^{-1} \kappa_1 \text{ and } L \geq 4 \geq 2\alpha. \tag{34}$$

Then

$$\frac{\kappa}{L} \mathcal{L}(\underline{x_1}, \dots, \underline{x_n}) - \frac{\kappa_1}{L} \mathcal{L}(\underline{y_1} \mid \dots \mid \underline{y_n}) \leq \kappa \sum_{l=1}^{n} \mathcal{L}(\underline{x_l}, \underline{y_l}) - \frac{\kappa_1}{2\,L} \mathcal{L}(\underline{x_1}, \underline{y_1}, \dots, \underline{x_n}, \underline{y_n}) \tag{35}$$

The estimate (35) is the only property of tree distances which we need in our bound for the fluctuation integral.

2.3.3 Estimate on the f-vertex

We proceed under the assumption (34). Return to (29). From the estimates (30) and (35), we claim it follows that:

$$\|F_L(V)_f\|_{L^{-1}\kappa} \leq \sum_{n=1}^{\infty} \left(L^{2-2\sigma}\, C_2\, C_3\right)^{n-1}$$
$$\sum_{f_1=1}^{\infty} \sum_{e_1=0}^{f_1-1} \binom{f_1}{e_1} C_1^{i_1} \, \|V_{f_1}\|_\kappa \cdots \sum_{f_n=1}^{\infty} \sum_{e_n=0}^{f_n-1} \binom{f_n}{e_n} C_1^{i_n} \, \|V_{f_n}\|_\kappa \, \delta_{f, e_1 + \cdots e_n}. \tag{36}$$

[5] For this reason, Gawedzki and Kupiainen use a different tree distance in [3].

To see this, note that for each vertex one chooses a point to anchor its tree. One then pulls out the vertex norms. The remaining integral over the anchors is estimated using the spared exponential decay,

$$\sup_{x_0} \int d^2x_1 \cdots \int d^2x_n \, \delta(x_0 - x_1) \exp\left(-\frac{\kappa_1}{2\alpha L}\mathcal{L}(x_1, \dots, x_n)\right) \leq \left(L^2 C_3\right)^{n-1}. \quad (37)$$

To obtain a bound on the (h, κ)-norm from this, we have to sum over the powers of fields. This yields the geometric series

$$\|F_L(V)\|_{L^{-\sigma}h, L^{-1}\kappa} \leq \sum_{n=1}^{\infty} \left(L^{2-2\sigma} C_2 C_3\right)^{n-1} \left(\|V\|_{L^{-\sigma}h + C_1, \kappa}\right)^n \quad (38)$$

which converges if $q\|V\|_{L^{-\sigma}h+C_1,\kappa} < 1$, where $q = L^{2-2\sigma} C_2 C_3$; then

$$\|F_L(V)\|_{L^{-\sigma}h, L^{-1}\kappa} \leq \frac{\|V\|_{L^{-\sigma}h+C_1,\kappa}}{1 - q\|V\|_{L^{-\sigma}h+C_1,\kappa}}. \quad (39)$$

This shows that the RG transform is well–defined on a ball of potentials that are analytic in the fields.

2.4 Estimate on R_L

Let h satisfy

$$h = L^{-\frac{\epsilon}{4}} h + C_1. \quad (40)$$

Both h and κ now depend on L. Let $V(\Psi)$ then be an element of the ball $B_r = \left\{V(\Psi) \in \mathbb{V}_{h,\kappa} \middle| \|V\|_{h,\kappa} \leq r\right\}$ with sufficiently small radius r. Then (26) and (39) imply together that $R_L : B_r \to B_{f_L(r)}$ with a flow of radii given by

$$f_L(r) = \frac{L^{1+3\epsilon} r}{1 - qr} \quad (41)$$

Unfortunately, this bound is not sufficient for an iteration of R_L because small potentials tend to grow. [6] This behavior indicates the necessity of renormalization. The factor L in (41) will be removed by restricting to a subspace of potentials with vanishing mass vertex.

3 Two point vertex

The scaling dimension of an $2f$–vertex is $2 - f\sigma$. Because all vertices with an odd number of fields f vanish, the lowest non-vanishing vertex is a two point vertex. Its scaling dimension is $1 + \epsilon$, which is also the scaling dimension of a local mass vertex. The two point vertex is the most relevant vertex of our flow. In this section, we will split it into a local and a non-local part. The non-local part will have an improved scaling dimension. The local part is zero for chirally invariant interactions.

[6] The largest eigenvalue of the linearized renormalization group is here $L^{1+\epsilon}$. The extra factor of $L^{2\epsilon}$ is due to the non-linear corrections.

3.1 Localization operator L

The localization operator amounts to a Taylor expansion with remainder in momentum space. For the purposes of this paper, a lowest order expansion of the two point vertex suffices. [7] In real space, we define $\mathbf{L}$ by the decomposition

$$V_2(\Psi) = \int \mathrm{d}X_1 \int \mathrm{d}X_2\, \Psi(X_1) \Big(\mathbf{L}(V_2)(X_1, X_2) + (\mathbf{1} - \mathbf{L})(V_2)(X_1, X_2)\Big)\, \Psi(X_2) \quad (42)$$

into a local part

$$\mathbf{L}(V_2)\Big((x_1, \lambda_1), (x_2, \lambda_2)\Big) = \delta(x_1 - x_2) \int \mathrm{d}^2 y_2\, V_2\Big((x_1, \lambda_1), (y_2, \lambda_2)\Big) \quad (43)$$

and a non-local part

$$\begin{aligned}&(\mathbf{1} - \mathbf{L})(V_2)\Big((x_1, \lambda_1), (x_2, \lambda_2)\Big) \\ &= \sum_{\mu=1}^{2} \int_0^1 \mathrm{d}t\, t^{-2}\, V_2\left((x_1, \lambda_1), \left(x_1 + \frac{x_2 - x_1}{t}\right), \lambda_2\right) \frac{x_2^\mu - x_1^\mu}{t} \frac{\partial}{\partial x_2^\mu}. \end{aligned} \quad (44)$$

The t-integral converges at $t = 0$ because the vertex decays exponentially fast at infinity. This splitting follows from a Taylor expansion with remainder term of the second field

$$\Psi(x_2, \lambda_2) = \Psi(x_1, \lambda_2) + \int_0^1 \mathrm{d}t\, (x_2 - x_1) \cdot (\partial\Psi)(x_1 + t(x_2 - x_1), \lambda_2) \quad (45)$$

around the position of the first one. After a change of integration variables, one obtains an expression of the form

$$\begin{aligned} V_2(\Psi) = &\int \mathrm{d}^2 x \sum_{\lambda_1, \lambda_2} \Psi(x, \lambda_1)\, m_{\lambda_1, \lambda_2}\, \Psi(x, \lambda_2) \\ &+ \sum_\mu \int \mathrm{d}X_1 \int \mathrm{d}X_2 \Psi(X_1)\, V_{2,\mu}(X_1, X_2)\, (\partial_\mu \Psi)(X_2) \end{aligned} \quad (46)$$

3.2 Redefinition of the norm $\|V_2\|_\kappa$

Let us represent the two point vertex as in (46). Then we may redefine its norm into $|||V_2|||_\kappa = \|\mathbf{L}V_2\| + \|(\mathbf{1} - \mathbf{L})V_2\|_\kappa$ with $\|\mathbf{L}V_2\| = \sum_{\lambda_1, \lambda_2} |m_{\lambda_1, \lambda_2}|$ and

$$\|(\mathbf{1} - \mathbf{L})V_2\|_\kappa = \kappa \sup_{x_0} \int \mathrm{d}X_1 \int \mathrm{d}X_2 \sum_\mu |V_{2,\mu}(X_1, X_2)| \exp\Big(\kappa\, \|x_1 - x_2\|\Big), \quad (47)$$

which is the old norm of the non-local part times κ. For the higher vertices, we use the old norm. We can now redo the above estimates with this redefined norm.

[7] In the case when $\epsilon = 0$, one has to expand the two point vertex to third order and the four point vertex to first order, as is done in [3] and [8]. The formulas are immediate generalizations of those presented here.

3.2.1 Estimate on S_L

The remainder term has an improved scaling dimension. S_L and $\mathbf{L}$ commute. Therefore, the local term scales according to $\|\mathbf{L}S_L V_2\| = L^{1+\epsilon}\|\mathbf{L}V_2\|$, while the non-local remainder scales as

$$\|(\mathbf{1}-\mathbf{L})S_L V_2\|_\kappa = L^\epsilon \, \|(\mathbf{1}-\mathbf{L})V_2\|_{L^{-1}\kappa} \tag{48}$$

because of its derivative field. In our model, the local mass term is zero because of the chiral symmetry (4). The net gain of the localization procedure is a factor of L^{-1} for the redefined norm, since

$$|||S_L V|||_{h,\kappa} = L^\epsilon \, h^2 \, |||V_2|||_{L^{-1}\kappa} + \sum_{n=2}^{\infty} L^{2-n(1-\epsilon)} \, h^{2n} \, |||V_{2n}|||_{L^{-1}\kappa}. \tag{49}$$

This gives the following refinement of (26). If $0 < \epsilon < \frac{2}{3}$, then

$$|||S_L V|||_{h,\kappa} \leq L^{3\epsilon} \, |||V|||_{L^{-\frac{\epsilon}{4}}h, L^{-1}\kappa} \tag{50}$$

3.2.2 Estimate on F_L

The norm of the non-local term can be bounded by the norm of the non-differentiated vertex. By definition

$$\|(\mathbf{1}-\mathbf{L})V_2\|_\kappa = \kappa \sup_{x_1} \int \mathrm{d}^2x_2 \sum_{\lambda_1,\lambda_2,\mu} \mathrm{e}^{\kappa\|x_1-x_2\|}$$
$$\left| \int_0^1 \mathrm{d}t \, t^{-2} \, V_2\left((x_1,\lambda_1), \left(x_1 + \frac{x_2-x_1}{t}, \lambda_2\right)\right) \frac{x_2^\mu - x_1^\mu}{t} \right| \tag{51}$$

it follows that

$$\|(\mathbf{1}-\mathbf{L})V_2\|_\kappa \leq \sup_{x_1} \int \mathrm{d}^2x_2 \sum_{\lambda_1,\lambda_2} \mathrm{e}^{\kappa\|x_1-x_2\|} \left| V_2\big((x_1,\lambda_1),(x_2,\lambda_2)\big) \right|$$
$$\kappa \sum_\mu |x_1^\mu - x_2^\mu| \int_0^1 \mathrm{d}t \, \mathrm{e}^{t\kappa\|x_1-x_2\|}. \tag{52}$$

For our convenience, we define $\|x\| = \sum_\mu |x^\mu|$. Then we have the promised estimate

$$\|(\mathbf{1}-\mathbf{L})V_2\|_\kappa \leq \|V_2\|_\kappa. \tag{53}$$

Consequently, we find the following estimate for the effective non-local two point vertex

$$\|(\mathbf{1}-\mathbf{L})S_L F_L(V)_2\|_\kappa = L^\epsilon \, \|(\mathbf{1}-\mathbf{L})F_L(V)_2\|_{L^{-1}\kappa} \leq L^\epsilon \, \|F_L(V)_2\|_{L^{-1}\kappa} \tag{54}$$

computed as the image of one renormalization group transformation. The local mass term is zero by the chiral invariance.

We now do the estimate of the fluctuation step exactly as in Section 2. This is possible because (47) is of the same form as the old norm up to a factor κ. For each factor V_2 we pick up a factor κ^{-1}. But we also get one derivative field for each factor V_2. Fortunately,

$$\left|\frac{\partial}{\partial x^\mu} C_L(X,Y)\right| \leq O(1)\, L^{-2\sigma-1}\, \exp\Big(- L^{-1}\, \|x-y\|\Big) \tag{55}$$

comes with an additional factor L^{-1}, which compensates the L-factor in κ^{-1}. The remaining constant is easily accommodated since it is of the order $O(1)$. We shift it into a modified cumulant bound. Thus C_1 and C_2 are now understood to be redefined such that the cumulant bound holds for the enlarged multiplet Ψ, which includes derivative fields.

3.3 Estimate on the massless renormalization group

Summing over powers of the field, we get

$$|||R_L(V)|||_{h,\kappa} \leq L^\epsilon\, h^2\, \|F_L(V)_2\|_{L^{-1}\kappa} + \sum_{n=2}^{\infty} L^{2-n(1-\epsilon)}\, h^{2n}\, \|F_L(V)_{2n}\|_{L^{-1}\kappa}. \tag{56}$$

The largest scale factor is now $L^{2\epsilon}$. For $0 < \epsilon < \frac{2}{3}$, it follows that

$$|||R_L(V)|||_{h,\kappa} \leq L^{3\epsilon} \sum_{n=1}^{\infty} \Big(L^{-\frac{\epsilon}{4}}\, h\Big)^{2n}\, \|F_L(V)_{2n}\|_{L^{-1}\kappa}. \tag{57}$$

As before, $h = L^{-\frac{\epsilon}{4}} h + C_1$. Then we can sum the series as above. The result is the estimate

$$|||R_L(V)|||_{h,\kappa} \leq L^{3\epsilon} \frac{|||V|||_{h,\kappa}}{1 - q|||V|||_{h,\kappa}} \tag{58}$$

with $q = L^{2-2\sigma}\, O(1)$. Thus, also in the massless renormalization group, small potentials tend to grow. But the pace is reduced.

In the following, we shall only work with $|||\cdot|||$; for simplicity, we denote it by the usual norm symbol $\|\cdot\|$.

4 Invariant ball

We turn our attention from points in the space of chirally invariant even potentials to parametrized continuous curves $V(\Psi|g)$, $g \in [0,\gamma]$, which are of the form

$$V(\Psi|g) = g\, \mathcal{O}_{GN}(\Psi) + g^{\frac{7}{4}}\, \mathcal{V}(\Psi|g). \tag{59}$$

Here $\mathcal{O}_{GN}(\Psi)$ denotes the normal ordered Gross-Neveu vertex and $\mathcal{V}(\Psi|g) = O(g^{\frac{1}{4}})$ denotes a second order correction to it. For any fixed g, the potentials of the type (59) form a linear space. We shall estimate the remainder in

$$\|\mathcal{V}\|_{\gamma,h,\kappa} = \sup_{g\in[0,\gamma]} \|\mathcal{V}(\cdot|g)\|_{h,\kappa}. \tag{60}$$

The additional parameter γ denotes the maximal admissible value of the coupling constant g in our estimates.

4.1 Step β-function

The linearization of R_L at the free field fixed point $V^\star(\Psi) = 0$ is the first term of the cumulant expansion

$$DR_L(V)(\Psi) = \int \mathrm{d}\mu_{C_L}(\Phi)\, V(S_L\Psi + \Phi) - \text{const.} \tag{61}$$

The normal ordered Gross-Neveu vertex is an eigenvector of the linearized renormalization group

$$DR_L(\mathcal{O}_{GN})(\Psi) = L^{2\epsilon}\, \mathcal{O}_{GN}(\Psi) \tag{62}$$

with eigenvalue $L^{2\epsilon}$. For $\epsilon > 0$, it is a relevant perturbation. We use the inverse of the eigenvalue in (62) to define our step β-function as the linear function

$$\beta_L(g) = L^{-2\epsilon}\, g. \tag{63}$$

4.2 Extended renormalization group

We then define an extended renormalization group transformation as the composition $T_L = \beta_L \circ R_L$ of a linear coupling transformation $\beta_L(V)(\Psi|g) = V(\Psi|\beta_L(g))$ and the renormalization group R_L. The additional step β-function turns the Gross-Neveu vertex into a fixed point

$$DT_L(g\mathcal{O}_{GN})(\Psi) = g\mathcal{O}_{GN}(\Psi) \tag{64}$$

of the linearized extended renormalization group. Our desire is a non-linear extension thereof. For this purpose, we consider the transformation of the second order correction

$$\mathcal{T}_L(\mathcal{V})(\Psi|g) = g^{-\frac{7}{4}}\, \beta_L S_L \mathcal{F}_L(\mathcal{V})(\Psi|g) = g^{-\frac{7}{4}} \left(T_L(V)(\Psi|g) - g\, \mathcal{O}_{GN}(\Psi) \right) \tag{65}$$

4.2.1 Estimate on β_L

The flow of the coupling constant yields an extra small factor. It will turn out to be sufficient to renormalize the theory. We have that

$$\|\mathcal{T}_L(\mathcal{V})\|_{\gamma,h,\kappa} = L^{-\frac{7\epsilon}{2}} \sup_{g\in[0,L^{-2\epsilon}\gamma]} g^{-\frac{7}{4}}\, \|S_L\mathcal{F}_L(\mathcal{V})(\cdot|g)\|_{h,\kappa} \tag{66}$$

Since $L > 1$ and $\epsilon > 0$, we have that $[0, L^{-2\epsilon}\gamma] \subset [0, \gamma]$ and therefore

$$\|\mathcal{T}_L(\mathcal{V})\|_{\gamma,h,\kappa} \leq L^{-\frac{7\epsilon}{2}} \sup_{g\in[0,\gamma]} g^{-\frac{7}{4}}\, \|S_L\mathcal{F}_L(\mathcal{V})(\cdot|g)\|_{h,\kappa}. \tag{67}$$

In the following, we do not need the small scale factors coming with terms of higher order than g^2.

4.2.2 Estimate on S_L

As a payoff of our general massless estimate (50) it follows that the right hand side of (67) itself can be further estimated by

$$\|\mathcal{T}_L(\mathcal{V})\|_{\gamma,h,\kappa} \leq L^{-\frac{\epsilon}{2}} \sup_{g\in[0,\gamma]} g^{-\frac{7}{4}} \|\mathcal{F}_L(\mathcal{V})(\cdot|g)\|_{L^{-\frac{\epsilon}{4}}h,L^{-1}\kappa} \tag{68}$$

The prefactor $L^{-\frac{\epsilon}{2}} < 1$ will become responsible for the contraction property.

4.2.3 Estimate on $\mathcal{F}_L$

In this renormalization group, we track the transformation of the non-linear corrections to a pure Gross–Neveu vertex. By its definition (65), the subtracted fluctuation step reads, in a selfexplanatory notation,

$$\mathcal{F}_L(\mathcal{V})(\Psi|g) = g^{\frac{7}{4}} \Big\langle \mathcal{V}(S_L\Psi + \Phi \cdot |g) \Big\rangle_{C_L}$$
$$+ \sum_{n=2}^{\infty} \frac{1}{n!} \Big\langle \prod_{i=1}^{n} \Big[g\, \mathcal{O}_{GN}(S_L\Psi + \Phi) + g^{\frac{7}{4}}\, \mathcal{V}(S_L\Psi + \Phi); \Big] \Big\rangle_{C_L}^{T}. \tag{69}$$

The estimate of (69) goes exactly as in Section 2. With the choice (40), the result is

$$\|\mathcal{F}_L(\mathcal{V})(\cdot|g)\|_{L^{-\frac{\epsilon}{4}}h,L^{-1}\kappa} \leq g^{\frac{7}{4}} \|\mathcal{V}(\cdot|g)\|_{h,\kappa}$$
$$+ \sum_{n=2}^{\infty} \Big(L^{2-2\sigma}\, C_2\, C_3 \Big)^{n-1} \Big(g\, \|\mathcal{O}_{GN}\|_{h,\kappa} + g^{\frac{7}{4}}\, \|\mathcal{V}(\cdot|g)\|_{h,\kappa} \Big)^{n} \tag{70}$$

When this estimate is plugged into (68), the factor $g^{-\frac{7}{4}}$ cancels, and

$$\|\mathcal{T}_L(\mathcal{V})\|_{\gamma,h,\kappa} \leq L^{-\frac{\epsilon}{2}} \sup_{g\in[0,\gamma]} \Big\{ \|\mathcal{V}(\cdot|g)\|_{h,\kappa}$$
$$+ \sum_{n=2}^{\infty} \Big(L^{2-2\sigma}\, C_2\, C_3\, g^{\frac{1}{4}} \Big)^{n-1} g^{\frac{3}{4}(n-2)} \Big(\|\mathcal{O}_{GN}\|_{h,\kappa} + g^{\frac{3}{4}}\, \|\mathcal{V}(\cdot|g)\|_{h,\kappa} \Big)^{n} \Big\} \tag{71}$$

We have chosen h and κ to depend on L. We now also choose γ to depend on L. We demand that $\gamma \leq 1$ be so small that

$$L^{2-2\sigma}\, C_2\, C_3\, \gamma^{\frac{1}{4}} \leq \frac{1}{2\, \|\mathcal{O}_{GN}\|_{h,\kappa}}. \tag{72}$$

With this, we have the following estimate on the extended renormalization group

$$\|\mathcal{T}_L(\mathcal{V})\|_{\gamma,h,\kappa} \leq L^{-\frac{\epsilon}{2}} \Big\{ \|\mathcal{V}\|_{\gamma,h,\kappa} + 2\, \|\mathcal{O}_{GN}\|_{h,\kappa} \sum_{n=2}^{\infty} \Big(\frac{1}{2} + \frac{\|\mathcal{V}\|_{\gamma,h,\kappa}}{2\, \|\mathcal{O}_{GN}\|_{h,\kappa}} \Big)^{n} \Big\}. \tag{73}$$

4.3 Invariant ball

The only parameter which has not been fixed yet is L. This last parameter in our map can be used to find an invariant ball of second order perturbations. Let

$$B = \left\{ \mathcal{V} \in \mathbb{V}_{\gamma,h,\kappa} \middle| \|\mathcal{V}\|_{\gamma,h,\kappa} \le \frac{\|\mathcal{O}_{GN}\|_{h,\kappa}}{2} \right\} \tag{74}$$

Let L be so large such that

$$L^{-\frac{\epsilon}{2}} \le \frac{1}{10}. \tag{75}$$

Then (73) implies that the ball of second order perturbations (74) is invariant under the extended renormalization group transformation $\mathcal{T}_L$.

5 Contraction property

The last property to be shown is that any pair of points in the invariant ball move closer under an extended renormalization group transformation.

5.1 Estimates on β_L and S_L

The treatment of β_L and S_L remains the same as in the previous section. The result is

$$\begin{aligned} &\|\mathcal{T}_L(\mathcal{V}_1) - \mathcal{T}_L(\mathcal{V}_2)\|_{\gamma,h,\kappa} \\ &\le L^{-\frac{\epsilon}{2}} \sup_{g\in[0,\gamma]} g^{-\frac{7}{4}} \|\mathcal{F}_L(\mathcal{V}_1)(\cdot|g) - \mathcal{F}_L(\mathcal{V}_2)(\cdot|g)\|_{L^{-\frac{\epsilon}{4}}h,L^{-1}\kappa} \end{aligned} \tag{76}$$

5.2 Estimate on $\mathcal{F}_L$

The difference on the right hand side of (76) leads to a cancellation of the $\mathcal{V}$-independent term, as is best seen from the formula

$$\begin{aligned} &\mathcal{F}_L(\mathcal{V}_1)(\Psi|g) - \mathcal{F}_L(\mathcal{V}_2)(\Psi|g) = g^{\frac{7}{4}} \left\langle \mathcal{V}_1(\cdot|g) - \mathcal{V}_2(\cdot|g) \right\rangle_{C_L} (\Psi) \\ &+ \sum_{n=2}^{\infty} \frac{1}{(n-1)!} \int_0^1 \mathrm{d}s \left\langle \left[g\,\mathcal{O}_{GN} + g^{\frac{7}{4}}\,\mathcal{V}_2 + s\,g^{\frac{7}{4}} \left(\mathcal{V}_1(\cdot|g) - \mathcal{V}_2(\cdot|g) \right); \right]^{n-1} \right. \\ &\left. ; g^{\frac{7}{4}} \left(\mathcal{V}_1(\cdot|g) - \mathcal{V}_2(\cdot|g) \right) \right\rangle^{T}_{C_L} (\Psi) \end{aligned} \tag{77}$$

In complete analogy to (70), we conclude that the following estimate holds

$$\begin{aligned} &\|\mathcal{F}_L(\mathcal{V}_1)(\Psi|g) - \mathcal{F}_L(\mathcal{V}_2)(\Psi|g)\|_{L^{-\frac{\epsilon}{4}}h,L^{-1}\kappa} \le g^{\frac{7}{4}} \|\mathcal{V}_1(\cdot|g) - \mathcal{V}_2(\cdot|g)\|_{h,\kappa} \\ &+ g^{\frac{7}{4}} \|\mathcal{V}_1(\cdot|g) - \mathcal{V}_2(\cdot|g)\|_{h,\kappa} \sum_{n=2}^{\infty} \left(L^{2-2\sigma} C_2 C_3 g^{\frac{1}{4}} \right)^{n-1} n\, g^{\frac{3}{4}(n-1)} \\ &\int_0^1 \mathrm{d}s \left(\|\mathcal{O}_{GN}\|_{h,\kappa} + g^{\frac{3}{4}} \|(1-s)\mathcal{V}_1(\cdot|g) + s\mathcal{V}_2(\cdot|g)\|_{h,\kappa} \right)^{n-1} \end{aligned} \tag{78}$$

On top of (and consistent with) the above choices of h, κ, γ, and L, we demand that γ be so small that $n\gamma^{\frac{3}{4}(n-1)} \leq \left(\frac{4}{3}\right)^2$ for all $n \in \{2, 3, 4, \dots\}$. Then we have that

$$\|\mathcal{T}_L(\mathcal{V}_1) - \mathcal{T}_L(\mathcal{V}_2)\|_{\gamma,h,\kappa} \leq L^{-\frac{\epsilon}{2}} \left\{1 + \left(\frac{4}{3}\right)^2 \sum_{n=2}^{\infty} \left(\frac{3}{4}\right)^n\right\} \|\mathcal{V}_1 - \mathcal{V}_2\|_{\gamma,h,\kappa} \tag{79}$$

But $L^{-\frac{\epsilon}{2}} \leq \frac{1}{10}$, so (79) implies that

$$\|\mathcal{T}_L(\mathcal{V}_1) - \mathcal{T}_L(\mathcal{V}_2)\|_{\gamma,h,\kappa} \leq \frac{1}{2} \|\mathcal{V}_1 - \mathcal{V}_2\|_{\gamma,h,\kappa}. \tag{80}$$

Thus our extended RG transformation $\mathcal{T}_L$ is indeed a contraction mapping on the ball B.

6 Conclusions

In this paper, we have constructed the renormalized Gross-Neveu trajectory as an invariant curve in the unstable manifold of the free field fixed point. We have chosen a parametrization, in which the renormalization group acts on the curve in a normal form. The normal form of super-renormalizable models is a linear step β-function. It can be used in models whose differential β-function

$$\dot{\beta}(g) = \partial_L \beta_L(g)|_{L=1} = \beta_1 g + \beta_2 g^2 + \beta_3 g^3 \cdots \tag{81}$$

has a non-vanishing coefficient $\beta_1 < 0$ (and is regular enough for the first coefficent to be leading). In our model $\beta_1 = -2\epsilon$.

In the non-deformation limit $\epsilon = 0$, the model stays renormalizable due to the sign of the second order correction. Its construction is slightly different from the super-renormalizable case. The normal form of the differential β-function is now cubic, that is, $\dot{\beta}(g) = \beta_2 g^2 + \beta_3 g^3$ with the well known universal constants β_2 and β_3. As L is increased, the coupling flows logarithmically rather than powerlike. Therefore, we cannot extract inverse powers of L from the flowing coupling. One deals with this situation in the usual way by imposing renormalization conditions the non-irrelevant vertices, namely the Gross-Neveu vertex and the wave function vertex. (The mass vertex is still forbidden.) But the infinite series of higher monomials in the fields ψ can be treated exactly as in this paper.

An interesting extension of the present work would be to gain complete control of the renormalized trajectory all the way from the ultraviolet to the expected infrared fixed point. In our model, this would require control of the large coupling limit.

Acknowledgements

We thank the *Forschungsinstitut für Mathematik* at the ETH Zürich for support and the referee for constructive remarks.

A Propagator properties

The decay properties follow immediately from the integral representation

$$\mathbf{C}_{1,L}(x,a,\sigma;x',a',\sigma') = \delta_{a,a'} \frac{\mathrm{i}}{4\pi\,\Gamma(1+\frac{\epsilon}{2})} \int_1^{L^2} \mathrm{d}\alpha\, \alpha^{\frac{\epsilon}{2}-1} \not{\partial}_{\sigma,\sigma'} \mathrm{e}^{-\frac{(x-x')^2}{4\alpha}}. \tag{82}$$

The Gram representation holds by the Fourier representation (12) of $\mathbf{C}_{1,L}$: with the spectral decomposition

$$\gamma_\mu = \sum_\rho \lambda_\rho |\mu,\rho\rangle\langle\mu,\rho|, \tag{83}$$

we have

$$(\gamma_\mu)_{\sigma,\sigma'} = \sum_\rho \lambda_\rho \langle\sigma|\mu,\rho\rangle\langle\mu,\rho|,\sigma'\rangle. \tag{84}$$

Thus

$$\varphi_L(x,a,\sigma)(\rho,p,\mu,c) \;=\; \delta_{a,c}\,\mathrm{e}^{-\mathrm{i}px}\,|\lambda_\rho p_\mu|^{\frac{1}{2}}\,\langle\sigma|\mu\rho\rangle f_L(p) \tag{85}$$

$$\tilde{\varphi}_L(x',a',\sigma')(\rho,p,\mu,c) \;=\; \delta_{c,a'}\,\mathrm{e}^{-\mathrm{i}px'}\,\frac{\lambda_\rho p_\mu}{|\lambda_\rho p_\mu|^{\frac{1}{2}}}\,\langle\sigma'|\mu\rho\rangle f_L(p) \tag{86}$$

with

$$f_L(p) = \left(\frac{\hat{\chi}(p)-\hat{\chi}(Lp)}{|p|^{1+\frac{\epsilon}{2}}}\right)^{1/2}. \tag{87}$$

The norms of φ and $\tilde{\varphi}$ are bounded by

$$(|\lambda_1|+|\lambda_2|)\int_{\mathbb{R}^2} \frac{\mathrm{d}^2p}{|p|^{1+\epsilon}}(\hat{\chi}(p)-\hat{\chi}(L\,p)) \le 2\int_{\mathbb{R}^2} \frac{\mathrm{d}^2p}{|p|^{1+\epsilon}}\hat{\chi}(p) \tag{88}$$

Because $\hat{\chi}(p) \le O(1)|p|^\epsilon \mathrm{e}^{-p^2}$ for $|p| \ge 1$, the integral converges at infinity. The integral over the unit disk is finite for $\epsilon < 1$.

References

[1] D. J. Gross and A. Neveu. Dynamical Symmetry Breaking in Asymptotically Free Field Theories. *Phys. Rev.*, D(10):3235–3253, 1974.

[2] P. K. Mitter and P. H. Weisz. Asymptotic Scale Invariance in a Massive Thirring Model with $U(n)$ Symmetry. *Phys. Rev. D*, 8(12), 1973.

[3] K. Gawedzki and A. Kupiainen. Gross-Neveu Model Through Convergent Perturbation Expansion. *Commun. Math. Phys.*, 102:1–30, 1985.

[4] A. Lesniewski. Effective Action for the Yukawa$_2$ Quantum Field Theory. *Commun. Math. Phys.*, 108:437–467, 1987.

[5] M. Salmhofer and Chr. Wieczerkowski. Positivity and Convergence in Fermionic Quantum Field Theory. *math-phys/9909002, to appear in Jour. Stat. Physics.*

[6] Chr. Wieczerkowski. The Renormalized ϕ_4^4-Trajectory by Perturbation Theory in the Running Coupling I: The Discrete Renormalization Group. *Nucl. Phys. B*, 488:441–465, 1997.

[7] Chr. Wieczerkowski. The Renormalized ϕ_4^4-Trajectory by Perturbation Theory in the Running Coupling II: The Continuous Renormalization Group. *Nucl. Phys. B*, 488:466–489, 1997.

[8] J. Feldman, J. Magnen, V. Rivasseau, and R. Sénéor. Massive Gross-Neveu Model: A Rigorous Perturbative Construction. *Phys. Rev. Lett.*, 54:1479–1481, 1985.

[9] D. Iagolnitzer and J. Magnen. Asymptotic Completeness and Multi-Particle Structure in Field Theories. *Commun. Math. Phys.*, 111:81–100, 1987.

[10] D. Iagolnitzer and J. Magnen. Bethe-Salpeter Kernel and Short Distance Expansion in the Massive Gross-Neveu Model. *Commun. Math. Phys.*, 111:81–100, 1987.

[11] C. Kopper, J. Magnen, and V. Rivasseau. Mass Generation in the Large N Gross-Neveu Model. *Commun. Math. Phys.*, 169:121–180, 1995.

[12] M. Disertori and V. Rivasseau. Continuous Fermionic Renormalization. *hep-th/9802145*, 1998.

[13] J. Feldman, H. Knörrer, E. Trubowitz. A Representation for Fermionic Correlation Functions. *Commun. Math. Phys.*, 195:465–493, 1998.

[14] M. Salmhofer. *Renormalization: An Introduction.* Texts and Monographs in Physics. Springer-Verlag, Berlin-Heidelberg-New York, 1999.

M P E J

MATHEMATICAL PHYSICS ELECTRONIC JOURNAL

ISSN 1086-6655
Volume 6, 2000

Paper 2
Received: Nov 19, 1999 Revised: Feb 10, 2000 Accepted:Feb 15, 2000
Editor: C.E. Wayne

Families of whiskered tori for a-priori stable/unstable Hamiltonian systems and construction of unstable orbits

Enrico Valdinoci

Department of Mathematics, The University of Texas at Austin, TX 78712-1082 (USA)
enrico@math.utexas.edu

Abstract

We give a detailed statement of a KAM theorem about the conservation of partially hyperbolic tori on a fixed energy level for an analytic Hamiltonian $H(I,\varphi,p,q) = h(I,pq;\mu) + \mu f(I,\varphi,p,q;\mu)$, where φ is a $(d-1)$−dimensional angle, I is in a domain of $\mathbb{R}^{d-1}$, p and q are real in a neighborhood 0, and μ is a small parameter. We show that invariant whiskered tori covering a large measure exist for sufficiently small perturbations. The associated stable and unstable manifolds also cover a large measure. Moreover, we show that there is a geometric organization to these tori. Roughly, the whiskered tori we construct are organized in smooth families, indexed by a Cantor parameter. The whole set of tori as well as their stable and unstable manifolds is smoothly interpolated. In particular, we emphasize the following items: sharp estimates on the relative measure of the surviving tori on the energy level, analyticity properties, including dependence upon parameters, geometric structures.

We apply these results to both "a-priori unstable" and "a-priori stable" systems. We also show how to use the information obtained in the KAM Theorem we prove to construct unstable orbits.

Contents

* * * * *

Acknowledgments

This work is part of my "tesi di laurea" [39] at Università di Roma 3 and the final version has been written at UT. Partially supported by CNR, INdAM, TARP. I would like to thank Profs. L. Chierchia and R. de la Llave for their help and encouragement. Comments of anonymous referees were useful. I am also indebted with L. Biasco for many helpful discussions. This paper is dedicated to my grandma, nonna Tina.

1 Introduction

Chains of whiskered invariant tori are the building blocks to prove *instability* of some Hamiltonian systems by a mechanism suggested in a celebrated example of Arnol'd [2]. Even if several precise formulations of Arnol'd diffusion have been proposed in the literature, all of them share that there are chains of invariant tori with hyperbolic directions so that the stable and unstable directions intersect. Then, orbits that follow these transition chains experience "large" changes in the actions.
In this paper, we want to make a detailed study of the survival of whiskered tori and prove results that we hope can be eventually used in the program outlined above.
In particular, we pay special attention to how the tori fit together. We not only prove measure theoretical statements about their abundance (which we show are optimal), but we also show that these tori are organized in a geometric manner. All the whiskered tori that we prove are invariant can be interpolated by a smooth family of whiskered tori (not all the members of the family are invariant). Moreover, we show that the set of invariant tori is organized in smooth families. This interpolation is constructed along the proof somewhat explicitly. We also construct a similar interpolation for the stable and unstable manifolds.
We also pay attention to the analyticity properties of the tori and of the transformation involved, including regularity with respect to the parameter of perturbation.

We will consider an unperturbed Hamiltonian $H(I,\varphi,p,q) = h(I,pq)$ with I in a domain[1] of $\mathbb{R}^{d-1}$, $\varphi \in \mathbb{T}^{d-1}$, p and q real in a neighborhood of the origin. Moreover, we will endow the phase space $\mathbb{R}^{d-1} \times \mathbb{T}^{d-1} \times \mathbb{R} \times \mathbb{R}$ with the standard symplectic structure $\omega = \sum_i dI_i \wedge d\varphi_i + dp \wedge dq$ so that I_i and φ_i are conjugate variables and so are p and q. We will refer to I as the action variables, φ as the angle variables and p, q as the hyperbolic variables.
The equations of motion of this unperturbed Hamiltonian are

$$\begin{aligned} \dot{I} &= -\partial_\varphi H \\ \dot{\varphi} &= \partial_I H \\ \dot{p} &= -\partial_q H \\ \dot{q} &= \partial_p H \end{aligned}$$

where the dot stands for the derivative with respect to time. We will denote by Φ_t the flow of these equations.
Note that fixing any I_0 in the domain of definition, the $d-1$ dimensional torus $\mathcal{T}_{I_0} = \{(I_0,\varphi,0,0),\ \varphi \in \mathbb{T}^{d-1}\}$ is invariant under the equations of motion. Moreover, the torus $\mathcal{T}_{I_0}$ is contained in the invariant manifold $\mathcal{S}_{I_0} = \{(I_0,\varphi,p,q)\}$, on which the motion is simply

$$\Phi^t(I_0,\varphi,p,q) = \left(I_0, \varphi + \omega t, pe^{-\lambda t}, qe^{\lambda t}\right)$$

with λ and ω depending only on the product of the hyperbolic variables p and q. If we call $\zeta = pq$, we have that ζ is an invariant of the motion and $\lambda(\zeta) = \partial_\zeta h(I_0,\zeta)$, $\omega(\zeta) = \partial_I h(I_0,\zeta)$.
The torus $\mathcal{T}_{I_0}$ has the following local stable and unstable manifolds

$$\begin{aligned} W^s_{I_0} &\equiv \{(I_0,\varphi,p,0)\,,\ \varphi \in \mathbb{T}^{d-1},\ |p| \le R\} \\ W^u_{I_0} &\equiv \{(I_0,\varphi,0,q)\,,\ \varphi \in \mathbb{T}^{d-1},\ |q| \le R\} \end{aligned}$$

and we will call them *whiskers*, following [2]. A formal definition of whiskered torus will be given in Definition 2.8 below.

We will prove that most of the above tori, together with their associated invariant manifolds, are preserved under perturbations. Roughly speaking, Theorem 3.1 below asserts that:

If the Hamiltonian

$$H(I,\varphi,p,q) = h(I,pq;\mu) + \mu f(I,\varphi,p,q;\mu) \tag{1.1}$$

is real analytic and "isoenergetically non-degenerate"[2] *and* $|\mu|$ *is sufficiently small, then there exists a smooth canonical transformation, close to the identity, analytic in the parameter* μ*, in the angles and in the hyperbolic variables, such that, on a suitable set, the new Hamiltonian depends only on the actions and on the product of the hyperbolic variables.*
In this way, (d − 1)−dimensional invariant tori and smoothly interpolated d−dimensional whiskers are obtained. Such whiskered tori fill the space with density at least $1 - O(\sqrt{|\mu|})$.

Here and in the sequel, when we refer to a KAM-type transformation as "smooth", we mean "C^∞ in the sense of Whitney" (see, for example, [40], [26], [32] and [7]). Anyhow, following [9], a direct and fully constructive extension is possible, making use of elementary "bump functions".

The above result applies directly to the so called "a-priori unstable" systems, in the terminology of [11]. Such systems are obtained perturbing Hamiltonians which possess separatrices: see Definition 2.2 below. A detailed result for whiskered tori in a-priori unstable systems is given in Theorem 5.2 below.

---

[1] In this paper, we use the word "domain" to denote the closure of an open, bounded, connected set.

[2] Which means that the matrix

$$\begin{pmatrix} \partial_I^2 h & \partial_I h \\ \partial_I h & 0 \end{pmatrix}$$

is nonsingular on the energy level. See Section 2.2.

We also apply our KAM Theorem to "a-priori stable" systems (see Definition 2.1 below), showing that such systems have the above mentioned geometry near simple resonances. Theorem 6.5 provides a detailed statement for whiskered tori in a-priori stable systems. Also, in Theorem 6.6, we prove that near the Diophantine resonances, that the density of the "holes" is *exponentially small* in ε, i.e. the persisting tori fill a suitable domain of the energy level with density at least $1 - O(e^{-O(1/\varepsilon^c)})$, where c is a suitable constant.

Theorem 3.1 is similar to Lemma 1 of [11]. The main differences between our Theorem 3.1 and [11] are listed below:

(*i*) We obtain a relation between the Diophantine constant γ and the size of the perturbation μ similar to the one obtained in the classic KAM Theorem (i.e. $\gamma = O(\sqrt{|\mu|})$: see, for example [29] and [32]). We also obtain some characterizations of the set of validity of the Theorem. As a consequence, we obtain estimates on the measure of the space covered by the surviving tori, proving that the "holes" on which the Theorem fails are not wider than $O(\sqrt{|\mu|})$. We will also exhibit a simple example which proves that these estimates are sharp.

(*ii*) We prove the analytical dependence upon the parameter μ for fixed energy rather than for fixed frequencies, as proved in [11].

(*iii*) We study the geometry of the interpolation between tori, observing a structure of "filaments".

The first KAM results about persistence of whiskered tori go back to [22] and [44], and normal forms for whiskered tori can be found in [18], [21], [14], [30]. The recent paper [35] also considers a KAM theory about partially hyperbolic tori; they obtain, with a different method, the relation between γ and μ mentioned in (*i*). Moreover, the statement in [35] is obtained for fixed frequencies instead of fixed energy. The proof in [35] follows the scheme in [32] and deals with the frequency–angle variables instead of the action–angles coordinates used in our paper. Actually, for our purposes, we need a slightly more detailed statement than the one in [35]: here (as well as in [12]) we need a "good" characterization for the set in which the KAM Theorem holds [namely, the relations of the type of (3.6)], in order to construct the diffusion [i.e., the unstable orbits built in Section 4].

The construction of the KAM tori and of the unstable orbits carried out in our paper are robust enough to apply to the so called "anisochronous" cases[3], in which the KAM tori are separated by gaps.

A KAM theory for systems with a degenerate integrable part is developed in [42], for the hyperbolic case, and in [41], for the elliptic case.

The question of the persistence of KAM tori of codimension one (or bigger) is also addressed in [28], Chapter V, Section 4. Results for whiskered tori of higher codimension can be found in [8]. See also [43], which constructs a KAM algorithm for elliptic low dimensional tori under a mild non-degeneracy condition on the small divisors, extending the results of [17], [33], [6].

Also, [25], [24] and [38] consider the creation of low dimensional whiskered tori in perturbation of integrable systems.

Some of the papers quoted above take into account the case in which the system is not analytic. With regard to the problem of the conservation of the invariant tori and their hyperbolic manifolds for smooth Hamiltonians, see also the very recent paper [23].

The persistence of KAM tori for reversible (instead of Hamiltonian) systems is discussed in [36].

With respect to the papers quoted above, the target of Theorem 3.1 here is to provide a more detailed description of the partially hyperbolic tori of codimension one and investigate the structure of their

[3]See [11] for the nomenclature. Also, the "gap-bridging" procedure that inspires our paper was introduced in [11]. The isochronous systems (see [20]), on the contrary, are systems with a fixed Diophantine frequency: such systems do not present gaps between the tori, and this fact makes the construction of the unstable orbits easier.

whiskers, in order to apply directly these results to Arnol'd diffusion. We pay special attention to the smooth interpolation of such manifolds and provide a very strong normal form, in order to describe *exactly* the motion of a $(d+1)-$dimensional neighborhood of any invariant torus, even if it is not possible to determine *all* the motions nearby. The existence of such a normal form is crucial in the construction of the unstable orbits presented here in Section 4.

The scheme of the present paper is the following. In Section 2 we recall common definitions such as the ones for "a-priori stable" and "a-priori unstable" systems, isoenergetic non-degeneracy and chain of whiskered tori. We also discuss a formula relating the determinant of the matrix of isoenergetic non-degeneracy with the function that implicitly defines the energy level, and we derive some characterizations of the isoenergetic non-degeneracy.
In Section 3 we state the KAM Theorem about the preservation of whiskered tori and we discuss an example showing the optimality of our estimate on the density of the preserved tori. We also briefly emphasize the geometric structure of "filaments" related to these tori.
In Section 4 we show that the whiskered tori built in Section 3 can be used to construct "unstable" orbits (i.e. trajectories that exhibit an excursion of order 1 in the action variables), provided that there exist a chain of such whiskered tori, in which the unstable whisker of each torus intersects transversally the stable whisker of the next one. We also remark that this procedure also allow to construct orbits "drifting towards infinity".
In Section 5 and 6 we apply the KAM Theorems of Section 3 to the a-priori unstable and stable systems, respectively.
Section 7 contains a detailed proof of the KAM Theorem which makes use of a Newton scheme. The Appendix collects some elementary Lemmas.

2 Preliminaries

2.1 A-priori stable and unstable systems

We recall here the terminology of [11].

Definition 2.1 *The Hamiltonian system $H(I,\varphi) = h(I) + \varepsilon f(I,\varphi;\varepsilon)$, in which h and f are real analytic for I in a domain of $\mathbb{R}^d$, $\varphi \in \mathbb{T}^d$, and ε is a small parameter, is called* **a-priori stable**.

Such a-priori stable systems are often called *nearly-integrable* since they are perturbations of completely integrable systems written in action-angle coordinates.

Definition 2.2 *The Hamiltonian system*

$$H(I,\varphi,p,q) = \mathcal{R}(I;\mu) + \mathcal{P}(I,p,q;\mu) + \mu f(I,\varphi,p,q;\mu) \tag{2.1}$$

in which $\mathcal{R}$, $\mathcal{P}$ and f are real analytic for I in a domain of $\mathbb{R}^{d-1}$, $\varphi \in \mathbb{T}^{d-1}$, p and q are real in the neighborhood of the origin, and μ is a small parameter, is called **a-priori unstable** *if*

$$\begin{aligned} &\partial_p\mathcal{P}(I,0,0;\mu) = \partial_p(\mathcal{P}+\mathcal{R})(I,0,0;\mu) = 0 = \partial_q\mathcal{P}(I,0,0;\mu) = \partial_q(\mathcal{P}+\mathcal{R})(I,0,0;\mu) \\ \text{and}\quad &\det\partial^2_{(p,q)}\mathcal{P} = \det\partial^2_{(p,q)}(\mathcal{R}+\mathcal{P}) \le -C < 0 \end{aligned} \tag{2.2}$$

when I, p and q vary in their own set of definition, and C is a positive constant, independent of μ.

Some authors refer to the a-priori unstable systems as "initially hyperbolic". Condition (2.2) means that $p = 0 = q$ is a hyperbolic equilibrium. An example of a-priori unstable system is obtained choosing $\mathcal{R}$ as free rotators

$$\mathcal{R} = \frac{1}{2}(I_1^2 + \ldots + I_{d-1}^2) \tag{2.3}$$

and $\mathcal{P}$ as a pendulum

$$\mathcal{P} = \frac{1}{2}p^2 + g^2(\cos q - 1) \tag{2.4}$$

where g is a constant.
As it turns out, a-priori stable systems also have partially hyperbolic orbits near simple resonances. In the distinction between a-priori stable and a priori unstable systems, a crucial role is played by the size of the Lyapunov exponent near hyperbolic equilibria. This exponent is of order one in the case of an a-priori unstable system because of (2.2), while it is of order $\sqrt{\varepsilon}$ near the simple resonances of a generic a-priori stable system. This will be clarified in Section 6.
To better understand the previous remark, the reader may check that the following example [to be compared with the previous (2.3)-(2.4)] is a-priori stable:

$$H(I,\varphi) = \frac{1}{2}(I_1^2 + \ldots + I_{d-1}^2) + \frac{1}{2}I_d^2 + \varepsilon(\cos\varphi_d - 1)\,,$$

where $\varepsilon > 0$ is a small parameter. The procedure of making use of an "independent" parameter in a singular-perturbation problem was already used in [31]. The use of such a procedure in our paper will be clarified in Section 6.

2.2 The isoenergetic non-degeneracy

In this subsection, we will denote by $\mathbb{F}$ either $\mathbb{R}$ or $\mathbb{C}$, and we will consider neighborhoods in $\mathbb{F}$. This is done to deal with both the real and the complex case at the same time.

Definition 2.3 *The (smooth) Hamiltonian $h(I)$, with I in a domain of $\mathbb{F}^d$, is called* **isoenergetically non-degenerate** *on the energy level $h = E$ if*

$$\det\begin{pmatrix} h'' & h' \\ (h')^T & 0 \end{pmatrix} \neq 0 \tag{2.5}$$

for any I in the domain of h such that $h(I) = E$.

Notational Remarks. In the rest of this paper, we will use the same notation for both column and row vectors: for instance, following [4] page 409, we write $\begin{pmatrix} h'' & h' \\ h' & 0 \end{pmatrix}$ instead of $\begin{pmatrix} h'' & h' \\ (h')^T & 0 \end{pmatrix}$.
The only place in which we denote row vectors with the symbol of transposition "$\cdot^T$" is Lemma A.1, in order to avoid confusion between $v \cdot w$ [i.e. the scalar product between $v = (v_1, \ldots, v_n)$ and $w = (w_1, \ldots, w_n)$] and vw^T [i.e. the matrix whose $(i,j)-$th entry is $v_i w_j$].
We will also denote $\omega \equiv h'$ and use the symbol "tilde" for the first $d-1$ components of a $d-$dimensional vector.
Also, we write $|\cdot|$ to denote a norm for real or complex vectors, and the sum of the absolute values of the components for integer vectors. We will not compute explicitly the constants appearing in the KAM proof, hence we do not fix explicitly the norm used in the finite-dimensional vector spaces, since they are, for our purposes, equivalent. On the other hand, a careful choice of the norms is necessary for a concrete and effective implementation of the scheme: see, for example, [10].

Proposition 2.4 *Consider the (smooth) Hamiltonian $h(I)$, with I in a domain of $\mathbb{F}^d$ and $d \geq 2$. Assume $\omega_d \neq 0$ on the energy level $h = E$. Denote by $I_d(\tilde{I})$ the function implicitly defined by $h(\tilde{I}, I_d(\tilde{I})) = E$. Then,*

$$\det \partial^2_{\tilde{I}} I_d = (-1)^d \omega_d^{-d-1} \det\begin{pmatrix} h'' & \omega \\ \omega & 0 \end{pmatrix}. \tag{2.6}$$

Proof. It follows from

$$\partial^2_{\tilde{I}} I_d = \frac{(\omega_d \partial^2_{\tilde{I}I_d} h - \partial^2_{I_d} h\tilde{\omega})\tilde{\omega}^T - \omega_d(\omega_d \partial^2_{\tilde{I}} h - \tilde{\omega}(\partial^2_{\tilde{I}I_d} h)^T)}{\omega_d^3}\,,$$

making use of Lemma A.1. □

Proposition 2.5 *Let $d \geq 2$. The following conditions are equivalent:*

(i) h is isoenergetically non-degenerate on the energy level $h = E$ for I in a suitable domain of $\mathbb{F}^d$.

(ii) $\omega \neq 0$ on the energy level $h = E$ and, assuming for example $\omega_d \neq 0$ and denoting by $I_d(\tilde{I})$ the function implicitly defined by $h(\tilde{I}, I_d(\tilde{I})) = E$ and

$$\alpha(\tilde{I}) \equiv \frac{\tilde{\omega}(\tilde{I}, I_d(\tilde{I}))}{\omega_d(\tilde{I}, I_d(\tilde{I}))}, \tag{2.7}$$

we have

$$\det \alpha' \neq 0 . \tag{2.8}$$

(iii) The following function G is a local diffeomorphism near $\sigma = 1$:

$$\begin{array}{rcl} G : \mathbb{F}^d \times \mathbb{F} & \longrightarrow & \mathbb{F}^d \times \mathbb{F} \\ (I;\sigma) & \longmapsto & (\sigma\omega(I),\, h(I)) . \end{array}$$

Proof. (i) and (iii) are equivalent because of the Implicit Function Theorem. The equivalence between (i) and (ii) follows from (2.6) and from $\partial_{\tilde{I}} I_d = -\alpha$. □

Writing the details in the proof of the previous Proposition, it is easy to obtain the following

Corollary 2.6 *Consider the (smooth) Hamiltonian $h(I)$, with I in a domain of $\mathbb{F}^d$ and $d \geq 2$. Assume $\omega_d \neq 0$ on the energy level $h = E$, and define $\alpha(\tilde{I})$ as in (2.7). Then,*

$$\det \alpha' = -(\omega_d)^{-d-1} \cdot \det \begin{pmatrix} h'' & \omega \\ \omega & 0 \end{pmatrix} . \tag{2.9}$$

Similar definitions and results hold for a "partially hyperbolic" Hamiltonian $h(I, pq)$, since the variable $\zeta = pq$ plays in this case only the role of a parameter. In particular, the condition of isoenergetic non-degeneracy becomes

$$\det \begin{pmatrix} \partial_I^2 h & \partial_I h \\ \partial_I h & 0 \end{pmatrix} \neq 0 , \tag{2.10}$$

for any I in a domain of $\mathbb{F}^d$ and p, q in a neighborhood of 0. With a slight abuse of notation, we will refer to both (2.5) and (2.10) with the term of "isoenergetic non-degeneracy". All through the paper, the isoenergetic non-degeneracy condition concerns only the derivatives with respect to the actions I and never the derivatives with respect to the action $\zeta \equiv pq$.

Proposition 2.5 can also be slightly modified to include hyperbolic variables as follows:

Proposition 2.7 *Let $d \geq 2$. The following conditions are equivalent:*

(i) $h(I, \zeta)$ is isoenergetically non-degenerate on the energy level $h = E$ for I in a domain of $\mathbb{F}^d$ and p, q in a neighborhood of 0.

(ii) $\omega \equiv \partial_I h \neq 0$ on the energy level $h = E$ and, assuming for example $\omega_d \neq 0$ and denoting by $I_d(\tilde{I}, \zeta)$ the function implicitly defined by $h(\tilde{I}, I_d(\tilde{I}, \zeta), \zeta) = E$ and

$$\alpha(\tilde{I}, \zeta) \equiv \frac{\tilde{\omega}(\tilde{I}, I_d(\tilde{I}, \zeta), \zeta)}{\omega_d(\tilde{I}, I_d(\tilde{I}, \zeta), \zeta)},$$

we have

$$\det \partial_I \alpha \neq 0 .$$

(iii) The following function G is a local diffeomorphism near $\sigma = 1$:

$$\begin{array}{rcl} G : \mathbb{F}^d \times \mathbb{F} \times \mathbb{F} & \longrightarrow & \mathbb{F}^d \times \mathbb{F} \times \mathbb{F} \\ (I;\sigma;\zeta) & \longmapsto & (\sigma\omega(I,\zeta),\, h(I,\zeta), \zeta) . \end{array}$$

2.3 Whiskered tori and transversality

Following [2], we give the following

Definition 2.8 *A torus $\mathcal{T}$ is called a* **whiskered torus** *for the flow Φ^t if it is a connected component of the intersection of two manifolds W^s and W^u invariant under Φ^t, such that $\forall \zeta^s \in W^s$, $\lim_{t\to\infty} \text{dist}\left(\Phi^t(\zeta^s), \mathcal{T}\right) = 0$ and $\forall \zeta^u \in W^u$, $\lim_{t\to-\infty} \text{dist}\left(\Phi^t(\zeta^u), \mathcal{T}\right) = 0$.*

In the KAM setting, the motion on these tori will be conjugated to an irrational (and, in fact, Diophantine) rotation, the trajectories on W^s will converge exponentially fast to $\mathcal{T}$ and the trajectories on W^u will diverge exponentially fast from $\mathcal{T}$. Such a motion will be described in detail in (3.15) below.

We call W^s and W^u the stable and unstable whisker, respectively. We will denote by $T_p\mathcal{M}$ the tangent space at p of the manifold $\mathcal{M}$.
If $V = \text{span}\{v_1, \ldots, v_i\}$ and $W = \text{span}\{w_1, \ldots, w_j\}$ are vector spaces, we set

$$V + W \equiv \text{span}\{v_1, \ldots, v_i, w_1, \ldots, w_j\} .$$

Definition 2.9 *Let $\mathcal{M}$ and $\mathcal{N}$ be submanifolds of the manifold $\mathcal{X}$. We say that $\mathcal{M}$ and $\mathcal{N}$ are* **transverse** *in the point p with respect to the ambient space $\mathcal{X}$ if $p \in \mathcal{M} \cap \mathcal{N}$ and $T_p\mathcal{M} + T_p\mathcal{N} = T_p\mathcal{X}$.*

In this paper, we will consider only transversality of whiskers with respect to a common, fixed energy level $h = E$. We now recall the standard definition of Diophantine vector:

Definition 2.10 *A vector $\omega \in \mathbb{R}^n$ is called $(\gamma, \tau)-$Diophantine if $|\omega \cdot n| \geq \gamma/|n|^\tau$ for any $n \in \mathbb{Z}^n - \{0\}$.*

We will use later the elementary fact that, if $\tau > n - 1$, the $(\gamma, \tau)-$Diophantine vectors fill the $n-$dimensional space with density $1 - O(\gamma)$. In the sequel, we will consider Diophantine vectors in the $(d-1)-$dimensional frequency space, so that in the rest of the paper $\tau > d - 2$ will be a fixed parameter.

3 A KAM Theorem about preservation of partially hyperbolic tori

In the sequel we will restrict to Hamiltonian systems whose number of degree of freedom is $d \geq 3$. This is motivated by the very well known fact that, in our setting, Arnol'd diffusion does not occur for autonomous Hamiltonian systems with less than 3 degrees of freedom (see for instance [1], [3]). Anyhow, it is easy to see that some of the results of this paper remain valid even in the case $d = 2$. But in the case $d = 2$ it is not possible to have transverse intersections between the unstable whisker of a torus in the chain with the stable whisker of the next torus, so that the procedure given in Section 4 can not work.

Theorem 3.1 *Fix $I^* \in \mathbb{R}^{d-1}$, $E \in \mathbb{R}$. Consider the Hamiltonian*

$$H(I, \varphi, p, q) = h(I, pq; \mu) + \mu f(I, \varphi, p, q; \mu) \tag{3.1}$$

with h and f real analytic in

$$\mathcal{O}_{\rho,\xi,R,\bar{\mu}} = \{(I, \varphi, p, q, \mu) \in \mathbb{C}^{(d-1)+(d-1)+1+1+1} \text{ s.t. } |I - I^*| \leq \rho,\ |\Im\varphi| \leq \xi,\ |p| \leq R,\ |q| \leq R,\ |\mu| \leq \bar{\mu}\}$$

and periodic in the angles φ. Denote $\zeta \equiv pq$ and assume that

$$\lambda_0 \equiv \inf_{|I-I^*|\leq\rho,\, |\zeta|\leq R^2,\, |\mu|\leq\bar{\mu}} |\partial_\zeta h| > 0$$

and that h is isoenergetically non-degenerate on the energy level $h(I, pq; \mu) = E$ when the variables vary in $\mathcal{O}_{\rho,\xi,R,\bar{\mu}}$.
Then, there exist R_∞, $0 < R_\infty \le R$, and a constant $\kappa_\star$ (eventually depending on d, τ, and the sizes of h and f), such that, for $|\mu| \le \mu_0 \equiv \kappa_\star \lambda_0^2 \le \bar{\mu}$, there exist:

(e1) a smooth canonical transformation Φ, close to the identity and real analytic (for a fixed action) in the angles, in the hyperbolic variables and in the parameter μ,

(e2) a function $h_\infty : \mathbb{R}^{(d-1)} \times \mathbb{R} \times \mathbb{R} \longrightarrow \mathbb{R}$ with the same smoothness as Φ,

(e3) a set $\Omega_\mu \subseteq \mathbb{R}^{(d-1)+1+1}$, with density at least $1 - \kappa_\star\sqrt{\mu_0}$,

such that, for $|\mu| \le \mu_0$,

$$\begin{aligned} &\partial^n \left(H \circ \Phi(I', \varphi', p', q')\right) = \partial^n h_\infty(I', p'q'; \mu)\,, \ \forall (I', p', q') \in \Omega_\mu,\ \varphi' \in \mathbb{T}^{d-1},\ n \in \mathbb{N}^{2d} \\ &h_\infty(I', p'q'; \mu) = E\,,\ \forall (I', p', q') \in \Omega_\mu\,. \end{aligned} \tag{3.2}$$

Moreover, setting $\zeta' \equiv p'q'$,

$$\begin{aligned} &\forall (I', p', q') \in \Omega_\mu,\ |\partial_{\zeta'} h_\infty| \ge \lambda_0/2\,, \\ &\forall (I', p', q') \in \Omega_\mu,\ \varphi' \in \mathbb{T}^{d-1},\ q' \ne 0,\ |\partial_{p'}(H \circ \Phi)| = |q'\,\partial_{\zeta'} h_\infty| > 0\,, \\ &\forall (I', p', q') \in \Omega_\mu,\ \varphi' \in \mathbb{T}^{d-1},\ p' \ne 0,\ |\partial_{q'}(H \circ \Phi)| = |p'\,\partial_{\zeta'} h_\infty| > 0\,. \end{aligned} \tag{3.3}$$

More precisely, if we denote

$$\mathcal{D}_\tau \equiv \{I \in \mathbb{R}^{d-1} \text{ s.t. } |I - I^*| \le \rho' \text{ and } \partial_I h(I, 0; 0) \text{ is } (\gamma_0, \tau) - \text{Diophantine}\} \tag{3.4}$$

with a suitable $0 < \rho' < \rho$, and a suitable γ_0 depending on μ_0 [i.e. $\gamma_0 = \kappa_\star\sqrt{\mu_0}$], then there exist:

$(E1)$ a function $\mathcal{I}^I_\infty(\zeta; \mu)$, with range in the action space, which is smooth in I, ζ, μ, and (for a fixed I) real analytic in ζ and μ, for $|\zeta| \le R^2_\infty$, $|\mu| \le \mu_0$, verifying $\mathcal{I}^I_\infty(0; 0) = I$,

$(E2)$ a function $\alpha^I_\infty(\zeta; \mu)$, with the same regularity as $\mathcal{I}^I_\infty$, that verifies

$$\alpha^I_\infty(0; 0) = 0 \qquad \textit{and} \qquad \sup_{|I - I^*| \le \rho',\, |\zeta| \le R^2_\infty,\, |\mu| \le \mu_0} |\alpha_\infty| \le c\rho \le \frac{1}{2}\,, \tag{3.5}$$

where c is a constant with the dimensions of the inverse of an action, such that:

$(P1)$ $\partial_{I'} h_\infty(\mathcal{I}^{\bar{I}}_\infty(\zeta; \mu), \zeta; \mu) = \partial_I h(\bar{I}, 0; 0) \cdot (1 + \alpha^{\bar{I}}_\infty(\zeta; \mu))$, $\forall \bar{I} \in \mathcal{D}_\tau$.

$(P2)$ The set Ω_μ in (e3) can be described in the following two ways:

$$\begin{aligned} \Omega_\mu \ &= \ \{(\mathcal{I}^{\bar{I}}_\infty(p'q'; \mu), p', q'),\ \bar{I} \in \mathcal{D}_\tau,\ |p'| \le R_\infty,\ |q'| \le R_\infty\} = \\ &= \ \{(I', p', q') \text{ s.t. } |p'| \le R_\infty,\ |q'| \le R_\infty,\ h_\infty(I', p'q'; \mu) = E \\ &\qquad \text{and } \exists \bar{I} \in \mathcal{D}_\tau \text{ s.t. } \partial_{I'} h_\infty(I', p'q'; \mu) = \partial_I h(\bar{I}, 0; 0) \cdot (1 + \alpha^{\bar{I}}_\infty(p'q'; \mu))\}\,. \end{aligned} \tag{3.6}$$

Furthermore,

$$\begin{aligned} \Omega_\mu \ \subseteq \ &\{(I', p', q'),\ I' \text{ in a neighborhood of } I^*, \\ &p' \text{ and } q' \text{ in a neighborhood of 0 s.t. } h_\infty(I', p'q'; \mu) = E \\ &\text{and } \partial_I h_\infty(I', p'q'; \mu) \text{ is } (\gamma_0/2, \tau)\text{–Diophantine}\} \end{aligned} \tag{3.7}$$

$$\begin{aligned} \Omega_\mu \ \supseteq \ &\{(I', p', q'),\ I' \text{ in a neighborhood of } I^*, \\ &p' \text{ and } q' \text{ in a neighborhood of 0 s.t. } h_\infty(I', p'q'; \mu) = E \\ &\text{and } \partial_I h_\infty(I', p'q'; \mu) \text{ is } (2\gamma_0, \tau)\text{–Diophantine}\}\,. \end{aligned} \tag{3.8}$$

$(P3)$ *Denoting by* Dens ^{E} *the* $(2d-1)-$*dimensional restriction of the Lebesgue density on the energy level* $\{(I',\varphi',p',q')$ s.t. $h_\infty(I',p'q';\mu)=E\}$ *and by* dens ^{E} *the* $d-$*dimensional restriction in the space of the actions and the hyperbolic variables of the Lebesgue density on the surface defined by the energy relation* $\{(I',p',q')$ s.t. $h_\infty(I',p'q';\mu)=E\}$, *we have*

$$\text{dens}\,^E\Omega_\mu \geq 1-\kappa_\star\sqrt{\mu_0}\,, \qquad \text{Dens}\,^E\Phi(\Omega_\mu\times\mathbb{T}^{d-1}) \geq 1-\kappa_\star\sqrt{\mu_0}\,. \tag{3.9}$$

$(P4)$ *We have the following equality of sets:*

$$\begin{aligned} \Omega^0_\mu &\equiv \{\mathcal{I}^{\bar I}_\infty(0;\mu),\ \bar I\in\mathcal{D}_\tau\} = \\ &= \{I\ \text{ s.t. }\ h_\infty(I,0;\mu)=E \text{ and} \\ &\qquad \exists\bar I\in\mathcal{D}_\tau \text{ s.t. } \partial_{I'}h_\infty(I,0;\mu)=\partial_I h(\bar I,0;0)\cdot(1+\alpha^{\bar I}_\infty(0;\mu))\}. \end{aligned} \tag{3.10}$$

Furthermore,

$$\begin{aligned} \Omega^0_\mu &\subseteq \{I,\ \text{ in a neighborhood of } I^* \text{ s.t.} \\ &\qquad h_\infty(I,0;\mu)=E \text{ and } \partial_I h_\infty(I,0;\mu) \text{ is } (\gamma_0/2,\tau)-\text{Diophantine}\} \end{aligned} \tag{3.11}$$

$$\begin{aligned} \Omega^0_\mu &\supseteq \{I,\ \text{ in a neighborhood of } I^*, \text{ s.t.} \\ &\qquad h_\infty(I,0;\mu)=E \text{ and } \partial_I h_\infty(I,0;\mu) \text{ is } (2\gamma_0,\tau)-\text{Diophantine}\}\,. \end{aligned} \tag{3.12}$$

$(P5)$ *Denoting by* dens $_E$ *the* $(d-2)-$*dimensional restriction of the Lebesgue density to the manifold defined by the energy relation* $h_\infty(I',0;\mu)=E$, *we have*

$$\text{dens}\,_E\Omega^0_\mu \geq 1-\kappa_\star\sqrt{\mu_0}\,. \tag{3.13}$$

We term the variables (I',φ',p',q') obtained with this procedure "normal coordinates", and refer to the Hamiltonian h_∞ as a "normal form", since the motion in the variables (I',φ',p',q') according to the Hamiltonian h_∞ is particularly simple. As a matter of fact, from (3.2) and (3.6), it follows that the tori (written in normal coordinates)

$$\mathcal{T}_{\bar I} = \{(\mathcal{I}^{\bar I}_\infty(0;\mu),\varphi',0,0),\ \varphi'\in\mathbb{T}^{d-1}\} \tag{3.14}$$

are invariant under the Hamiltonian flow of h_∞.
Moreover, it follows that $\mathcal{T}_{\bar I}$ is contained in the manifold with boundary

$$\mathcal{S}_{\bar I} = \{(\mathcal{I}^{\bar I}_\infty(p'q';\mu),\varphi',p',q'),\ \varphi'\in\mathbb{T}^{d-1},\ |p'|\leq R_\infty,\ |q'|\leq R_\infty\}\,,$$

which is locally invariant and on which the motion is simply:

$$\Phi^t_{h_\infty}(\mathcal{I}^{\bar I}_\infty(p'q';\mu),\varphi',p',q') = (\mathcal{I}^{\bar I}_\infty(p'q';\mu),\varphi'+\omega_\infty t,p'e^{-\lambda_\infty t},q'e^{\lambda_\infty t})\,, \tag{3.15}$$

provided that $|p'e^{-\lambda_\infty(I',p'q')t}|$, $|q'e^{\lambda_\infty(I',p'q')t}|\leq R_\infty$, where $\omega_\infty\equiv\partial_{I'}h_\infty$ and $\lambda_\infty\equiv\partial_{\xi'}h_\infty$ depend only on $\xi'\equiv p'q'$, $\bar I$ and μ. In particular, the whiskers are (locally) parameterized as

$$\begin{aligned} W^s_{\bar I} &= \{(\mathcal{I}^{\bar I}_\infty(0;\mu),\varphi',p',0)\,,\ \varphi'\in\mathbb{T}^{d-1}\ |p'|\leq R_\infty\} \text{ and} \\ W^u_{\bar I} &= \{(\mathcal{I}^{\bar I}_\infty(0;\mu),\varphi',0,q')\,,\ \varphi'\in\mathbb{T}^{d-1}\ |q'|\leq R_\infty\}\,. \end{aligned} \tag{3.16}$$

We propose the name of *fan* to call sets of the type $\Omega_\mu\times\mathbb{T}^{d-1}$, which collects the tori, their whiskers, and their normal hyperbolic trajectories.

The method of proof used here yields a very strong normal form, since (3.15) describes exactly the motion of a $(d+1)$–dimensional neighborhood of the torus (even if it does not determine *all* the motions near the torus). This normal form is at the basis of the construction of the unstable orbits presented here in Section 4 (as well as in [11] and in [12]). Another fundamental ingredient in the construction of such unstable trajectories will be the fact that Diophantine properties (or, more generally, rationally independence) of the "old frequency" $\partial_I h$ are preserved for the "new frequency" $\partial_{I'} h_\infty$, according to $(P1)$ of Theorem 3.1.

We remark the fact that the hypotheses of [12] can be readily derived[4] from the conclusions of our KAM Theorem. Namely, hypothesis (ii) of [12] follows from (3.2), (3.3), $(P1)$ and (3.15); hypothesis (iii) of [12] follows from (3.14) and (3.16). We also remind that for "isochronous" systems a much stronger normal form holds. See [20].

We also remark that, leaving out the hyperbolic variables p and q, our proof also establishes the classic KAM Theorem for Lagrangian tori in isoenergetically non-degenerate systems. Moreover, the same result as Theorem 3.1 holds for Hamiltonians depending on several small parameters $\mu^{(1)}, \ldots, \mu^{(n)}$: the proof would remain the same, denoting $\mu \equiv (\mu^{(1)}, \ldots, \mu^{(n)})$ and considering it as a vector.

The proof of Theorem 3.1 is deferred to Section 7.

We now derive from Theorem 3.1 a KAM result for Hamiltonians depending on two parameters ε and μ, in which the parameter ε plays the role of a fixed singular-perturbation parameter, while the dependence on μ will be uniform. We will apply the following Corollary 3.2 in the a-priori stable setting, in which the Lyapunov exponent is not bounded from zero uniformly in the parameter. In reference to this, see Lemma 6.3 below.

Corollary 3.2 *Fix $I^* \in \mathbb{R}^{d-1}$, $E \in \mathbb{R}$. Consider the Hamiltonian*

$$H(I, \varphi, p, q) = h(I, pq; \varepsilon) + f(I, \varphi, p, q; \varepsilon, \mu) \tag{3.17}$$

with h and f real for any real value of $(I, \varphi, p, q, \varepsilon, \mu)$, analytic, for any fixed ε, $|\varepsilon| \le \bar{\varepsilon}$, in

$$\begin{aligned} \mathcal{O}_{\rho,\xi,R,\bar{\mu}} \equiv\ & \{(I, \varphi, p, q, \mu) \in \mathbb{C}^{(d-1)+(d-1)+1+1+1} \text{ s.t. } |I - I^*| \le \rho,\ |\Im\varphi| \le \xi, \\ & |p| \le R,\ |q| \le R,\ |\mu| \le \bar{\mu}\}, \end{aligned}$$

and periodic in the angles φ. Fix $\varepsilon \in \mathbb{R}$, $|\varepsilon| \le \bar{\varepsilon}$, and denote $\zeta \equiv pq$. Assume that there exists a constant $C > 0$ such that

$$\sup_{\mathcal{O}_{\rho,\xi,R,\bar{\mu}} \times \{|\varepsilon| \le \bar{\varepsilon}\}} |h| \le C \quad \text{and} \quad \sup_{\substack{|I-I^*| \le \rho,\ |\Im\varphi| \le \xi \\ |p| \le R,\ |q| \le R,\ |\varepsilon| \le \bar{\varepsilon}}} |f| \le C\mu, \qquad \forall |\mu| \le \bar{\mu}$$

and that

$$\lambda_0 = \lambda_0(\varepsilon) \equiv \inf_{|I-I^*| \le \rho,\ |\zeta| \le R^2} |\partial_\zeta h| > 0.$$

Assume also that h is isoenergetically non-degenerate on the energy level $h(I, pq; \varepsilon) = E$ when the variables vary in $\mathcal{O}_{\rho,\xi,R,\bar{\mu}} \times \{|\varepsilon| \le \bar{\varepsilon}\}$.
Then the analogous statement of Theorem 3.1 holds, with the constant $\kappa_\star$ eventually depending on d, τ, C, but independent of ε.

Now we will show that the estimates (3.9) and (3.13) are optimal, using an example which is an extension of one of [29]. Consider

$$H(I_1, I_2, \varphi_1, \varphi_2, p, q; \mu) \equiv h(I_1, I_2, pq) + \mu(\cos\varphi_1 - 1)$$

[4]Beware of some slight changes in the notation with respect to [12]: for instance, the transformation Φ is called C in [12], and the normal variables (I', φ', p', q') here correspond to $(\vec{A}', \vec{\psi}, p, q)$ in [12]. For a comparison with the notations in [11], see footnote 4 of [12].

$$\text{with} \qquad h(I_1, I_2, pq) \equiv \frac{I_1^2}{2} + I_2 + pq\,, \tag{3.18}$$

where $\mu > 0$ is a small parameter. If $(I_1, I_2) \in \mathbb{R}^2$ is sufficiently close to the origin, h is isoenergetically non-degenerate. The unperturbed system has the invariant tori $\mathcal{T}_{(I_1,I_2)} = \{(I_1, I_2, \varphi_1, \varphi_2),\ (\varphi_1, \varphi_2) \in \mathbb{T}^2\}$. We will show that the tori destroyed by the perturbation have measure $\approx \sqrt{\mu}$: actually, if $\mathcal{X}$ is the set of (I_1, φ_1) enclosed inside the separatrices of the pendulum $I_1^2/2 + \mu(\cos\varphi_1 - 1)$, we have that there is no surviving KAM torus in $\mathcal{X}$, hence the measure of the holes is

$$\begin{aligned} & \operatorname{meas}_E\{(I_1, \varphi_1) \in \mathcal{X},\ \varphi_2 \in \mathrm{S}^1,\ h(I_1, I_2, 0) = E\} = \\ = \ & \operatorname{meas}_E\{(I_1, \varphi_1) \in \mathcal{X},\ \varphi_2 \in \mathrm{S}^1,\ I_2 = E - I_1^2/2\} = \\ = \ & \int_{(I_1,\varphi_1)\in\mathcal{X},\,\varphi_2\in\mathrm{S}^1} \sqrt{1 + I_1^2}\, dI_1\, d\varphi_1\, d\varphi_2 \geq \int_{(I_1,\varphi_1)\in\mathcal{X},\,\varphi_2\in\mathrm{S}^1} dI_1\, d\varphi_1\, d\varphi_2 = 2\pi \operatorname{meas}\mathcal{X} \geq 4\pi^2\sqrt{\mu}\,. \end{aligned}$$

In the KAM settings, the measure of the surviving tori is usually large for small values of the perturbation, but a surprising exception can be found in [37].

3.1 The filaments

We will observe that the interpolation between the tori preserved in Theorem 3.1 presents a structure of "filaments". As stated in (3.14), these tori are interpolated by the smooth function $\mathcal{I}^I_\infty(\zeta;\mu)$, i.e. the preserved invariant tori correspond to $\mathcal{I}^{\bar{I}}_\infty(0;\mu)$ with $\bar{I}$ in the Diophantine set $\mathcal{D}_\tau$. Propositions 3.3 and 3.4 will prove that, fixing μ and letting $\bar{I}$ vary in $\mathcal{D}_\tau$, the curves $\mathcal{I}^{\bar{I}}_\infty(\zeta;\mu)$ obtained in this way do not have self-intersections, and different curves do not intersect, so they can be seen as filaments, side by side:

Proposition 3.3 *Fixed μ, R_∞ sufficiently small, we have:*

$$\mathcal{I}^I_\infty(\zeta;\mu) = \mathcal{I}^I_\infty(\zeta';\mu)\,,\ |\zeta|, |\zeta'| \leq R_\infty \iff \zeta = \zeta'\,.$$

Proof. First notice that $\partial_\zeta \mathcal{I}^I_\infty(\zeta;\mu) \neq 0$. If it were zero, differentiating $h_\infty\left(\mathcal{I}^I_\infty(\zeta;\mu), \zeta;\mu\right) = E$, one would get that the Lyapunov exponent is zero, in contradiction with our assumption. Then apply the Inverse Function Theorem. □

Proposition 3.4 *Fixed μ sufficiently small, let $\sigma : \mathbb{R} \longrightarrow \mathbb{R}^{d-1}$ be a smooth curve in the action space with $\sigma'(0) \neq 0$. Then, $\Psi(s,\zeta) \equiv \mathcal{I}^{\sigma(s)}_\infty(\zeta;\mu)$ is injective near $s = 0 = \zeta$.*

Proof. We will show that $\operatorname{Ran}\partial_{(s,\zeta)}\mathcal{I}^{\sigma(s)}_\infty(\zeta;\mu)\Big|_{s=0,\zeta=0,\mu=0} = 2$. Then apply the Inverse Function Theorem. By contradiction, if there were $(a,b) \in \mathbb{R}^2 - \{0\}$ such that

$$0 = a\partial_s\mathcal{I}^{\sigma(s)}_\infty(\zeta;\mu) + b\partial_\zeta\mathcal{I}^{\sigma(s)}_\infty(\zeta;\mu)\Big|_{s=0,\zeta=0,\mu=0}\,. \tag{3.19}$$

Differentiating $h_\infty(\mathcal{I}^{\sigma(s)}_\infty(\zeta;\mu), \zeta;\mu) = E$ we have:

$$\begin{aligned} (\partial_I h_\infty) \circ (\mathcal{I}^{\sigma(s)}_\infty(\zeta;\mu), \zeta;\mu) \cdot \partial_\zeta\mathcal{I}^{\sigma(s)}_\infty(\zeta;\mu) + (\partial_\zeta h_\infty) \circ (\mathcal{I}^{\sigma(s)}_\infty(\zeta;\mu), \zeta;\mu) &= 0 \\ (\partial_I h_\infty) \circ (\mathcal{I}^{\sigma(s)}_\infty(\zeta;\mu), \zeta;\mu) \cdot \partial_s\mathcal{I}^{\sigma(s)}_\infty(\zeta;\mu) &= 0 \end{aligned}$$

so that, multiplying (3.19) by $(\partial_I h_\infty) \circ (\mathcal{I}^{\sigma(s)}_\infty(\zeta;\mu), \zeta;\mu)$, we get:

$$b\,(\partial_\zeta h_\infty) \circ (\mathcal{I}^{\sigma(0)}_\infty(0;0), 0;0) = 0$$

that, for the non-vanishing of the Lyapunov exponent, implies $b = 0$. Then, in order that $(a,b) \neq (0,0)$, it must be $a \neq 0$; so from (3.19) and $\mathcal{I}^I_\infty(0;0) = I$, we get $0 = \partial_s\mathcal{I}^{\sigma(s)}_\infty(\zeta;\mu)\Big|_{s=0,\zeta=0,\mu=0} = \sigma'(0)$, contradicting the hypothesis. □

4 Existence of unstable orbits

Following [12], we show now that the invariant partially hyperbolic tori, whose existence is ensured by Theorem 3.1, can be easily used to construct orbits with an excursion of order one in the actions, provided that the unstable whisker of each torus intersects *transversally* the stable whisker of the next one. The construction of such an unstable orbit can be seen as a generalization of the one in [2] and §23 of [5]. We also extend this procedure, via an elementary argument of point-set topology, showing the existence of an orbit "drifting towards infinity", as stated in (4.1); see also [16]. Inclination Lemmas and diffusion paths are also considered in Section 4 of [19].

The proof presented here is essentially "topological", in the sense that it makes use only of the continuous dependence on the initial data, so that it covers also cases in which the Hamiltonian has less smoothness.

The problem of showing the validity of the hypothesis of transverse intersection of the whiskers is not addressed in this paper. This assumption is established in [11] for a-priori unstable systems, under suitable regularity conditions and non-degeneracy of the perturbation.

Theorem 4.1 *Consider a chain of whiskered tori* $\{\mathcal{T}_{\bar I_j}\}_{\substack{j\in\mathbb{N}\\ 1\le j\le N}}$ *as in* (3.14), *with whiskers* $\{W^s_{\bar I_j}\}_{\substack{j\in\mathbb{N}\\ 1\le j\le N}}$ *and* $\{W^u_{\bar I_j}\}_{\substack{j\in\mathbb{N}\\ 1\le j\le N}}$ *as in* (3.16), *with flow in local coordinates as in* (3.15). *If* $W^u_{\bar I_j}$ *intersects transversally* $W^s_{\bar I_{j+1}}$ *with respect to the* $E-$*energy level for* $j = 1, \ldots, N-1$, *then there exists an open set of points of the phase space arbitrarily close to the first torus* $\mathcal{T}_{\bar I_1}$ *that evolves under the flow arbitrarily close to the* N*th torus* $\mathcal{T}_{\bar I_N}$ *in a finite time.*
In particular, if there exists a constant c *independent of* μ *such that* $|\bar I_1 - \bar I_N| \ge c$, *then the orbit constructed here exhibits an instability of order one in the action variables.*

Also, if the system admits a sequence of whiskered tori $\{\mathcal{T}_{\bar I_j}\}_{j\in\mathbb{N}}$, *verifying the same assumptions as above, with the property that*

$$\lim_{j\longrightarrow\infty} |\bar I_j| = \infty\,,$$

then, there exists an orbit $(I(t), \varphi(t), p(t), q(t))$ *such that*

$$\limsup_{t\longrightarrow\infty} |I(t) - I(0)| = \infty\,. \tag{4.1}$$

Proof. Consider a neighborhood U_i of the i-th torus in which the normal form above holds; the condition of transversality assures the existence of a piece of the stable manifold of the $(i+1)-$th torus lying in U_i. It is not difficult to see that this piece of manifold contains a curve at constant $q' = q_0'$

$$\Gamma_i(p) = \{(I_i'(p'), \varphi_i'(p'), p', q_0')\,,\ |p'| \le r^*\}$$

such that $\Gamma_i(0) \in W_i^u \cap W_{i+1}^s \cap U_i$ and the evolution at time t (in local coordinates) of the point $(I_i'(p'), \varphi_i'(p'), p', q_0')$ is simply given by

$$(I_i'(p'),\ \varphi_i'(p') + \omega(p')t,\ p'e^{-\lambda_i(p')t},\ q_0 e^{\lambda_i(p')t})\,,$$

where

$$\omega_i(p') \equiv \partial_{I'} h_\infty(I', p'q'; \mu)\Big|_{\{I'=I_i'(0), q'=q_0\}}\,, \qquad \lambda_i(p') \equiv \partial_{\zeta'} h_\infty(I', p'q'; \mu)\Big|_{\{I'=I_i'(0), q'=q_0\}}$$

and $\zeta' \equiv p'q'$. Hence, using also the irrationality of $\omega_i(p')$, one can see that, given any neighborhood B_i of a point in $W_i^s \cap U_i$, there exists p_i^*, $0 < p_i^* < r^*$, and a finite time t_i^* such that the backward evolution of $\Gamma_i(p_i^*)$ at time t_i^* lies inside B_i. By continuity, there exists a small neighborhood B_i^* of $\Gamma_i(p_i^*)$ whose backward evolution at time t_i^* is contained in B_i.
This process can be iterated torus after torus, choosing B_{i+1} as the evolution of B_i^* in the neighborhood U_{i+1} of the $(i+1)-$th torus, leading to an unstable orbit. This proves the first claim of this result.

For the drift towards infinity, notice that, as below, one can construct a sequence of closed ball $\{B_j\}_{j\in\mathbb{N}}$ and a sequence of times $\{t_j\}_{j\in\mathbb{N}}$ such that

(i) each B_j is in a small neighborhood of $\mathcal{T}_{\bar{I}_j}$
(ii) $\Phi^{-t_{j-1}}(B_j) \subseteq B_{j-1}$, $\forall j \in \mathbb{N}$.

Hence, defining $\mathcal{B}_j \equiv \Phi^{-t_1-\ldots-t_{j-1}}(B_j)$ for any $j \in \mathbb{N}$, $j \geq 2$, we have that $\mathcal{B}_j \subseteq \mathcal{B}_{j-1} \subseteq B_1$. A well known argument shows that the intersection of all the $\mathcal{B}_j$'s is not empty. So if $\eta \in \bigcap \mathcal{B}_j$, the orbit $(I(t), \varphi(t), p(t), q(t)) = \Phi^t(\eta)$ has the desired property (4.1).

For further details about the construction of such unstable orbits see [12]. □

5 Whiskered tori for a-priori unstable systems

The following Lemma provides a good "normal form" for the unperturbed part of an a-priori unstable system:

Lemma 5.1 *Let $H(I,\varphi,p,q) = \mathcal{R}(I;\mu) + \mathcal{P}(I,p,q;\mu)$ be analytic for I in a domain of $\mathbb{R}^{d-1}$, p and q in a neighborhood of the origin. Fixed $\mu \in \mathbb{R}$, assume that*

$$\partial_p \mathcal{P}(I,0,0;\mu) = 0 = \partial_q \mathcal{P}(I,0,0;\mu)$$
$$\det \partial^2_{(p,q)} \mathcal{P}(I,0,0;\mu) < 0 \, .$$

Then, there exists a canonical transformation $(I,\varphi,p,q) \longleftrightarrow (I',\varphi',p',q')$, real analytic for I' in a suitable domain of $\mathbb{R}^{d-1}$, $\varphi' \in \mathbb{T}^{d-1}$, p' and q' in a suitable neighborhood of the origin, sending H into the new Hamiltonian $h^*(I', p'q')$, depending only on the actions I' and on the product of the hyperbolic variables $\zeta' \equiv p'q'$. This transformation does not affect the action variables, i.e. $I'(I,\varphi,p,q) = I$. Of course, this transformation preserves the Lyapunov exponent, which in our case implies

$$\partial_\zeta h^*(I',0) = \sqrt{-\det \partial^2_{(p,q)} \mathcal{P}(I',0,0)} \, . \tag{5.1}$$

Furthermore, $\forall n \in \mathbb{N}^{d-1}$,

$$\partial^n_{I'} h^*|_{\zeta=0} = \partial^n_{I'} H|_{p=q=0} \, . \tag{5.2}$$

Also, if there exist $\bar{\mu} > 0$ such that, for any $|\mu| \leq \bar{\mu}$,

$$\det \partial^2_{(p,q)} \mathcal{P}(I,0,0;\mu) \leq -C \, , \tag{5.3}$$

for a suitable positive constant C, independent of μ, then there exists $\mu_0 > 0$ such that the above transformation depends analytically on μ, for $|\mu| \leq \mu_0$.

Proof. See [27], where the convergence of the Birkhoff series is shown, or Appendix A3 of [11], in which a KAM algorithm is used. See also [13] for a more general approach. □

Remark. In case condition (5.3) is not fulfilled, the above transformation may experience a very drastic loss of regularity in the parameter. This can be understood looking at equations (A3.9) and (A3.42) of [11], or just considering the following example. We claim that there is no canonical transformation $p = p(p',q';\mu)$, $q = q(p',q';\mu)$, *continuously* depending on the parameter μ, that sends the Hamiltonian $p^2/2 + \mu^2(\cos q - 1)$ into $\mu p'q' + (p'q')^2 G(p'q';\mu)$, with G depending continuously on μ. Arguing by contradiction, we would obtain

$$\frac{1}{2}\Big(p(p',q';0)\Big)^2 = (p'q')^2 G(p'q';0) \, ,$$

which implies $|p(p',q';0)| = |p'q'|\sqrt{2\,G(p'q';0)}$. Then,

$$\partial_{p'} p(0,0;0) = \lim_{p' \to 0} \frac{p(p',0;0) - p(0,0;0)}{p'} = 0 \, ,$$

and analogously $\partial_{q'}p(0,0;0) = 0$. Hence

$$\partial_{p'}p(0,0;0)\partial_{q'}q(0,0;0) - \partial_{q'}p(0,0;0)\partial_{p'}q(0,0;0) = 0\,,$$

which contradicts that the transformation is symplectic. □

Here is the application of the KAM Theorem to the a-priori unstable systems:

Theorem 5.2 *Let*

$$H(I,\varphi,p,q) = \mathcal{R}(I;\mu) + \mathcal{P}(I,p,q;\mu) + \mu f(I,\varphi,p,q;\mu)$$

be an a-priori unstable Hamiltonian according to Definition 2.2. Assume that $H_0 \equiv \mathcal{R}+\mathcal{P}$ is isoenergetically non-degenerate on the level surface $H_0 = E$. Then $\exists \mu_0 > 0$ such that, if $|\mu| \leq \mu_0$, the energy level is filled by whiskered tori with density at least $1 - O(\sqrt{\mu_0})$. More precisely:

There exist μ_0 and R_∞, $0 < \mu_0 \leq \bar{\mu}$ and $0 < R_\infty \leq R$, such that, for $|\mu| \leq \mu_0$, there exist:

(e1) *a smooth canonical transformation Φ, close to the identity and real analytic (for a fixed action) in the angles, in the hyperbolic variables and in the parameter μ,*

(e2) *a function $h_\infty : \mathbb{R}^{(d-1)} \times \mathbb{R} \times \mathbb{R} \longrightarrow \mathbb{R}$ with the same smoothness as Φ,*

(e3) *a set $\Omega_\mu \subseteq \mathbb{R}^{(d-1)+1+1}$, with density at least $1 - O(\sqrt{\mu_0})$,*

such that, fixed $|\mu| \leq \mu_0$,

$$\begin{array}{l} \partial^n\left(H \circ \Phi(I',\varphi',p',q')\right) = \partial^n h_\infty(I',p'q';\mu)\,,\ \forall (I',p',q') \in \Omega_\mu,\ \varphi' \in \mathbb{T}^{d-1},\ n \in \mathbb{N}^{2d} \\ h_\infty(I',p'q';\mu) = E\,,\ \forall (I',p',q') \in \Omega_\mu\,. \end{array} \tag{5.4}$$

Moreover, setting $\zeta' \equiv p'q'$,

$$\begin{array}{l} \forall (I',p',q') \in \Omega_\mu,\ \varphi' \in \mathbb{T}^{d-1},\ q' \neq 0,\ \ |\partial_{p'}(H\circ\Phi)| = |q'\,\partial_{\zeta'}h_\infty| > 0\,, \\ \forall (I',p',q') \in \Omega_\mu,\ \varphi' \in \mathbb{T}^{d-1},\ p' \neq 0,\ \ |\partial_{q'}(H\circ\Phi)| = |p'\,\partial_{\zeta'}h_\infty| > 0\,. \end{array}$$

More precisely: one can find a ball $B \subset \mathbb{R}^{d-1}$ in the actions and a set $\mathcal{D}_\tau$ of the form

$$\mathcal{D}_\tau \equiv \{I \in B \text{ and } \partial_I H_0(I,0;0) \text{ is } (\gamma_0,\tau) - \text{diophantine}\}$$

with a suitable γ_0 [i.e. $\gamma_0 = O(\sqrt{\mu_0})$], such that there exist

(E1) *a function $\mathcal{I}^I_\infty(\zeta;\mu)$, with range in the action space, which is smooth in I, ζ, μ, and (for a fixed I) real analytic in ζ and μ, for $|\zeta| \leq R^2_\infty$, $|\mu| \leq \mu_0$, verifying $\mathcal{I}^I_\infty(0;0) = I$,*

(E2) *a function $\alpha^I_\infty(\zeta;\mu)$ with the same regularity as $\mathcal{I}^I_\infty$, that verifies $\alpha^I_\infty(0;0) = 0$,*

such that:

(P1) $\partial_{I'}h_\infty(\mathcal{I}^{\bar{I}}_\infty(\zeta;\mu),\zeta;\mu) = \partial_I H_0(\bar{I},0;0)\cdot(1+\alpha^{\bar{I}}_\infty(\zeta;\mu))$, $\forall \bar{I} \in \mathcal{D}_\tau$.

(P2) *The set Ω_μ in (e3) can be described in the following two ways:*

$$\begin{array}{rcl} \Omega_\mu & = & \{(\mathcal{I}^{\bar{I}}_\infty(p'q';\mu),p',q'),\ \bar{I} \in \mathcal{D}_\tau,\ |p'| \leq R_\infty,\ |q'| \leq R_\infty\} = \\ & = & \{(I',p',q') \text{ s.t. } |p'| \leq R_\infty,\ |q'| \leq R_\infty,\ h_\infty(I,p'q';\mu) = E \\ & & \text{and } \exists \bar{I} \in \mathcal{D}_\tau \text{ s.t. } \partial_{I'}h_\infty(I,p'q';\mu) = \partial_I H_0(\bar{I},0;0)\cdot(1+\alpha^{\bar{I}}_\infty(p'q';\mu))\}\,. \end{array} \tag{5.5}$$

(P3) *Denoting by Dens^E [resp., by dens^E] the $(2d-1)$-dimensional restriction of the Lebesgue density on the energy level $\{(I',p',q') \text{ s.t. } h_\infty(I',p'q';\mu) = E\}$ [resp., the d-dimensional restriction in the space of the actions of the Lebesgue density on the energy level], we have*

$$\mathrm{dens}^E\Omega_\mu \geq 1 - O(\sqrt{\mu_0})$$

$$\text{Dens}\,^E\Phi(\Omega_\mu \times \mathbb{T}^{d-1}) \geq 1 - O(\sqrt{\mu_0}) \tag{5.6}$$

(P4) We have the following equality of sets:

$$\begin{aligned} \Omega_\mu^0 &\equiv \{\mathcal{I}_\infty^{\bar{I}}(0;\mu),\ \bar{I} \in \mathcal{D}_\tau\} = \\ &= \{I \text{ s.t. } h_\infty(I,0;\mu) = E \\ &\quad \text{and } \exists \bar{I} \in \mathcal{D}_\tau \text{ s.t. } \partial_{I'} h_\infty(I,0;\mu) = \partial_I H\Big|_{\substack{I=\bar{I},\mu=0\\ p=0=q}} \cdot (1 + \alpha_\infty^{\bar{I}}(0;\mu))\}\,. \end{aligned} \tag{5.7}$$

(P5) Denoting by $\text{dens}\,_E$ *the* $(d-2)-$*dimensional restriction of the Lebesgue density to the manifold defined by the energy relation* $h_\infty(I',0;\mu) = E$,

$$\text{dens}\,_E\Omega_\mu^0 \geq 1 - O(\sqrt{\mu_0})\,. \tag{5.8}$$

Finally, one can take $\mu_0 = O\Big(\inf(-\det \partial^2_{(p,q)}\mathcal{P})\Big)$.

Proof. Using Lemma 5.1, we obtain the new Hamiltonian

$$\mathcal{H}(I',\varphi',p',q') = h^\star(I',p'q';\mu) + \mu f^\star(I',\varphi',p',q';\mu)\,.$$

Notice that by (5.2) the matrices of isoenergetic non-degeneracy of $h^\star$ and H_0 agree in the origin of the hyperbolic coordinates; and by (5.1) $\partial_{\zeta'} h^\star > 0$, where $\zeta' \equiv p'q'$. Therefore, Theorem 3.1 can be applied. □

6 Whiskered tori for a-priori stable systems

In this section, ε will be a strictly positive, fixed, small parameter. Our target will be to look at an a-priori stable system near a simple resonance and recognize that these systems (under extremely mild conditions) are "hyperbolic in the first order". In this way we will be able to apply the previous results to the a-priori stable case too.

We note that this implies that the $d-$dimensional resonant tori break down for generic perturbations, creating $(d-1)-$dimensional whiskered tori. The mechanism of such a breakdown was considered, without measure estimates, in [38] and [25].

Lemma 6.1 *Consider the function*

$$h^{[0]}(I,p,q;\varepsilon) = h(I,p;\varepsilon) + \varepsilon f(I,p,q;\varepsilon) \tag{6.1}$$

with h *and* f *real analytic for* (I,p) *in a domain of* $\mathbb{R}^{d-1} \times \mathbb{R}$ *and* $q \in \mathrm{S}^1$. *Assume that there exists* $(\bar{I},\bar{p}) \in \mathbb{R}^{d-1} \times \mathbb{R}$, *verifying* $\partial_p h(\bar{I},\bar{p};0) = 0$ *and* $\partial_p^2 h(\bar{I},\bar{p};0) \neq 0$. *Assume that the function* $\bar{f}(q) \equiv f(\bar{I},\bar{p},q;0)$ *has a non singular critical point, i.e. there exists* $\bar{q}$ *such that* $\partial_q \bar{f}(\bar{q}) = 0$ *and* $\partial_q^2 \bar{f}(\bar{q}) \neq 0$. *Then, two functions exist* $p(I;\varepsilon)$ *and* $q(I;\varepsilon)$, *real analytic for* I *near* $\bar{I}$ *and* ε *small, with* $p(\bar{I};0) = \bar{p}$, $q(\bar{I};0) = \bar{q}$, *such that*

$$\partial_p h^{[0]}(I,\,p(I;\varepsilon),\,q(I;\varepsilon);\varepsilon) = 0 = \partial_q h^{[0]}(I,\,p(I;\varepsilon),\,q(I;\varepsilon);\varepsilon)\,. \tag{6.2}$$

Moreover, if $\bar{f}$ *has a non singular maximum and a nonsingular minimum, we can make the previous choice of* $\bar{q}$ *in order to verify*

$$\partial_p^2 h(\bar{I},\bar{p};0)\,\partial_q^2 f(\bar{I},\bar{p},\bar{q};0) < 0\,. \tag{6.3}$$

Proof. Apply the Implicit Function Theorem to

$$\mathcal{L}(p,q,I,\varepsilon) \equiv \Big(\partial_p h^{[0]}(I,p,q;\varepsilon),\ \partial_q f(I,p,q;\varepsilon)\Big)$$

near $I=\bar{I}$, $p=\bar{p}$, $q=\bar{q}$ and $\varepsilon=0$. □

The following Lemma sets the equilibria found in (6.2) in the origin:

Lemma 6.2 *Consider the Hamiltonian system* (6.1), *under the same assumptions as the previous Lemma. Let $\varphi \in \mathbb{T}^{d-1}$ denote the angles conjugated to the actions I. Then, the canonical transformation associated to the generating function*

$$\mathcal{G}(I^{[1]},p^{[1]},\varphi,q) = \Big(q - q(I^{[1]};\varepsilon)\Big)\, p^{[1]} + p(I^{[1]};\varepsilon)\,\sin\Big(q - q(I^{[1]};\varepsilon)\Big) + \varphi\cdot I^{[1]}$$

sends the Hamiltonian (6.1) *into a new Hamiltonian*

$$h^{[1]}(I^{[1]},p^{[1]},q^{[1]};\varepsilon)\,,\quad \textit{verifying}\quad \partial_{p^{[1]}} h^{[1]}(I^{[1]},0,0;\varepsilon) = 0 = \partial_{q^{[1]}} h^{[1]}(I^{[1]},0,0;\varepsilon)\,. \tag{6.4}$$

Proof. Straightforward check. □

We now inspect the hyperbolic structure of the above $h^{[1]}$ near $I=\bar{I}$ and $p=0=q$, showing that, for ε small enough, $h^{[1]}$ inherits such a hyperbolic structure from the one of $h^{[0]}$ stated in (6.3). In detail:

Lemma 6.3 *Let $h^{[1]}$ be the Hamiltonian obtained from $h^{[0]}$ in the previous Lemma. Define*

$$\lambda(I,p,q;\varepsilon) \equiv \sqrt{-\det \partial^2_{(p^{[1]},q^{[1]})} h^{[1]}(I,p,q;\varepsilon)}\,. \tag{6.5}$$

Then,

$$\Big(\lambda(I,0,0;\varepsilon)\Big)^2 = -\varepsilon(\partial_p^2 h\, \partial_q^2 f) - \varepsilon^2 \det \partial^2_{(p,q)} f\,,$$

with the functions on the right hand side evaluated in $p=p(I;\varepsilon)$ and $q=q(I;\varepsilon)$. In particular, if ε is small enough, $\lambda(\bar{I},0,0;\varepsilon)$ is real and positive, and $|\Re\lambda(I,p,q;\varepsilon)| \ge c_*\sqrt{\varepsilon}$, for a suitable constant c_*, for any I in a suitable neighborhood of $\bar{I}$ and p and q near 0.

Proof. Straightforward check. □

Lemma 6.4 *Consider the system* (6.1). *Assume that $(\bar{I},\bar{p}) \in \mathbb{R}^{d-1}\times\mathbb{R}$ verifies $\partial_p h(\bar{I},\bar{p};0)=0$ and $\partial_p^2 h(\bar{I},\bar{p};0)\neq 0$. Assume that the function $\bar{f}(q)\equiv f(\bar{I},\bar{p},q;0)$ has a non singular maximum and a non singular minimum. Then, there exists a canonical transformation $(I,\varphi,p,q)\longleftrightarrow(I^{[2]},\varphi^{[2]},p^{[2]},q^{[2]})$, defined for $p^{[2]}$ and $q^{[2]}$ in a neighborhood of* 0, *$I^{[2]}$ in a neighborhood of $\bar{I}$ and $\varphi^{[2]}\in\mathbb{T}^{d-1}$, with new Hamiltonian $h^{[2]}(I^{[2]},\zeta^{[2]};\varepsilon)$ verifying*

$$|\partial_{\zeta^{[2]}} h^{[2]}(I^{[2]},p^{[2]}q^{[2]};\varepsilon)| = |\lambda(I^{[2]},p^{[2]},q^{[2]};\varepsilon)| \ge c_*\sqrt{\varepsilon} \tag{6.6}$$

for a suitable constant c_*, for any $I^{[2]}$ in a suitable neighborhood of $\bar{I}$, $p^{[2]}$ and $q^{[2]}$ near 0, *where we defined $\zeta^{[2]}\equiv p^{[2]}q^{[2]}$ and $\lambda(I,p,q;\varepsilon)$ is defined in* (6.5). *Furthermore,*

$$\partial^n_{I^{[2]}} h^{[2]}(I^{[2]},0;\varepsilon) = \partial^n_I(h+\varepsilon f)(I^{[2]},\, p(I^{[2]};\varepsilon),\, q(I^{[2]};\varepsilon);\varepsilon)\,,\ \ \forall n\in\mathbb{N}^{d-1}\,. \tag{6.7}$$

Proof. First apply Lemma 6.2 to obtain a Hamiltonian like (6.4), and recall also Lemma 6.3. Then apply Lemma 5.1. □

The next theorem will show the existence of whiskered tori near simple resonances for a-priori stable systems. It will follow via Corollary 3.2, applying the previous Lemmas, where $(J_1,\ldots,J_d)$ and $(\psi_1,\ldots,\psi_d)$ in the next statement will correspond respectively to $(I_1,\ldots,I_{d-1},p)$ and $(\varphi_1,\ldots,\varphi_{d-1},q)$ of the Lemmas above. This is done making use of a classical result in perturbation theory, namely the Averaging Theorem (see, for instance, §5 of [3] and §52 of [4]).

Theorem 6.5 *Fix* $\nu \in \mathbb{N}$, $\nu \geq 2$. *Consider the system* $H(J,\psi) = h(J) + \varepsilon f(J,\psi;\varepsilon)$, *with* h *and* f *real analytic for* J *in a domain of* $\mathbb{R}^d$ *and* $\psi \in \mathbb{T}^d$. *Assume that* h *is isoenergetically non-degenerate on the energy level* $h = E$ *with respect to the first* $(d-1)$ *action variables. Let* $\bar{J}$ *be such that* $\partial_{J_d} h(\bar{J}) = 0$, $\partial^2_{J_d} h(\bar{J}) \neq 0$ *and let* $\partial_{(J_1,\dots J_{d-1})} h(\bar{J})$ *be rationally independent. Set*

$$\mathcal{F}_{\bar{J}}(x) \equiv \frac{1}{\operatorname{meas} \mathbb{T}^{d-1}} \int_{\mathbb{T}^{d-1}} f(\bar{J}, \psi_1, \dots, \psi_{d-1}, x; 0)\, d\psi_1 \dots d\psi_{d-1}\,.$$

Assume that $\mathcal{F}_{\bar{J}}$ *has nonsingular maximum and minimum. Then, a suitable subset (depending on* ν*) of the energy level near* $\bar{J}$*, is filled by whiskered invariant tori with density at least* $1 - O(\varepsilon^{\nu/2})$*, provided that* ε *is small enough; more precisely, the tori [resp., the fan*[5]*] fill the space, near* $\bar{J}$*, with* $(2d-3)-$*dimensional density [resp.,* $(2d-1)-$*dimensional density] at least* $1 - O(\varepsilon^{\nu/2})$*.*
More precisely: there exist

(i) a smooth canonical transformation $(J,\psi) = \Phi(I',\varphi',p',q')$, *with* $I' \in \mathbb{R}^d$, $p', q' \in \mathbb{R}$, $\varphi' \in \mathbb{T}^{d-1}$,

(ii) a smooth function $h_\infty : \mathbb{R}^{d-1} \times \mathbb{R} \longmapsto \mathbb{R}$,
(iii) a set $\Omega_{\varepsilon,\nu} \subset \mathbb{R}^{(d-1)+1+1}$, *with density at least* $1 - O(\varepsilon^{\nu/2})$,

such that:

$$\partial^n (H \circ \Phi(I',\varphi',p',q')) = \partial^n h_\infty(I',p'q';\varepsilon)\,,\ \forall (I',p',q') \in \Omega_{\varepsilon,\nu},\ \varphi' \in \mathbb{T}^{d-1}, n \in \mathbb{N}^{2d}$$
$$h_\infty(I',p'q';\varepsilon) = E\,,\ \forall (I',p',q') \in \Omega_{\varepsilon,\nu}\,.$$

In the coordinates (I',φ',p',q')*, the above mentioned tori are given by*

$$\mathcal{T}(I') = \{(I',\varphi',0,0)\,,\ \varphi' \in \mathbb{T}^{d-1}\}\,,$$

for I' *in a suitable set* $\Omega^0_{\varepsilon,\nu}$*, whose density is at least* $1 - O(\varepsilon^{\nu/2})$*. The corresponding (local) whiskers are*

$$W^s(I') = \{(I',\varphi',p',0)\,,\ \varphi' \in \mathbb{T}^{d-1}\ |p'| \leq R_\infty\}$$
$$W^u(I') = \{(I',\varphi',0,q')\,,\ \varphi' \in \mathbb{T}^{d-1}\ |q'| \leq R_\infty\}\ ,$$

for a suitable $R_\infty > 0$.
Furthermore: for any $I' \in \Omega^0_{\varepsilon,\nu}$ *there exists a smooth function* $\mathcal{I}_{I',\varepsilon,\nu} : \mathbb{R} \longrightarrow \mathbb{R}^{d-1}$ *such that* $\mathcal{I}_{I',\varepsilon,\nu}(0) = I'$ *and*

$$\Omega_{\varepsilon,\nu} = \{(\mathcal{I}_{I',\varepsilon,\nu}(p'q'),p',q')\ I' \in \Omega^0_{\varepsilon,\nu}\ |p'| \leq R_\infty\,,\ |q'| \leq R_\infty\}\ .$$

Moreover, setting $\zeta' \equiv p'q'$ *and* $\lambda_\infty \equiv \partial_{\zeta'} h_\infty$*, we have that* $|\lambda_\infty| \geq c^* \sqrt{\varepsilon}$*, for a suitable constant* $c^* > 0$ *and for any* $(I',p',q') \in \Omega_{\varepsilon,\nu}$*. Also,* $\omega_\infty \equiv \partial_{I'} h_\infty$ *is a* $(\gamma,\tau)-$*Diophantine vector with* $\gamma = O(\varepsilon^{\nu/2})$ *for any* $(I',p',q') \in \Omega_{\varepsilon,\nu}$*.*
Finally, for any $(I',p',q') \in \Omega_{\varepsilon,\nu}$ *and for any* $\varphi' \in \mathbb{T}^{d-1}$*,*

$$\Phi^t_{h_\infty}(I',\varphi',p',q') = (I',\ \varphi' + \omega_\infty(I',p'q')\,t,\ p'e^{-\lambda_\infty(I',p'q')t},\ q'e^{\lambda_\infty(I',p'q')t})\ ,$$

provided that $|p'e^{-\lambda_\infty(I',p'q')t}|$, $|q'e^{\lambda_\infty(I',p'q')t}| \leq R_\infty$.

Proof. Making use of the Averaging Theorem, we can find a canonical transformation, close to the identity for small ε, sending the Hamiltonian $H(J,\psi)$ of the hypothesis into $H^\flat(I,\varphi,p,q) = h(I,p) + \varepsilon f^\flat(I,p,q;\varepsilon) + O(\varepsilon^\nu)$. Such a transformation is defined in a suitable neighborhood of $\bar{J}$ (which is small if ν is big). Moreover

$$f^\flat(I,p,q;0) = \frac{1}{\operatorname{meas} \mathbb{T}^{d-1}} \int_{\mathbb{T}^{d-1}} f(I,p,\psi_1,\dots,\psi_{d-1},q;0)\, d\psi_1 \dots d\psi_{d-1}\,.$$

[5]Recall the notation of the fan at page 10.

Then, use Lemma 6.4 and Corollary 3.2 with $\mu \equiv \varepsilon^\nu$. Notice also that it is important that Theorem 3.1 contains a quantitative estimate on how small μ_0 is. In particular, it must be smaller than $\kappa_* \lambda_0^2$, and this estimate is satisfied if $\lambda \approx \sqrt{\epsilon}$, and $\mu \approx \epsilon^\nu$, with $\nu \geq 2$. □

The statement of the previous Theorem can be sharpened considering Diophantine simple resonances and optimizing the choice of ν as done in the Nekhoroshev theory:

Theorem 6.6 *Consider the system* $H(J,\psi) = h(J) + \varepsilon f(J,\psi;\varepsilon)$*, under the same assumptions as Theorem 6.5. Assume also that* $\partial_{(J_1,\ldots J_{d-1})} h(\bar{J})$ *is* (γ,τ)*−Diophantine. Then, if* ε *is small enough, a neighborhood of* $\bar{J}$ *in the energy level is filled by whiskered invariant tori with density at least* $1 - O(e^{-O(1/\varepsilon^c)})$*, where* $c > 0$ *is a suitable constant. More precisely, the tori [resp., the fan.] fill the space, near* $\bar{J}$*, with* $(2d-3)$*−dimensional density [resp.,* $(2d-1)$*−dimensional density] at least* $1 - O(e^{-O(1/\varepsilon^c)})$

Proof. Following the notations of [34], we set $\Lambda \equiv \{n = (n_1, \ldots, n_d) \in \mathbb{Z}^d \text{ s.t. } n_1 = \ldots = n_{d-1} = 0\}$, and

$$K \equiv c_1 \left(\frac{\gamma^2}{\varepsilon}\right)^{1/(2\tau+2)}, \qquad \alpha \equiv \frac{\gamma}{2K^\tau}, \qquad r \equiv c_2 \frac{\alpha}{K \sup|h''|}, \tag{6.8}$$

where the c_i's are suitable constants, chosen so that the hypotheses of the "Normal Form Lemma" of [34], page 192, are verified. Applying it to $H(J,\psi)$, it leads to the new Hamiltonian $H^\flat(I,\varphi,p,q) = h(I,p) + \varepsilon f^\flat(I,p,q;\varepsilon) + f_*(I,\varphi,p,q)$, with

$$f^\flat(I,p,q;0) = \frac{1}{\text{meas}\,\mathbb{T}^{d-1}} \int_{\mathbb{T}^{d-1}} f(I,p,\psi_1,\ldots,\psi_{d-1},q;0)\, d\psi_1 \ldots d\psi_{d-1},$$

and the size of f_* is controlled by $\varepsilon\, e^{-O(1/\varepsilon^c)}$. Then, as in the proof of the previous Theorem, apply Lemma 6.4 and Theorem 4.1. □

Notice that, in the proof of the previous Theorem, an explicit dependence of the constants with respect to the size of the domain of analyticity can be easily carried out. Namely, if the strip of analyticity in the angles ψ has width ξ, then the "Normal Form Lemma" bounds the size of f_* by $\varepsilon \exp(-K\xi/6)$, where K is defined in (6.8). Related measure estimates for elliptic equilibria can be found in [15] and in Section 4.1.5 of [7].

7 Proof of the KAM Theorem about partially hyperbolic tori

Proof of Theorem 3.1. The proof presented here makes use of a Newton-type algorithm, that will provide a sequence of canonical transformations converging on a suitable Cantor set. The general step of the algorithm can be summarized as follows:

Defining recursively suitable quantities as in (7.27)–(7.35)*, and assuming condition* (7.36) *[which is fulfilled by* $\gamma_0 = O(\sqrt{\mu_0})$*], there exists a sequence of canonical changes of variables* Φ_j*, converging in a suitable Cantor set, transforming the Hamiltonian* (3.1) *into* $H_j = h_j + f_j$*, with* h_j *depending only on the actions and on the product of the hyperbolic variables, and* $\sup_{V_j} |f_j| \leq \theta_j$*, where* V_j *is a sequence of sets, converging to a Cantor set, and* θ_j *converges to zero super-exponentially fast.*
Also, the set V_j *can be written as follows:*

$$\begin{aligned} V_j \;=\; & \Big\{(I,\varphi,p,q;\mu) \in \mathbb{C}^{(d-1)+(d-1)+1+1+1} \text{ s.t. } |p| \leq R_j,\ |q| \leq R_j,\ |\Im\varphi| \leq \xi_j,\ |\mu| \leq \mu_0, \\ & \textit{and there exists } \bar{I} \in \mathcal{D}_\tau \textit{ st } |I - \mathcal{I}_j^{\bar{I}}(pq;\mu)| \leq \tilde{\rho}_j \Big\}, \end{aligned}$$

where $\mathcal{D}_\tau$ *is defined in* (3.4)*, the quantities* R_j*,* $\tilde{\rho}_j$ *and* ξ_j *are defined in* (7.27)–(7.35)*, and* $\mathcal{I}_j$ *and* α_j *are functions defined via the Implicit Function Theorem by the relations*

$$\begin{aligned} \partial_I h_j(\mathcal{I}_j^I(\zeta;\mu),\zeta;\mu) &= \partial_I h(I,0;0) \cdot (1+\alpha_0^I(\zeta;\mu)) \cdot \ldots \cdot (1+\alpha_j^I(\zeta;\mu)) \\ h_j(\mathcal{I}_j^I(\zeta;\mu),\zeta;\mu) &= E\,. \end{aligned}$$

The fact that the KAM tori are of codimension (not higher than) one, i.e. p and q are (at most) one-dimensional, is crucial, in this argument, for the estimate on the small divisors.

In order to have dimensional estimates, we introduce a constant c with the dimensions of the inverse of an action. This is done only to have "dimensional" estimates: the reader who does not find it useful may set $c = 1$ in the sequel. In this way the matrix of isoenergetic non-degeneracy becomes

$$\mathcal{U}_0 \equiv \begin{pmatrix} \partial_I^2 h & c\omega \\ c\omega & 0 \end{pmatrix}, \quad \text{where } \omega \equiv \partial_I h.$$

In the sequel, we will often make use of the following easy relation: for $\delta < 1$

$$\sum_{j\geq 0} e^{-j\delta} = \frac{e^\delta}{e^\delta - 1} \leq \frac{e^\delta}{\delta} \leq \frac{e}{\delta}. \tag{7.1}$$

Also, we will use that, if $a \geq 0$, $0 < \delta < 1$, then there exist two constants C and C' (depending only on d and a) such that

$$\begin{aligned} \sum_{n\in\mathbb{Z}^d} |n|^a e^{-|n|\delta} &\leq C\,\delta^{-(d+a)} \\ \sum_{\substack{n\in\mathbb{Z}^d \\ |n|\geq N}} |n|^a e^{-|n|\delta} &\leq C'\,\delta^{-(d+a)}\,e^{-N\delta/2}. \end{aligned} \tag{7.2}$$

Now we start the iterative process. The first step is slightly different from the other ones, since we need to build the first couple of functions $\mathcal{I}_0^I$ and α_0^I as follows.

THE FIRST STEP. Set $h_0(I, pq; \mu) \equiv h(I, pq; \mu)$, $f_0(I, \varphi, p, q; \mu) \equiv \mu f(I, \varphi, p, q; \mu)$. Consider $\rho_0 \leq \rho/4$, $R_0 \leq R/4$ and $\mu_0 \leq \bar\mu/4$ small enough: then, via the Implicit Function Theorem, thanks to the isoenergetic non-degeneracy, we can find two functions $\mathcal{I}_0^I(\zeta;\mu)$ and $\alpha_0^I(\zeta;\mu)$, real analytic in $|\zeta| \leq R_0^2$ and $|\mu| \leq \mu_0$, verifying

$$\begin{aligned} &\mathcal{I}_0^I(0;0) = I, && \partial_I h_0(\mathcal{I}_0^I(\zeta;\mu), \zeta; \mu) = \omega_0(I)\,(1 + \alpha_0^I(\zeta;\mu)), && h_0(\mathcal{I}_0^I(\zeta;\mu), \zeta; \mu) = E, \\ &|\mathcal{I}_0^I(\zeta;\mu) - I| \leq \rho_0 \quad \text{and} && |\alpha_0^I(\zeta;\mu)| \leq c\rho_0, && \end{aligned} \tag{7.3}$$

where $\omega_0(I) \equiv \partial_I h_0(I, 0; 0)$. Let θ_0, A_0, B_0 and L_0 be such that $\sup|f_0| \leq \theta_0$, $\sup|\partial_I^2 h_0| \leq A_0$, $\sup|\mathcal{U}_0^{-1}| \leq B_0$, and $\sup|\partial_I h_0| \leq L_0$, where the sup is done over $\mathcal{O}_{\rho,\xi,R,\mu_0}$. Obviously, we may choose $\theta_0 = O(\mu_0)$.
For any real analytic $F(I, \varphi, p, q; \mu)$ we will write the Taylor–Fourier expansion

$$F(I, \varphi, p, q; \mu) = \sum_{\substack{k,j\in\mathbb{N} \\ n\in\mathbb{Z}^{d-1}}} F_{kjn}(I;\mu)\, p^k q^j e^{in\cdot\varphi}.$$

Also, without loss of generality, we may assume that $\forall |I - I^*| \leq 2\rho_0$, $|p| \leq 2R_0$, $|q| \leq 2R_0$, $|\mu| \leq 2\mu_0$, we have that $|\Re \partial_\zeta h_0(I, pq; \mu)| \geq \lambda_0/2$.

THE ITERATIVE SCHEME. Fix N_0 suitably large (see (7.7) below) and $\tilde\rho_0$ suitably small (see (7.11) below). Also define $\xi_0 \equiv \xi/2$ and fix δ_0, $0 < \delta_0 < \min\{1, \xi_0/4\}$. Denote f_{kjn}^0 the Taylor–Fourier terms of f_0. Set

$$\chi_0(I', \varphi', p', q'; \mu) \equiv \sum_{\substack{k,j\in\mathbb{N},\, n\in\mathbb{Z}^{d-1} \\ |k-j|+|n|>0,\, |n|\leq N_0}} \frac{-f_{kjn}^0(I';\mu)}{(k-j)\partial_\zeta h_0(I', p'q'; \mu) - i\partial_I h_0(I', p'q'; \mu)\cdot n}\,(p')^k (q')^j e^{in\cdot\varphi'} \tag{7.4}$$

defined on the set

$$V^0_{\tilde\rho_0,R_0,\xi_0,\mu_0} \equiv \Big\{(I,\varphi,p,q;\mu)\in\mathbb{C}^{(d-1)+(d-1)+1+1+1} \text{ s.t. } |p|\le R_0,\ |q|\le R_0,\ |\Im\varphi|\le\xi_0,\ |\mu|\le\mu_0,$$
$$\text{and there exists } \bar I\in\mathcal{D}_\tau \text{ st } |I-\mathcal{I}_0^{\bar I}(pq;\mu)|\le\tilde\rho_0\Big\}\,.$$

In the definition of $V^0_{\tilde\rho_0,R_0,\xi_0,\mu_0}$, the index "0" high above refers to the index "0" of $\mathcal{I}_0^I$. We now consider the Lie transform $(I,\varphi,p,q)\equiv\Phi^1_{\chi_0}(I',\varphi',p',q')$. From (7.4):

$$\{h_0,\chi_0\} = -f_0(I',\varphi',p',q';\mu)+ + \sum_{\substack{k,j\in\mathbb{N}\\ n\in\mathbb{Z}^{d-1},\ |n|>N_0}} f^0_{kjn}(I';\mu)\,(p')^k(q')^j\,e^{in\cdot\varphi'}+\sum_{k\in\mathbb{N}} f^0_{kk0}(I';\mu)\,(p'q')^k\,. \tag{7.5}$$

The next c_i's in this section stand for suitable constants (that can be explicitly determined by the algorithm). Set $\gamma_0^*\equiv\min\{\gamma_0,\lambda_0\}$. From (7.2):

$$\sup_{V^0_{\tilde\rho_0/2,R_0e^{-\delta_0},\xi_0-\delta_0,\mu_0}} |\textstyle\sum_{\substack{k,j\in\mathbb{N},\ n\in\mathbb{Z}^{d-1}\\ |n|>N_0}} f^0_{kjn}(I';\mu)\,(p')^k(q')^j\,e^{in\cdot\varphi'}|\le$$
$$\le c_1\theta_0\delta_0^{-(d+2)}\,e^{-N_0\delta_0/2}=\theta_0^2\delta_0^{-(d+2)}\,Ec^2\,(\gamma_0^*)^{-2}\,, \tag{7.6}$$

where we have chosen

$$N_0\equiv\frac{2}{\delta_0}\log\frac{c_1(\gamma_0^*)^2}{c^2E\theta_0}\,. \tag{7.7}$$

Furthermore, from (7.5), (7.1) and (7.2):

$$\sup_{V^0_{\tilde\rho_0/2,R_0e^{-\delta_0},\xi_0-\delta_0,\mu_0}} |\{h_0,\chi_0\}| = \sup_{V^0_{\tilde\rho_0/2,R_0e^{-\delta_0},\xi_0-\delta_0,\mu_0}} |\sum_{\substack{k,j\in\mathbb{N},\ n\in\mathbb{Z}^{d-1}\\ |k-j|+|n|>0\\ |n|\le N_0}} f^0_{kjn}(I';\mu)\,(p')^k(q')^j\,e^{in\cdot\varphi}|\le$$
$$\le c_1\,\theta_0\,\delta_0^{-d-2}\,. \tag{7.8}$$

Estimates on the small divisors. Assume that

$$\rho_0\le\min\left\{\frac{1}{3c},\ \frac{\lambda_0}{8cL_0N_0}\right\}\,. \tag{7.9}$$

Define[6] $A_0^*\equiv A_0$. Assume also that

$$\gamma_0\le 2A_0^*N_0^{\tau+1}\min\{\rho_0/2,\ R_0^2\}\,. \tag{7.10}$$

Define

$$\tilde\rho_0\equiv\frac{\gamma_0}{2A_0^*N_0^{\tau+1}}\le\min\left\{\frac{\rho_0}{2},\ R_0^2,\ \frac{\gamma_0}{2A_0N_0^{\tau+1}},\ \frac{\lambda_0}{8A_0N_0}\right\} \tag{7.11}$$

the inequality above following from (7.10). Now, $\forall\bar I\in\mathcal{D}_\tau$ and $n\in\mathbb{Z}^{d-1}-\{0\}$,

$$|\partial_I h_0\left(\mathcal{I}_0^{\bar I}(\zeta;\mu),\zeta;\mu\right)\cdot n|=|(1+\alpha_0^{\bar I}(\zeta;\mu))\,\omega_0(\bar I)\cdot n|\ge(1-c\rho_0)\,|\omega_0(\bar I)\cdot n|\ge\frac{2}{3}\frac{\gamma_0}{|n|^\tau}\,. \tag{7.12}$$

Thus, if $|I-\mathcal{I}_0^{\bar I}(\zeta;\mu)|\le\tilde\rho_0$,

$$|\partial_I h_0(I,\zeta;\mu)\cdot n| \ge |\partial_I h_0\left(\mathcal{I}_0^{\bar I}(\zeta;\mu),\zeta;\mu\right)\cdot n|-|\partial_I h_0\left(\mathcal{I}_0^{\bar I}(\zeta;\mu),\zeta;\mu\right)-\partial_I h_0(I,\zeta;\mu)|\,N_0\ge$$

[6] One needs the dummy definition of A_0^* just to make the notation uniform with the $j-$th step of the algorithm, in which one will set $A_j^*\equiv\max\{A_0,A_j\}$.

$$\begin{aligned}
&\geq \frac{2\gamma_0}{3\,|n|^\tau} - A_0\tilde\rho_0 N_0 \geq \\
&\geq \frac{\gamma_0}{6\,|n|^\tau}, \qquad \forall n\in\mathbb{Z}^{d-1}-\{0\}.
\end{aligned} \tag{7.13}$$

Besides, since $\omega_0(\bar I)$ is real, if $k\neq j$ and $|I-\mathcal{I}_0^{\bar I}(\zeta;\mu)|\leq\tilde\rho_0$:

$$\begin{aligned}
&|i\partial_I h_0(I,\zeta;\mu)\cdot n+\partial_\zeta h_0(I,\zeta;\mu)\,(k-j)|\geq \\
\geq\ &|\Re[i\partial_I h_0\left(\mathcal{I}^{\bar I}(\zeta;\mu),\zeta;\mu\right)\cdot n+\partial_\zeta h_0(I,\zeta;\mu)\,(k-j)]|-A_0\tilde\rho_0N_0\geq \\
\geq\ &|\Re[i\left(1+\alpha_0^{\bar I}(\zeta;\mu)\right)\,\omega_0(\bar I)\cdot n+\partial_\zeta h_0(I,\zeta;\mu)\,(k-j)]|-\lambda_0/8= \\
=\ &|\Re[i\alpha_0^{\bar I}(\zeta;\mu)\,\omega_0(\bar I)\cdot n+\partial_\zeta h_0(I,\zeta;\mu)\,(k-j)]|-\lambda_0/8\geq \\
\geq\ &|k-j|\,|\Re\,\partial_\zeta h_0(I,\zeta;\mu)|-|\Re[i\alpha_0^{\bar I}(\zeta;\mu)\,\omega_0(\bar I)\cdot n]|-\lambda_0/8\geq \\
\geq\ &|\Re\,\partial_\zeta h_0(I,\zeta;\mu)|-|i\alpha_0^{\bar I}(\zeta;\mu)\,\omega_0(\bar I)\cdot n|-\lambda_0/8\geq\lambda_0/4\,.
\end{aligned} \tag{7.14}$$

The estimate on the small divisors in χ_0 is thus given by (7.13) and (7.14). These inequalities also show the convergence of the series defining χ_0 on $V^0_{\tilde\rho_0,R_0,\xi_0,\mu_0}$.

Estimates on the Lie transform. From the estimates on the small denominators and (7.2), it follows that

$$\sup_{V^0_{\tilde\rho_0,R_0e^{-\delta_0/2},\xi_0-\delta_0/2,\mu_0}}|\chi_0|\leq c_2\,\frac{\theta_0}{\gamma_0^*}\,\delta_0^{-\kappa_0} \tag{7.15}$$

so that, by the Cauchy Estimate:

$$\begin{aligned}
\sup_{V^0_{\tilde\rho_0/2,R_0e^{-\delta_0},\xi_0-\delta_0,\mu_0}}\ \sup_{1\leq i\leq d-1}|\partial_{I'_i}\chi_0| &\leq c_3\,\frac{\theta_0}{\gamma_0^*\,\tilde\rho_0}\,\delta_0^{-\kappa_1} \\
\sup_{V^0_{\tilde\rho_0/2,R_0e^{-\delta_0},\xi_0-\delta_0,\mu_0}}\ \sup_{1\leq i\leq d-1}|\partial_{\varphi'_i}\chi_0| &\leq c_3\,\frac{\theta_0}{\gamma_0^*}\,\delta_0^{-\kappa_1} \\
\sup_{V^0_{\tilde\rho_0/2,R_0e^{-\delta_0},\xi_0-\delta_0,\mu_0}}|\partial_{p'}\chi_0| &\leq c_3\,\frac{\theta_0}{\gamma_0^*\,R_0}\,\delta_0^{-\kappa_1} \\
\sup_{V^0_{\tilde\rho_0/2,R_0e^{-\delta_0},\xi_0-\delta_0,\mu_0}}|\partial_{q'}\chi_0| &\leq c_3\,\frac{\theta_0}{\gamma_0^*\,R_0}\,\delta_0^{-\kappa_1}\,,
\end{aligned} \tag{7.16}$$

where κ_i's denote suitable constants (depending only on d and τ). Hence, using Lemma A.3, $\forall|t|\leq 3$,

$$\Phi^t_{\chi_0}(V^0_{\tilde\rho_0/4,R_0e^{-4\delta_0},\xi_0-4\delta_0,\mu_0})\subseteq V^0_{\tilde\rho_0/3,R_0e^{-3\delta_0},\xi_0-3\delta_0,\mu_0}\subseteq V^0_{\tilde\rho_0,R_0e^{-\delta_0},\xi_0-\delta_0,\mu_0} \tag{7.17}$$

provided that

$$c_5\,\frac{\theta_0}{\gamma_0^*\,\tilde\rho_0}\,\delta_0^{-\kappa_2}\leq 1\,. \tag{7.18}$$

Estimates on the new Hamiltonian. Define

$$\begin{aligned}
h_0^\dagger(I',\varphi',p',q';\mu) &\equiv \int_0^1(1-t)\,\{\{h_0,\chi_0\},\,\chi_0\}\circ\Phi^t_{\chi_0}(I',\varphi',p',q';\mu)\,dt \\
f_0^\dagger(I',\varphi',p',q';\mu) &\equiv \int_0^1\{f_0,\chi_0\}\circ\Phi^t_{\chi_0}(I',\varphi',p',q';\mu)\,dt \\
f_0^*(I',p'q';\mu) &\equiv \sum_{k\in\mathbb{N}}f^0_{kk0}(I';\mu)(p'q')^k \\
h_1(I',p'q';\mu) &\equiv h_0(I',p'q';\mu)+f_0^*(I',p'q';\mu) \\
f_1(I',\varphi',p',q';\mu) &\equiv h_0^\dagger(I',\varphi',p',q';\mu)+f_0^\dagger(I',\varphi',p',q';\mu)+\sum_{\substack{k,j\in\mathbb{N},\ n\in\mathbb{Z}^{d-1}\\|n|>N_0}}f^0_{kjn}(I';\mu)p'^k q'^j\,e^{in\cdot\varphi'} \\
H_1(I',\varphi',p',q') &\equiv H\circ\Phi^1_{\chi_0}(I',\varphi',p',q')
\end{aligned}$$

Using Lemma A.5 (at the first order for f_0 and at the second order for h_0) one has

$$\begin{aligned} h_0 \circ \Phi^1_{\chi_0} &= h_0 + \{h_0, \chi_0\} + h_0^\dagger \\ f_0 \circ \Phi^1_{\chi_0} &= f_0 + f_0^\dagger \,. \end{aligned}$$

This implies, by (7.5), that

$$H_1(I',\varphi',p',q') = h_1(I',p'q';\mu) + f_1(I',\varphi',p',q';\mu)\,. \tag{7.19}$$

By Lemma A.6, (7.15) and (7.8), making use of (7.17) to control the domains, we obtain:

$$\sup_{V^0_{\tilde\rho_0/4, R_0 e^{-4\delta_0}, \xi_0-4\delta_0,\mu_0}} |h_0^\dagger| \le \sup_{V^0_{\tilde\rho_0/3, R_0 e^{-3\delta_0}, \xi_0-3\delta_0,\mu_0}} |\{\{h_0,\chi_0\},\chi_0\}| \le c_6 \frac{\theta_0^2}{\gamma_0^* \tilde\rho_0} \delta_0^{-\kappa_3} \tag{7.20}$$

$$\sup_{V^0_{\tilde\rho_0/4, R_0 e^{-4\delta_0}, \xi_0-4\delta_0,\mu_0}} |f_0^\dagger| \le c_6 \frac{\theta_0^2}{\gamma_0^* \tilde\rho_0} \delta_0^{-\kappa_3}\,. \tag{7.21}$$

Hence, by (7.6)

$$\sup_{V^0_{\tilde\rho_0/4, R_0 e^{-4\delta_0}, \xi_0-4\delta_0,\mu_0}} |f_1| \le c_7 \frac{\theta_0^2}{\gamma_0^* \tilde\rho_0} \delta_0^{-\kappa_4} \equiv \theta_1 \tag{7.22}$$

Then, setting $\rho_1 \equiv \tilde\rho_0/8$, $R_1 \equiv R_0\, e^{-4\delta_0}$, $\xi_1 \equiv \xi_0 - 4\delta_0$, we obtain a new Hamiltonian like (7.19) with $\sup_{V^0_{2\tilde\rho_1,R_1,\xi_1,\mu_0}} |f_1| \le \theta_1$.

By the Implicit Function Theorem, we obtain two functions $\mathcal{I}_1^I(\zeta;\mu)$ and $\alpha_1^I(\zeta;\mu)$, real analytic for $|\zeta| \le R_1^2$ and $|\mu| \le \mu_0$ verifying

$$|\mathcal{I}_1^I(\zeta;\mu) - \mathcal{I}_0^I(\zeta;\mu)| \le \rho_1/2 \tag{7.23}$$

$$|\alpha_1^I(\zeta;\mu)| \le c\rho_1/2 \tag{7.24}$$

$$\begin{aligned} \partial_I h_1\left(\mathcal{I}_1^I(\zeta;\mu),\zeta;\mu\right) &= \partial_I h_0\left(\mathcal{I}_0^I(\zeta;\mu),\zeta;\mu\right)\cdot\left(1+\alpha_1^I(\zeta;\mu)\right) = \\ &= \omega_0(I)\cdot\left(1+\alpha_0^I(\zeta;\mu)\right)\cdot\left(1+\alpha_1^I(\zeta;\mu)\right) \end{aligned} \tag{7.25}$$

$$h_1\left(\mathcal{I}_1^I(\zeta;\mu),\zeta;\mu\right) = E\,. \tag{7.26}$$

By construction $V^1_{\rho_1,R_1,\xi_1,\mu_0} \subseteq V^0_{2\rho_1,R_1,\xi_1,\mu_0}$, so that $\sup_{V^1_{\rho_1,R_1,\xi_1,\mu_0}} |f_1| \le \theta_1$, and we can iterate the previous arguments (writing the appropriate index instead of the index 0), from (7.4) onwards.

ITERATION OF THE ALGORITHM. Set $\gamma^* \equiv \min\{\gamma_0, \lambda_0/2\}$. Fix a suitable $\ell > 1$, and define recursively:

$$\delta_j \equiv \frac{\delta_{j-1}}{\ell} = \frac{\delta_0}{\ell^j} \tag{7.27}$$

$$N_j \equiv \frac{2}{\delta_j} \log \frac{c_1(\gamma_j^*)^2}{c^2 E \theta_j} \tag{7.28}$$

$$\tilde\rho_j \equiv \frac{\gamma_j^*}{2A_j^* N_j^{\tau+1}} \tag{7.29}$$

$$\rho_{j+1} \equiv \frac{\tilde\rho_j}{8} \tag{7.30}$$

$$R_{j+1} \equiv R_j\, e^{-4\delta_j} = R_0\, e^{-4\sum_{i=0}^j \delta_i} \tag{7.31}$$

$$\xi_{j+1} \equiv \xi_j - 4\delta_j = \xi_0 - 4\sum_{i=0}^j \delta_i \tag{7.32}$$

$$\theta_{j+1} \equiv c_7 \frac{\theta_j^2}{\gamma_j^* \, \tilde{\rho}_j} \, \delta_j^{-\kappa_4} \tag{7.33}$$

$$\varepsilon_j \equiv \frac{c^2 E \theta_j}{c_1 (\gamma_j^*)^2} \tag{7.34}$$

$$\gamma_j^* \equiv \min\{\gamma_{j-1}^*, \, \lambda_j\} \, . \tag{7.35}$$

Obviously

$$N_j = \frac{2}{\delta_j} \log \frac{1}{\varepsilon_j} \, .$$

Iterating the scheme, one obtains

$$H_i(I, \varphi, p, q) = h_i(I, pq; \mu) + f_i(I, \varphi, p, q; \mu)$$

with

$$\sup_{V_i} |f_i| \leq \theta_i$$

where

$$V_i \equiv V^i_{\tilde{\rho}_i, R_i, \xi_i, \mu_0} \, .$$

In order to apply recursively the algorithm above, one has to check that the following conditions are satisfied at the general $\bar{\jmath}$–th step of the scheme:

$(C1)$ The sup [resp., the inf] over $V_{\bar{\jmath}}$ of a quantity involving only $h_{\bar{\jmath}}$ (or its derivatives up to a suitable order) is less or equal than the double of the corresponding sup [resp., greater or equal than the half of the corresponding inf] over V_0 of the corresponding quantity with index 0 (e.g.: $\lambda_{\bar{\jmath}} \equiv \inf_{V_{\bar{\jmath}}} |\partial_\zeta h_{\bar{\jmath}}| \geq \lambda_0/2$, etc.),

$(C2)$ The matrix $\mathcal{U}_{\bar{\jmath}}$ is nonsingular on $V_{\bar{\jmath}}$,

$(C3)$ $\gamma_j \leq 2 A_{\bar{\jmath}}^* N_{\bar{\jmath}}^{\tau+1} \min\{\rho_{\bar{\jmath}}/2, \, R_{\bar{\jmath}}^2\}$ and $\rho_j \leq \lambda_j/(8 c L_j N_j)$,

$(C4)$ There exists a constant C^* such that $\varepsilon_{\bar{\jmath}} \leq (C^* \, \Lambda_0^\tau \, \varepsilon_0)^{2^{\bar{\jmath}}}$.

To prove $(C1)$–$(C4)$, we will assume the following **main condition**:

$$K_1 \left(\log \frac{c_1 (\min\{\gamma_0, \lambda_0\})^2}{c^2 E \theta_0} \right)^{K_2} \frac{c^2 E \theta_0}{(\min\{\gamma_0, \lambda_0\})^2} \leq 1 \tag{7.36}$$

where K_1 and K_2 are suitable constants. We remark [see (7.42) below] that this condition is satisfied choosing $\gamma_0 = O(\sqrt{\theta_0})$, for θ_0 small enough, $\theta_0 \leq O(\lambda_0^2)$.
The proof of $(C1)$–$(C4)$ is by induction, assuming them true for $i = 1, \ldots, \bar{\jmath} - 1$. In these pages, k_i's will stand for suitable constants.
First notice that, by definition of $h_{\bar{\jmath}}$, the relation $\lambda_{\bar{\jmath}} \geq \lambda_{\bar{\jmath}-1}/2$ follows, and so $\gamma_{\bar{\jmath}}^* \geq \gamma_{\bar{\jmath}-1}^*/2$.
Notice also that $\varepsilon_i \geq \varepsilon_{i-1}^2$, $\forall 1 \leq i \leq \bar{\jmath}$, so

$$\varepsilon_i \geq \varepsilon_{i-1}^2 \geq \ldots \geq \varepsilon_0^{2^i} \, , \; \forall 1 \leq i \leq \bar{\jmath} \, . \tag{7.37}$$

Therefore, defining $\Lambda_0 \equiv \log(1/\varepsilon_0)$,

$$N_i \leq \frac{2^{i+1} \ell^i}{\delta_0} \log \frac{1}{\varepsilon_0} = \frac{2^{i+1} \ell^i}{\delta_0} \Lambda_0 \, , \; \forall 1 \leq i \leq \bar{\jmath} \tag{7.38}$$

Making use of the inductive hypothesis, this implies that

$$\tilde{\rho}_i \geq \frac{\gamma_0}{k_1^i \, A_0^* \, \Lambda_0^{\tau+1}} \, , \; \forall 1 \leq i \leq \bar{\jmath} - 1 \, . \tag{7.39}$$

Thus

$$\theta_{\bar{j}} \le \frac{k_2^{\bar{j}} \theta_{\bar{j}-1}^2 \, A_0 \, \Lambda_0^{\tau+1}}{(\gamma_{\bar{j}-1}^*)^2} \, .$$

Then,

$$\varepsilon_{\bar{j}} \le k_3^{\bar{j}} \, A_0 \Lambda_0^{\tau+1} \, c^{-2} E^{-1} \, \varepsilon_{\bar{j}-1}^2 \, . \tag{7.40}$$

Iterating (7.40), one gets $(C4)$. Furthermore, it is easy to see that $\sup_{V_j} |\mathcal{U}_j - \mathcal{U}_{j-1}| \le B_0^{-1} 3^{-j}$, hence

$$\sup_{V_{\bar{j}}} |\mathcal{U}_{\bar{j}} - \mathcal{U}_0| \le \sum_{i=1}^{\bar{j}} \sup_{V_{\bar{j}}} |\mathcal{U}_i - \mathcal{U}_{i-1}| \le B_0^{-1} \sum_{i \ge 1} 3^{-i} = \frac{1}{2B_0} \, .$$

This implies $(C2)$, via Lemma A.2. Incidentally, we have also proved that $B_{\bar{j}} \le 2B_0$. The other relations in $(C1)$ follow in the same way. Also, the already proved $(C4)$ implies that

$$N_{\bar{j}} \ge \frac{2^{\bar{j}+1} \ell^{\bar{j}}}{\delta_0} \log \frac{1}{C^* \, \Lambda_0 \, \varepsilon_0} \, . \tag{7.41}$$

then, recalling (7.38), one obtains $(C3)$.

Passage to the limit. From (7.23):

$$\sup_{|\zeta| \le R_\infty^2, \, |\mu| \le \mu_0} |\mathcal{I}_{j+m}^I - \mathcal{I}_j^I| \le \sum_{i=j}^{j+m-1} \sup_{|\zeta| \le R_\infty^2, \, |\mu| \le \mu_0} |\mathcal{I}_{i+1}^I - \mathcal{I}_i^I| \le \sum_{i=j}^{j+m-1} \rho_i \le \sum_{i \ge j} \frac{\rho_0}{4^i} \, ,$$

showing the uniform convergence of $\mathcal{I}_j^I$ to a suitable $\mathcal{I}_\infty^I$ for $|\zeta| \le R_\infty^2$ and $|\mu| \le \mu_0$.
Also, if we set

$$\alpha_\infty^I(\zeta; \mu) \equiv \prod_{j=0}^{\infty} \left(1 + \alpha_j^I(\zeta; \mu)\right) - 1 \, , \;\; \forall |\zeta| \le R_\infty^2, \; |\mu| \le \mu_0 \, ,$$

using the fact that $|\alpha_j| \le c\rho_j \le c\rho_0/4^j$ it is easy to prove that the above product converges uniformly and that $|\alpha_\infty| \le c\rho$.

Via iteration of (7.15), the convergence of the transformation $\Phi_j \equiv \Phi_{\chi_j}^1 \circ \ldots \circ \Phi_{\chi_0}^1$ readily follows. Since the convergences are uniform for complex $|\mu| \le \mu_0$, $|p| \le R_\infty$, $|q| \le R_\infty$, $|\Im \varphi| \le \xi_\infty$, we obtain the claimed analyticity in the angles, in the hyperbolic variables and in the parameter μ.

CHARACTERIZATION OF THE SETS Ω_μ AND Ω_μ^0 OF VALIDITY OF THE THEOREM [see (3.6) and (3.10)] AND MEASURE OF THE PRESERVED TORI. Let $\bar{x}$ such that if $x \ge \bar{x}$ then $c_1 K_1 \, (\log x)^{K_2} \, x^{-1} \le 1$. Set

$$\gamma_0 \equiv \sqrt{\frac{c^2 \, E \, \bar{x} \, \theta_0}{c_1}} = O(\sqrt{\theta_0}) = O(\sqrt{|\mu_0|}) \, . \tag{7.42}$$

Then, the KAM condition (7.36) is satisfied.
Since h_∞ is isoenergetically non-degenerate for I and ζ sufficiently close to I^* and 0 respectively and for μ_0 small enough, without loss of generality, we may assume that $\inf_{I \in B, \, |\zeta| \le R_\infty^2, |\mu| \le \mu_0} |\partial_{I_{d-1}} h_\infty| > 0$, where B stands for a suitable $(d-1)$–dimensional domain.
Then, if we denote by the symbol "tilde" the projection onto the first $d-2$ components and by ι the function implicitly defined by $h_\infty(\tilde{I}, \iota(\tilde{I}), 0; \mu) = E$, Proposition 2.5 implies that

$$\mathcal{A}(\tilde{I}) \equiv \frac{\partial_{\tilde{I}} h_\infty(\tilde{I}, \iota(\tilde{I}), 0; \mu)}{\partial_{I_{d-1}} h_\infty(\tilde{I}, \iota(\tilde{I}), 0; \mu)} \tag{7.43}$$

is a diffeomorphism. This proves that

$$I = \mathcal{I}_\infty^{\bar{I}}(0;\mu) \iff h_\infty(I,0;\mu) = E \text{ and } \partial_I h_\infty(I,0;\mu) = \omega_0(\bar{I}) \cdot (1 + \alpha_\infty^{\bar{I}}(0;\mu)) . \tag{7.44}$$

And this implies (3.10) and (3.11).
We now prove that the function $\mathcal{I}_\infty$ is locally invertible on the E−energy level, i.e., there exist a suitable $\rho' > 0$ such that

$$\begin{aligned}&\text{for any } I,\ |I - I^*| \le \rho',\ h_\infty(I,0;\mu) = E, \text{ there exists a unique } \bar{I},\ |\bar{I} - I^*| \le \rho \\ &\text{such that } h_0(\bar{I},0;0) = E \text{ and } I = \mathcal{I}_\infty^{\bar{I}}(0;\mu).\end{aligned} \tag{7.45}$$

To show this, recall Proposition 2.5 and consider the local diffeomorphism

$$G(I,\sigma) \equiv (\sigma \partial_I h_0(I,0;0), h_0(I,0;0)) . \tag{7.46}$$

Given I, let $\bar{I}$ be the component in the actions of $G^{-1}(\partial_I h_\infty(I,0;\mu), E)$. From (7.43) it follows that $\mathcal{A}(\bar{I}) = \mathcal{A}(\mathcal{I}_\infty^{\bar{I}}(0;\mu))$, proving the existence in (7.45). The uniqueness follows from the fact that $\mathcal{I}_\infty^{\bar{I}}(0;\mu) = \mathcal{I}_\infty^{\bar{y}}(0;\mu)$ and $h_0(\bar{I},0;0) = E = h_0(\bar{y},0;0)$ imply $G(\bar{I}, 1 + \alpha_\infty^{\bar{I}}(0;\mu)) = G(\bar{y}, 1 + \alpha_\infty^{\bar{y}}(0;\mu))$. Then, (3.12) readily follows from (7.45). The other characterizations of the set of validity of the Theorem and the related estimates on the measure can be proved with similar arguments. □

* * * * *

Appendix

A Some technical Lemmas

A.1 Some linear algebra

Lemma A.1 *Consider $A \in \mathrm{Mat}(n \times n)$. Let a, b, c, d be n−dimensional column vectors and $\alpha, \beta, \gamma \in \mathbb{R}$, with $\beta \neq 0 \neq \gamma$. Then*

$$\det \begin{pmatrix} A & a & b \\ c^T & \alpha & \gamma \\ d^T & \beta & 0 \end{pmatrix} = (-\gamma\beta)^{-n+1} \det\left((\gamma a - \alpha b)d^T - \beta(\gamma A - bc^T)\right) . \tag{A.1}$$

Proof. We have:

$$\begin{aligned}
\det \begin{pmatrix} A & a & b \\ c^T & \alpha & \gamma \\ d^T & \beta & 0 \end{pmatrix} &= \gamma^{-n} \det \begin{pmatrix} \gamma A & \gamma a & \gamma b \\ c^T & \alpha & \gamma \\ d^T & \beta & 0 \end{pmatrix} = \gamma^{-n} \det \begin{pmatrix} (\gamma A \ \ \gamma a \ \ \gamma b) - b(c^T \ \ \alpha \ \ \gamma) \\ c^T \ \ \alpha \ \ \gamma \\ d^T \ \ \beta \ \ 0 \end{pmatrix} = \\
&= \gamma^{-n} \det \begin{pmatrix} \gamma A - bc^T & \gamma a - \alpha b & 0 \\ c^T & \alpha & \gamma \\ d^T & \beta & 0 \end{pmatrix} = \\
&= (\gamma\beta)^{-n} \det \begin{pmatrix} \beta(\gamma A - bc^T) & \gamma a - \alpha b & 0 \\ \beta c^T & \alpha & \gamma \\ \beta d^T & \beta & 0 \end{pmatrix} = \\
&= (\gamma\beta)^{-n} \det \begin{pmatrix} \begin{pmatrix} \beta(\gamma A - bc^T) \\ \beta c^T \\ \beta d^T \end{pmatrix} - \begin{pmatrix} \gamma a - \alpha b \\ \alpha \\ \beta \end{pmatrix} d^T & \begin{matrix} \gamma a - \alpha b \\ \alpha \\ \beta \end{matrix} & \begin{matrix} 0 \\ \gamma \\ 0 \end{matrix} \end{pmatrix} = \\
&= (\gamma\beta)^{-n} \det \begin{pmatrix} \beta(\gamma A - bc^T) - (\gamma a - \alpha b)d^T & \gamma a - \alpha b & 0 \\ \beta c^T - \alpha d^T & \alpha & \gamma \\ 0 & \beta & 0 \end{pmatrix} =
\end{aligned}$$

$$= (\gamma\beta)^{-n}\cdot(-1)^{2n+3}\gamma\det\begin{pmatrix}\beta(\gamma A-bc^T)-(\gamma a-\alpha b)d^T & \gamma a-\alpha b\\ 0 & \beta\end{pmatrix} =$$
$$= (\gamma\beta)^{-n}\cdot(-1)^{2n+3}\gamma\cdot(-1)^{2n+2}\beta\det\left(\beta(\gamma A-bc^T)-(\gamma a-\alpha b)d^T\right).$$

proving (A.1). □

A.2 Perturbations of nonsingular matrices

Lemma A.2 *Let M, N be square matrices of the same order. If M is nonsingular and $|N| < 1/|M^{-1}|$, then $(M+N)$ is nonsingular too. Moreover*

$$|(M+N)^{-1}| \le \frac{|M^{-1}|}{1-|M^{-1}|\,|N|}\,.$$

In particular, if M is nonsingular and $|N| \le 1/(2\,|M^{-1}|)$, then $(M+N)$ is nonsingular too and $|(M+N)^{-1}| \le 2\,|M^{-1}|$.

Proof. We have that $M^{-1}\sum_{k\ge 0}(-1)^k(M^{-1}N)^k = (M+N)^{-1}$. □

A.3 Estimates on Lie transforms

Lemma A.3 *Let V be a domain of $\mathbb{C}^d\times\mathbb{C}^d$. Fixed $r=(r_1,\ldots,r_{2d})\in\mathbb{R}^{2d}$, with $r_i\ge 0$, we call*

$$V_r \equiv \{z=(z_1,\ldots,z_{2d})\in\mathbb{C}^{2d} \text{ s.t. } \exists w=(w_1,\ldots,w_{2d})\in V \text{ s.t. } |z_i-w_i|\le r_i\}\,.$$

Assume that $\chi(x,y)$ is real analytic on V_r. Fixed $r'\in\mathbb{R}^{2d}$, $0\le r_i'<r_i$, then $\Phi^t_\chi(V_{r'})\subseteq V_r$ provided that $|t|\le t_0$ with

$$t_0\max\left\{\frac{\sup_{V_r}|\partial_{y_i}\chi|}{r_i-r_i'}\,,\ \frac{\sup_{V_r}|\partial_{x_i}\chi|}{r_{i+d}-r_{i+d}'}\,,\ i=1,\ldots,d\right\}\le 1\,. \tag{A.2}$$

Proof. If $(x,y)\in V_{r'}$, then $\exists(\bar x,\bar y)\in V$ such that $|x_i-\bar x_i|\le r_i'$ and $|y_i-\bar y_i|\le r_{i+d}'$, $1\le i\le d$. Set $(x(t),y(t))\equiv\Phi^t_\chi(x,y)$. If the thesis were false, there would exist $\bar t$, $|\bar t|<t_0$, which is the time of "first exit" from V_r. Explicitly, $(x(t),y(t))\in V_r$ for all $|t|\le|\bar t|$ and $(x(\bar t),y(\bar t))\in\partial V_r$. But

$$|x_i(\bar t)-x_i| = |x_i(\bar t)-x_i(0)| \le \sup_{|t|\le|\bar t|}|\dot x_i(t)|\,|\bar t| < \sup_{V_r}|\partial_{y_i}\chi|\,t_0 \le r_i-r_i' \tag{A.3}$$

that[7] shows $|x_i(\bar t)-\bar x_i|\le|x_i(\bar t)-x_i|+|x_i-\bar x_i| < r_i-r_i'+r_i' = r_i$. In the same way one sees $|y_i(\bar t)-\bar y_i|<r_{i+d}$. So $(x(\bar t),y(\bar t))\in\operatorname{Int}V_r$, in contrast with $(x(\bar t),y(\bar t))\in\partial V_r$. □

We denote by $L_\chi H\equiv\{H,\chi\}$ the Poisson operator and by L^j_χ the operator L_χ applied for j times.

Lemma A.4 *For all $m\in\mathbb{N}$*

$$\frac{d^m}{dt^m}\left(H\circ\Phi^t_\chi(x,y)\right) = (L^m_\chi H)\circ\Phi^t_\chi(x,y)\,.$$

Proof. Induction over m. □

Lemma A.5 *Assume the hypotheses over r, r', χ, t_0 in Lemma A.3. Fix $k\in\mathbb{N}$, $k\ge 1$. Then*

$$H\circ\Phi^t_\chi(x,y) = \sum_{j=0}^{k-1}\frac{t^j}{j!}L^j_\chi H + t^k R_k(x,y;t)\,,$$

[7] In (A.3) we assumed $\sup_{V_r}|\partial_{y_i}\chi|\ne 0$; if not, obviously $|x_i(\bar t_i)-x_i| = 0 < r_i-r_i'$, and the argument goes on in the same way.

with

$$R_k(x,y;t) \;=\; \int_0^1 \frac{(1-s)^{k-1}}{(k-1)!}\frac{d^k}{d\tau^k}\left(H\circ\Phi_\chi^\tau(x,y)\right)\Big|_{\tau=ts} ds \tag{A.4}$$

$$\sup_{V_{r'}}|R_k| \;\le\; \frac{\sup_{V_r}|H|}{(t_0-|t|)^k}, \quad \forall|t|<t_0 . \tag{A.5}$$

Proof. Set $\mathcal{F}(t) \equiv H\left(\Phi_\chi^t(x,y)\right)$. By the Taylor expansion one has

$$\mathcal{F}(t) = \sum_{j=0}^{k-1}\frac{t^j}{j!}\frac{d^j\mathcal{F}}{dt^j}(0) + \int_0^1 \frac{(1-s)^{k-1}}{(k-1)!}\frac{d^j\mathcal{F}}{dt^j}(ts)\,t^k\,ds$$

so that (A.4) follows from Lemma A.4. Then, by the Cauchy Estimate and Lemma A.3,

$$\begin{aligned}
\sup_{V_{r'}}|R_k| \;&\le\; \sup_{(x,y)\in V_{r'},\,|\tau|\le|t|}\left|\frac{d^k}{d\tau^k}\left(H\circ\Phi_\chi^\tau(x,y)\right)\right|\int_0^1\frac{(1-s)^{k-1}}{(k-1)!}ds \le \\
&\le\; \frac{1}{k!}\cdot k!\cdot\frac{\sup_{(x,y)\in V_{r'},\,|\tau|\le t_0}\left|H\circ\Phi_\chi^\tau(x,y)\right|}{(t_0-|t|)^k} \le \\
&\le\; \frac{\sup_{V_r}|H|}{(t_0-|t|)^k},
\end{aligned}$$

proving (A.5). □

Lemma A.6 *Assume the hypotheses over r, r' and χ in Lemma A.3. Assume also that χ is real analytic on V_R with $R_i > r_i$ and that H is real analytic in V_r. Then,*

$$\sup_{V_{r'}}|L_\chi^j H| \le j!\,\sup_{V_r}|H|\,(\sup_{V_R}|\chi|)^j\left(\max\left\{\frac{1}{(r_i-r_i')(R_{i+d}-r_{i+d})},\ \frac{1}{(r_{i+d}-r_{i+d}')(R_i-r_i)}\right\}\right)^j . \tag{A.6}$$

Proof. If we set

$$t_0 \equiv \frac{1}{\max\left\{\frac{\sup_{V_r}|\partial_{y_i}\chi|}{r_i-r_i'},\ \frac{\sup_{V_r}|\partial_{x_i}\chi|}{r_{i+d}-r_{i+d}'},\ i=1,\dots,d\right\}}$$

we have that t_0 verifies (A.2). By Lemmas A.4 and A.3, one gets by the Cauchy Estimate

$$\begin{aligned}
\sup_{V_{r'}}|L_\chi^j H| \;&=\; \sup_{V_{r'}}\left|\frac{d^j}{dt^j}\Big|_{t=0} H\circ\Phi_\chi^t\right| \le \frac{j!}{t_0^j}\sup_{V_{r'},\,|t|\le t_0}|H\circ\Phi_\chi^t| \le \\
&\le\; \frac{j!}{t_0^j}\sup_{V_r}|H| = j!\,\sup_{V_r}|H|\left(\max\left\{\frac{\sup_{V_r}|\partial_{y_i}\chi|}{r_i-r_i'},\ \frac{\sup_{V_r}|\partial_{x_i}\chi|}{r_{i+d}-r_{i+d}'}\right\}\right)^j \le \\
&\le\; j!\,\sup_{V_r}|H|\,(\sup_{V_R}|\chi|)^j\left(\max\left\{\frac{1}{(r_i-r_i')(R_{i+d}-r_{i+d})},\ \frac{1}{(r_{i+d}-r_{i+d}')(R_i-r_i)}\right\}\right)^j ,
\end{aligned}$$

proving (A.6). □

References

[1] Arnol'd, V.I., *Small denominators and problems of stability of motion in classical and celestial mechanics*, in Russian Mathematical Survey 18(6), 1963, p.91.

[2] Arnol'd, V.I., *Instability of Dynamical Systems with several degrees of freedom*, in Soviet Mathematics – Doklady 5(3), 1964, p.581.

[3] Arnol'd, V.I. ed., *Dynamical Systems III – Encyclopaedia of Mathematical Sciences, Vol. 3*, New York, Springer–Verlag, 1985.

[4] Arnol'd, V.I., *Mathematical methods of Classical Mechanics* New York, Springer–Verlag, 1989.

[5] Arnol'd, V.I. – Avez, A., *Ergodic problems of classical mechanics*, New York, Benjamin, 1968.

[6] Bourgain, J., *On Melnikov's persistency problem*, in Math. Res. Lett., 4(4), 1997, p.445.

[7] Broer, H.W. – Huitema, G.B. – Sevryuk, M.B., *Quasi-Periodic Motions in Families of Dynamical Systems*, New York, Springer–Verlag, 1996.

[8] Cheng, C. – Wang, S., *The surviving of lower-dimensional tori from a resonant torus of Hamiltonian systems*, in Journal of Diff. Equations, 155(2), 1999, p.311.

[9] Chierchia, L., *A direct method for constructing solutions of Hamilton-Jacobi Equation*, in Meccanica, 25(4), 1990, p.246.

[10] Chierchia, L. – Celletti, A., *Construction of analytic KAM surfaces and effective stability bounds*, in Comm. Math. Phys., 118(1), 1988, p.119.

[11] Chierchia, L. – Gallavotti, G., *Drift and diffusion in phase space*, in Ann. Inst. H. Poincaré Phys. Théor., 60(1), 1994, p.1.

[12] Chierchia, L. – Valdinoci, E., *A note on the construction of Hamiltonian trajectories along heteroclinic chains*, in Forum Math., 12(2), 2000, p.247.

[13] DeLatte, D., *On normal forms in Hamiltonian dynamics, a new approach to some convergence questions*, in Ergodic Theory Dynam. Systems, 15(1), 1995, p.49.

[14] Delshams, A. – Gelfreich, V. – Jorba, A. – Seara, T.M., *Exponentially small splitting of separatrices under fast quasiperiodic forcing*, in Comm. Math. Phys., 189(1), 1997, p.35.

[15] Delshams, A. – Gutiérrez, P., *Estimates on invariant tori near an elliptic equilibrium point of a Hamiltonian system*, in Journal of Diff. Equations, 131(2), 1996, p.277.

[16] Delshams, A. – de la Llave, R. – Seara, T.M., *A geometric approach to the existence of orbits with unbounded energy in generic periodic perturbations by a potential of generic geodesic flows of* $\mathbb{T}^2$, Comm. Math. Phys., 209(2), 2000, p.353.

[17] Eliasson, L.H., *Perturbations of stable invariant tori for Hamiltonian systems*, in Ann. Scuola Norm. Sup. Pisa Cl. Sci. IV, 15(1), 1989, p.115.

[18] Eliasson, L.H., *Biasymptotic solutions of perturbed integrable Hamiltonian systems*, in Bol. Soc. Brasil. Mat. (N.S.), 25(1), 1994, p.56.

[19] Fontich E. – Martin P., *Arnold diffusion in perturbations of analytic integrable Hamiltonian systems*, mp_arc@math.utexas.edu, #98-319.

[20] Gallavotti, G., *Hamilton–Jacobi's equation and Arnold's diffusion near invariant tori in a priori unstable isochronous systems*, mp_arc@math.utexas.edu, #97-555.

[21] Gentile, G., *A proof of existence of whiskered tori with quasi-flat homoclinic intersections in a class of almost integrable Hamiltonian systems*, Forum Math. 7(6), 1995, p.709.

[22] Graff, S.M., *On the conservation of Hyperbolic Invariant Tori for Hamiltonian Systems*, in Journal of Diff. Equations, 15, 1974, p.1.

[23] Huang, D. – Zengrong, L., *On the persistence of lower-dimensional invariant hyperbolic tori for smooth Hamiltonian systems*, to appear in Nonlinearity.

[24] Jorba, A. – de la Llave, R. – Zou, M., *Lindstedt series for lower dimensional tori*, in Hamiltonian systems with three or more degrees of freedom, NATO Adv. Sci. Inst. Ser. C Math. Phys. Sci. 533, Dordrecht, Kluwer Acad. Publ., 1999, p.151.

[25] de la Llave, R. – Wayne, C.E., *Whiskered and low dimensional tori in nearly-integrable Hamiltonian systems*, preprint, 1996.

[26] Malgrange, B., *Ideals of differentiable functions*, London, Oxford University Press, 1967.

[27] Moser, J., *The Analytic Invariants of an Area-Preserving Mapping near a Hyperbolic Fixed Point*, in Comm. on Pure and Applied Mathematics, 9, 1956, p.673.

[28] Moser, J., *Stable and random motions in dynamical systems*, Annals of Mathematics Studies, 77, Princeton University Press, 1973.

[29] Neishtadt, A.I., *Estimates in the Kolmogorov Theorem on Conservation of Conditionally Periodic Motions*, in Journal of Applied Mathematics and Mechanics, 45(6), 1982, p.766.

[30] Niederman, L., *Dynamic around a chain of simple resonant tori in nearly integrable Hamiltonian systems*, preprint, mp_arc@math.utexas.edu, #97-142.

[31] Poincaré, H., *Les méthodes nouvelles de la mécanique céleste*, Gauthier Villars, Paris, 1899.

[32] Pöschel, J., *Integrability of Hamiltonian Systems on Cantor Sets*, in Comm. on Pure and Applied Mathematics, 35(5), 1982, p.653.

[33] Pöschel, J., *On elliptic lower-dimensional tori in Hamiltonian systems*, in Math. Zeitschrift, 202(4), p.559.

[34] Pöschel, J., *Nekhoroshev estimates for quasi-convex Hamiltonian systems*, in Math. Zeitschrift, 213(2), 1993, p.187.

[35] Rudnev, M. – Wiggins, S., *KAM Theory Near Multiplicity One Resonant Surfaces*, in Journal of Nonlinear Science, 7, 1997, p.177.

[36] Sevryuk, M.B., *Lower-dimensional tori in reversible systems*, in Chaos, 1(2), 1991, p.160.

[37] Sevryuk, M.B., *Invariant sets of degenerate Hamiltonian systems near equilibria*, mp_arc@math.utexas.edu, #98-684.

[38] Treshchev, D.V., *The mechanism of destruction of resonance tori of Hamiltonian Systems*, in Math. USSR Sbornik, 68(1), 1991, p.181.

[39] Valdinoci, E., *Tori di transizione nella teoria KAM*, tesi di laurea at Università di Roma 3, mp_arc@math.utexas.edu, #98-152.

[40] Whitney, H., *Analytical extension of differentiable functions defined on closed set*, in Trans. A.M.S., 36, 1934, p.63.

[41] Xu, J., *Persistence of elliptic lower-dimensional invariant tori for small perturbation of degenerate integrable Hamiltonian systems*, in J. Math. Anal. Appl., 208(2), 1997, p.372.

[42] You, J., *A KAM theorem for hyperbolic-type degenerate lower-dimensional tori in Hamiltonian systems*, in Comm. Math. Phys., 192(1), 1998, p.145.

[43] You, J., *Perturbations of lower-dimensional tori for Hamiltonian systems*, in Journal of Diff. Equations, 152(1), 1999, p.1.

[44] Zehnder, E., *Generalized implicit function theorems with applications to some small divisor problems II*, in Comm. Pure Appl. Math., 29(1), 1976, p.49.

M P E J

MATHEMATICAL PHYSICS ELECTRONIC JOURNAL

ISSN 1086-6655
Volume 6, 2000

Paper 3
Received: Oct 28, 1999, Revised: Mar 21, 2000, Accepted: Mar 22, 2000
Editor: G. Benettin

Computer–Assisted Proofs for Fixed Point Problems in Sobolev Spaces

Alain Schenkel[1,†], Jan Wehr[2,‡], and Peter Wittwer[3,†]

[1] Department of Mathematics, Helsinki University, P.O. Box 4, 00014 Helsinki, Finland
[2] Interdisciplinary Center for Mathematical and Computer Modeling, Warsaw University, Pawinskiego 5a, Warszawa 02 106, Poland [*]
[3] Département de Physique Théorique, Université de Genève, CH-1211 Genève 4, Switzerland

Abstract. In this paper we extend the technique of computer–assisted proofs to fixed point problems in Sobolev spaces. Up to now the method was limited to the case of spaces of analytic functions. The possibility to work with Sobolev spaces is an important progress and opens up many new domains of applications. Our discussion is centered around a concrete problem that arises in the theory of critical phenomena and describes the phase transition in a hierarchical system of random resistors. For this problem we have implemented in particular the convolution product based on the Fast Fourier Transform (FFT) algorithm with rigorous error estimates.

Key words: computer–assisted proofs, constructive analysis in Sobolev spaces, phase transitions in random media, discrete convolutions are convolutions of splines.

[†] The research of these authors was supported in part by the Fonds National Suisse.
[‡] The research of this author was supported in part by the NSF Probability and Mathematical Statistics grant 9706915.
[*] Permanent address: Department of Mathematics, University of Arizona, Tucson AZ 85721, USA.

Contents

1. Introduction

In this paper we study certain aspects of a model that describes the conductivity in a disordered material. A disordered material is often modeled in statistical mechanics by what is known as a network of random resistors. One would like to be able to describe, for example, what happens in a d–dimensional cubic lattice where each link represents a resistor whose resistance is a random variable. One of the quantities of interest is the effective conductivity of such a network. This conductivity could be defined, for instance, by taking the limit as L goes to infinity of the conductivity $\sigma(L)$ measured on a cube of side length L.

More generally, a network of random resistors is defined by giving a graph and a sequence Σ_{ij} of random variables with $0 \leq \Sigma_{ij} \leq \infty$. Σ_{ij} describes the conductivity of the link in the graph that connects the vertices i and j. For simplicity one considers in general families of independent identically distributed random variables. Such models have been widely studied over the last decades, see for example [Z, K1, Be1, K2, Be2, BW, SW, BSW, K3, Be3, BO, SS] and for a purely probabilistic approach to the problem [Bl, EB, W1-2].

The model which we study in this paper belongs to a class of models where one permits the links to be perfect insulators, that is $q \equiv P(\Sigma_{ij} = 0) > 0$, but for which on the other hand $\Sigma_{ij} < \infty$. This situation is interesting, as it presents the phenomenon of percolation: For q close to one, the links that are conducting form disconnected finite sets and the effective conductivity of the network is trivially zero. This is not any longer the case for q close to zero, where the conductivity depends on the resistivity of the connecting links. As a consequence, a phase transition occurs in the effective conductivity for a certain critical value of the parameter q. The introduction of the parameter q permits therefore, through the mechanism of percolation, to obtain a model that models a critical phenomenon.

The study of phase transitions has progressed enormously with the arrival of renormalization group methods. Different types of renormalization schemes have been proposed to describe the phase transition on regular lattices of random resistors, in particular transformations of the Migdal–Kadanoff type [K3, BO, Bl], and so–called exact renormalization schemes on hierarchical lattices [SW, Be3, SS].

Some of the difficulties that have to be overcome in a renormalization group study disappear when one considers hierarchical models. Moreover, these models typically seem to provide good approximations to more complicated systems [BO].

In Section 1.1 we describe the hierarchical lattice of random resistors for which we have studied the phase transition in this paper. This lattice has been proposed by [SW] as an approximation to the square lattice in two dimensions. In Section 1.2, we state our main result.

1.1. The Model

The hierarchical network that we study in this paper is constructed recursively as indicated in Fig. 1.2.

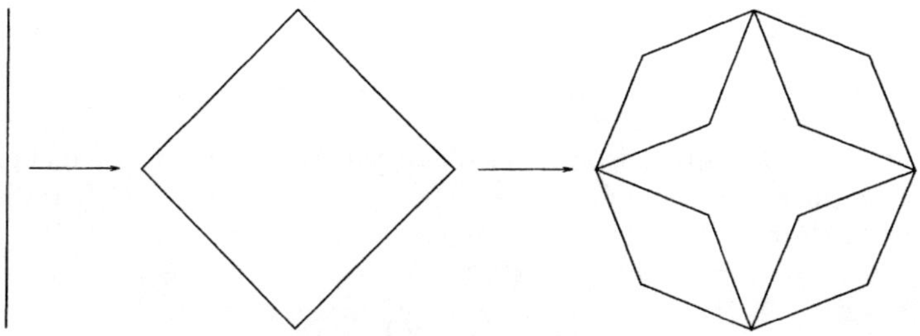

Fig. 1.1: The hierarchical lattice at order 2.

Consider the mapping on graphs that consists of replacing every link by two pairs of links. If we start with a graph consisting of one link connecting two sites, then after applying this operation we end up with four sites and after n applications we end up with a graph of 4^n links. For $n \geq 1$, the network of random resistors that we consider is obtained after n iterations of the procedure outlined above, and consists of resistors with conductivities described by 4^n independent copies $\Sigma_0^{(k)}$ of a random variable Σ_0. The random variables $\Sigma_0^{(k)}$ are therefore *i.i.d.*.

The choice of a hierarchical geometry permits to give a simple formulation of the effective conductivity of the network, when measured between the two vertices of the initial link, in terms of a map. Indeed, using the composition laws for conductors connected in series and in parallel, the conductivity of a circuit that consists of four resistors with conductivities σ_1, σ_2, σ_3, σ_4 arranged in a loop is

$$\sigma = D_c(\sigma_1, \sigma_2, \sigma_3, \sigma_4) \equiv \frac{1}{\frac{1}{\sigma_1} + \frac{1}{\sigma_2}} + \frac{1}{\frac{1}{\sigma_3} + \frac{1}{\sigma_4}}. \tag{1.1}$$

If the conductivities σ_1, σ_2, σ_3, σ_4 are random, this is also the case for the effective conductivity σ. Therefore, by applying this nonlinear average to each of the loops of the hierarchical network of order n, we obtain the random variables that describe the conductivities of the links of the network at the level $n-1$. Since the random variables that were given on the network of order n were *i.i.d.*, the 4^{n-1} random variables that are obtained for the network of order $n-1$ are also *i.i.d.*. They are independent copies of the random variable given by

$$\Sigma_1 = D_c(\Sigma_0^{(1)}, \Sigma_0^{(2)}, \Sigma_0^{(3)}, \Sigma_0^{(4)}).$$

By applying successively the nonlinear average D_c, one can therefore go up the hierarchy to compute the effective conductivity Σ_n of the network of order n. We note that by applying one more time the average D_c to four independent copies of Σ_n, we obtain the conductivity Σ_{n+1} of the hierarchical network of order $n+1$ made up from resistors with conductivities given by 4^{n+1} independent copies of the random variable Σ_0.

We are interested in the limit as $n \to \infty$ of the sequence Σ_n that we have just defined. This limit corresponds to the effective conductivity of our hierarchical network in the infinite volume limit. It is not difficult to see that our model is an approximation to the renormalization on a simple square lattice in $d = 2$ dimensions. In [SW] a detailed discussion of this approximation can be found. See also [BO]. The questions that arise naturally are the following. First of all, a phenomenon of self–averaging should lead to a deterministic effective conductivity for the infinite network. Next, it is interesting to know in what way this conductivity depends on the conductivity of each link of the network, that is on the distribution of the initial random variable Σ_0. Finally, once the convergence of the effective conductivity is established, one can study the fluctuations of Σ_n around this limit.

A certain amount of information can easily be obtained by studying the parameter

$$p_n = P(\Sigma_n > 0). \tag{1.2}$$

Recall that we permit the links to be perfect insulators with a nonzero probability. Here, $p_0 \equiv p$ is the probability that a resistor of the original (infinite) network is not broken. From (1.1) it is not difficult to see that the conductivity of a diamond circuit is nonzero with probability

$$\tilde{p} = g(p) \equiv 1 - (1 - p^2)^2, \tag{1.3}$$

where p is the probability that each of the for resistors has a nonzero conductivity. The function g is characteristic for percolation problems [G]. The function g is increasing on the interval $[0, 1]$ and has, in addition to the fixed points at zero and one, a unique unstable fixed point p_c in the interval $(0, 1)$. The value of p_c is,

$$p_c = \frac{\sqrt{5} - 1}{2}. \tag{1.4}$$

There are therefore three cases. If $p < p_c$, it follows immediately that $p_n \to 0$ as $n \to \infty$. Hence, the effective conductivity of the network is zero with probability one in this case. For $p > p_c$, one has $\lim_{n\to\infty} p_n = 1$. This means that with probability one there is a path made from resistors with nonzero conductivity that connects the two sites of the lattice for $n = 0$. In other words, the percolation threshold of the network is given by p_c.

We note that this does not imply that the effective conductivity is nonzero. However, it has been proved in [W1] that for $p > p_c$ the sequence Σ_n converges with probability one to a constant $\sigma^*(p)$, and in [Sh] that this constant is strictly positive. Therefore, the percolation threshold corresponds exactly to the phase transition of the effective conductivity.

At the critical point $p = p_c$, one has $p_n = p_c$ for all n. This means that the probability $P(\Sigma_n = 0)$ is invariant. In the following we are going to be interested in the part of the distribution of Σ_n supported on $(0, \infty)$. An argument based on the study of the expectation of Σ_n shows, however, that the effective conductivity of the network in the infinite volume limit is still zero in this case. Indeed, if we denote by $E(X)$ the expectation of a random variable X, one has

$$E(\Sigma_{n+1}) = E\big(D_c(\Sigma_n^{(1)}, \ldots, \Sigma_n^{(4)})\big) = E\Big(\frac{2}{\frac{1}{\Sigma_n^{(1)}} + \frac{1}{\Sigma_n^{(2)}}}\Big),$$

where we have used independence of the random variables. Since, in addition we have the following inequality between the arithmetic mean and the geometric mean,

$$\frac{2}{\frac{1}{x} + \frac{1}{y}} \leq \frac{x+y}{2},$$

and since the left hand side of this expression is equal to zero if x or y are zero, we can bound $E(\Sigma_{n+1})$ by the expectation of the average of $\Sigma_n^{(1)}$ and $\Sigma_n^{(2)}$ over the set where the random variables are strictly positive. Therefore, one obtains

$$E(\Sigma_{n+1}) \leq p_n E(\Sigma_n).$$

Since $p_n = p_c$ for all n, we have $E(\Sigma_n) \to 0$ as $n \to \infty$, and the sequence Σ_n converges to zero in probability.

The goal of this paper is to describe how the Σ_n converge to zero at the critical point, that is to describe the fluctuations of the Σ_n around their limiting value. A numerical study [SW] of the probability densities of the random variables Σ_n indicates that if one normalizes Σ_n with an appropriate factor μ_n that fixes the expectation of the random variable $\mu_n \Sigma_n$ at the expectation of Σ_0, the sequence $\mu_n \Sigma_n$ converges in distribution to a multiple of a (universal) random variable Σ_*, and

$$\lim_{n \to \infty} \frac{\mu_{n+1}}{\mu_n} = \lambda^* \approx 1.756,$$

independently of the choice of Σ_0. This means that the fluctuations of the effective conductivity of the network present a certain universality in the limit $n \to \infty$: the limiting probability densities for different initial random variables Σ_0 distinguish themselves only by a change of scale. Furthermore, the probability density for the positive values of Σ_* decays faster than exponentially at zero and infinity. Therefore, at the critical point, the behavior of the fluctuations distinguishes itself from the supercritical case for which a perturbative computation [SS] indicates that the sequence of properly normalized random variables Σ_n converges to a normal distribution (a proof of this fact will appear in [WW]). Also, since $\lambda^* < 2$, conductance fluctuations can be thought of as anomalously large compared to the supercritical case.

In this work we address the question of the existence of a positive real number λ^* and a random variable Σ_* such that

$$\Sigma_* = \lambda^* D_c(\Sigma_*^{(1)}, \ldots, \Sigma_*^{(4)}). \tag{1.5}$$

We note that the map D_c is homogeneous of degree one and the random variables $\lambda\Sigma_*$ are therefore solutions of (1.5) for all $\lambda > 0$. The number λ^* gives the dynamics of the renormalization group D_c at the critical point on the whole of the set $\lambda\Sigma_*$ and is related to the critical exponent t that describes the phase transition of the effective conductivity,

$$\sigma^*(p) \sim (p - p_c)^t, \quad p > p_c,$$

through the formula, cf. [SW],

$$t = \frac{\log \lambda^*}{\log g'(p_c)}.$$

1.2. Main Result

In order to study the fluctuations of the effective conductivity of our hierarchical network at the critical point, we work in the framework of functional analysis. The part of the distribution of the random variables Σ_n that interests us is the one that is supported on $(0, \infty)$. One assumes that this part of the distribution of Σ_0 is absolutely continuous with respect to Lebesgue measure and one derives the functional equation for the probability densities that correspond to the nonlinear average (1.1). It turns out to be simpler to work with the resistivities instead of the conductivities. One considers therefore the random variables Υ for the resistivity given by

$$\Upsilon = \frac{1}{\Sigma}.$$

If σ is the density of the random variable Σ, then the density ρ of Υ is given by

$$\rho(x) = T(\sigma)(x) \equiv \frac{1}{x^2}\sigma\Big(\frac{1}{x}\Big). \tag{1.6}$$

The average D_c for the conductivities can be rewritten for the resistivities as

$$D_r(r_1, \ldots, r_4) = \frac{1}{\frac{1}{r_1+r_2} + \frac{1}{r_3+r_4}}. \tag{1.7}$$

The functional equation for the density $\mathcal{D}_r(\rho)$ of the average (1.7) of four independent copies of a random variable Υ with density ρ is therefore given in terms of the map T and the convolution operator. Indeed, the probability density ρ of a sum of two random variables with densities ρ_1 and ρ_2 is given by the convolution of ρ_1 and ρ_2, that is by

$$\rho(x) = (\rho_1 * \rho_2)(x) = \int_0^x \rho_1(y)\rho_2(x-y)\,dy. \tag{1.8}$$

Therefore, it follows from (1.7) that

$$\mathcal{D}_r(\rho) = T\big(T(\rho * \rho) * T(\rho * \rho)\big). \tag{1.9}$$

One observes that, formally, the Dirac–densities $\delta(x-a)$ are fixed points of this transformation for all $a \geq 0$. They correspond to the limiting densities in the non–critical cases.

In order to obtain an equation for the probability densities with support on $(0, \infty)$ we determine the contributions of the four resistors to the finite value of $D_r(r_1, \ldots r_4)$. By inspecting (1.7), one observes that they are of two types: either $r_1 + r_2 = \infty$ and $r_3 + r_4 < \infty$ (with the corresponding symmetric case $r_1 + r_2 < \infty$ and $r_3 + r_4 = \infty$), or all of the four resistivities are finite. At the critical point p_c, one determines easily that the probability of the first case is given by

$$c_1 \equiv 2p_c(1-p_c^2) = 0.763..., \tag{1.10}$$

whereas the probability of the second case is

$$c_2 \equiv p_c^3 = 0.236.... \tag{1.11}$$

Obviously $c_1 + c_2 = 1$. Therefore, the operator that acts on the probability densities and corresponds to the finite part of the map D_r on the random variables is given at the critical point by

$$\mathcal{D}(\rho) = c_1(\rho * \rho) + c_2 T\big(T(\rho * \rho) * T(\rho * \rho)\big). \tag{1.12}$$

In order to rewrite the fixed point problem (1.5) in terms of the probability densities, one uses that for $\lambda > 0$ the probability density of a random variable Υ/λ is given by $S_\lambda \rho$, where ρ is the probability density of Υ and where S_λ is the operator that changes the scale,

$$S_\lambda f(x) \equiv \lambda f(\lambda x)\,, \quad \lambda > 0. \tag{1.13}$$

Therefore the fixed point problem (1.5) can be rewritten as

$$\rho^* = S_{\lambda^*} \mathcal{D}(\rho^*). \tag{1.14}$$

The proof of the existence of a real $\lambda^*$and a function ρ^* satisfying (1.14) which we present here is constructive. In particular, we will be capable of providing explicit bounds on λ^* as well as an approximation to the function ρ^*. The graph of the approximation that we have obtained this way is represented in Fig. 1.2.

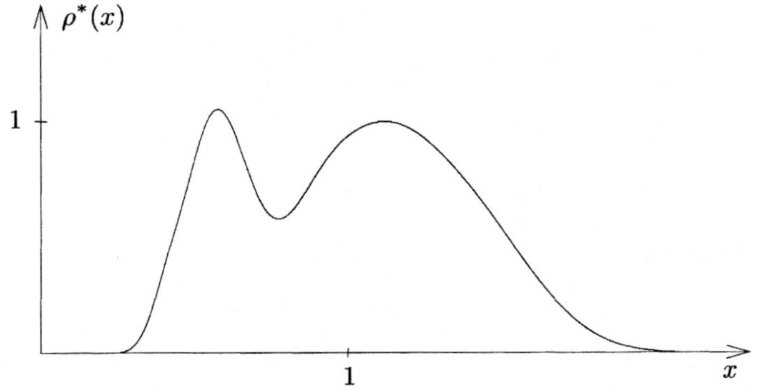

Fig. 1.2: The fixed point ρ^*.

The operator (1.12) was studied by mathematically rigorous, computer–assisted, constructive analysis. Before stating with precision the result that is proved in this paper, we define the function spaces with which we work. We first define the notation that will be used later.

Notation. We denote by $\mathbb{R}_+$ the set of nonnegative real numbers. The set of positive real numbers will be denoted by $\mathbb{R}_+^*$. For an interval $I \subseteq \mathbb{R}$, we denote by $\mathcal{C}^n(I)$ the set of functions that are n times continuously differentiable on I. The derivative of a function f of one variable will be denoted by f'. For a positive function μ, we denote by $L^1(\mathbb{R}_+, \mu(x)\, dx)$ the space of functions defined on $\mathbb{R}_+$ and integrable with respect to the measure $\mu(x)\, dx$. Finally $W_1^1(\mathbb{R}_+, \mu(x)\, dx)$ is the Sobolev space of functions of $L^1(\mathbb{R}_+, \mu(x)\, dx)$ with one distributional derivative in $L^1(\mathbb{R}_+, \mu(x)\, dx)$. For $r \geq 0$ and x in a metric space, $B_r(x)$ will denote the open ball of radius r centered at x.

Definition 1.1. *For α, $\beta \geq 0$ and functions $w_{\alpha\beta}$ given by*

$$w_{\alpha\beta}(x) = \exp\Bigl(\frac{\alpha}{x} + \beta x\Bigr), \tag{1.15}$$

we define $\mathcal{B}_{\alpha\beta}$ to be the Banach space $L^1(\mathbb{R}_+, w_{\alpha\beta}(x)\, dx)$. We denote the norm of $f \in \mathcal{B}_{\alpha\beta}$ by $\|f\|_{\alpha\beta}$, that is,

$$\|f\|_{\alpha\beta} = \int_0^\infty w_{\alpha\beta}(x)\, |f(x)|\, dx. \tag{1.16}$$

We furthermore define

$$\mathcal{B} = \bigcap_{\substack{\alpha>0\\ \beta>0}} \mathcal{B}_{\alpha\beta}. \tag{1.17}$$

Remark 1.2. It is clear that we have the inclusion $\mathcal{B}_{\sigma\tau} \subseteq \mathcal{B}_{\alpha\beta}$ for $\sigma \geq \alpha$ and $\tau \geq \beta$. The inclusion is strict unless $\sigma = \alpha$ and $\tau = \beta$.

We will also need the following definitions of the mass and expectation.

Definition 1.3. *We define the mass $M(f)$ of a function $f \in L^1(\mathbb{R}_+)$ by*

$$M(f) = \int_0^\infty f(x)\, dx. \tag{1.18}$$

If $f \in L^1(\mathbb{R}_+, (1+|x|)\, dx)$, we define the expectation $E(f)$ by

$$E(f) = \int_0^\infty x f(x)\, dx. \tag{1.19}$$

Finally, we will need to exclude functions f such that $M(f)E(f) = 0$, i.e., functions in

$$\mathcal{H} = \{f \in L^1(\mathbb{R}^+, (1+|x|)\, dx) \mid M(f)E(f) = 0\}. \tag{1.20}$$

We can now state the main result of this paper.

Theorem 1.4. *There exists a real number λ^* and a function $f^* \in \mathcal{B}\backslash\mathcal{H}$ that satisfy the equation*

$$f^* = S_{\lambda^*}\mathcal{D}(f^*).$$

In addition f^* has the following two properties

$$(1)\ M(f^*) = 1,$$
$$(2)\ f^* \in \mathcal{C}^\infty(\mathbb{R}^+).$$

Note that this theorem does not imply that the fixed point f^* is a probability density, since f^* is not necessarily a positive function. While we see strong numerical evidences for positivity of the fixed point f^*, we have no proof of this fact.

Before terminating this section, we summarize in the following lemma some of the properties satisfied by the maps that are contained in (1.14).

Lemma 1.5. *The maps S_λ form a multiplicative group, i.e., $S_1 = I$, and $S_{\lambda_1}S_{\lambda_2} = S_{\lambda_1\lambda_2}$. Moreover, for f and g integrable functions one has*

$$S_\lambda(f * g) = S_\lambda f * S_\lambda g, \tag{1.21}$$
$$S_\lambda T = T S_{1/\lambda}. \tag{1.22}$$

If $f,\ g \in L^1(\mathbb{R}_+, (1+|x|)\, dx)$, then the mass and expectation satisfy the following identities

$$\begin{aligned} M(f) &= M(S_\lambda f) = M(Tf)\,, \\ M(f * g) &= M(f)M(g)\,, \\ E(S_\lambda f) &= \frac{1}{\lambda}E(f)\,, \\ E(f * g) &= M(f)E(g) + E(f)M(g)\,. \end{aligned} \tag{1.23}$$

Even though the proof of Theorem 1.4 needs in part a computer for its proof, the properties (1) and (2) follow directly from the existence of the fixed point. The regularity of the fixed point will be established later, cf. Proposition 2.3, whereas property (1) is proved in the following lemma.

Lemma 1.6. *Let $f \in L^1(\mathbb{R}_+)\backslash\mathcal{H}$ and let $\lambda > 0$ arbitrary. If f satisfies $f = S_\lambda \mathcal{D}(f)$, then $M(f) = 1$.*

Proof. Using the relations (1.23) together with $M(f) \neq 0$, one computes from the identity $M(f) = M(S_\lambda \mathcal{D}(f))$ that

$$1 = c_1 M(f) + c_2 M(f)^3.$$

Using the monotonicity of the function $x \mapsto c_2 x^3 + c_1 x - 1$, one verifies that if $c_1 + c_2 = 1$, then the only zero is given by $x_0 = 1$. Therefore, $M(f) = 1$.

■

1.3. Computer–Assisted Proofs

The rest of the proof of Theorem 1.4 is the main part of this paper. It is based on a very large number of inequalities proved rigorously with the help of a computer. The use of a computer for proving theorems in analysis has become standard by now. This method, which allows to do constructive functional analysis on a computer, has been developed by O.E. Lanford in his seminal paper [L1], and has then been generalized by [EKW2]. This technique of proof has since then been applied to problems of various origin. See for example [BS, C, CC, dlL, EKW1-2, EW1-2, FL, FS, KP, KSW, KW1-7, L1-3, LR1-3, M, R, Se1-2, St].

The proofs constructed up to now have in common that they all deal with spaces of analytic functions. One important novelty of the work presented here is that a proof is constructed for function spaces of L^1–type. The basic ideas underlying the proof remain the same, but the generalization to L^1 spaces uses approximation methods which are typical in numerical analysis, and we will explain how to control discretization errors in this context.

A computer–assisted proof is complete once the program has come to an end without a "domain error".

2. Organization of the Proof

In order to prove existence of a fixed point f^* for the map $S_{\lambda^*}\mathcal{D}$, we will rely on the contraction mapping principle. The following argument shows, however, that $S_{\lambda^*}\mathcal{D}$ cannot even be hyperbolic due to the presence of a symmetry. Recall that the nonlinear average D_r is homogeneous of order 1. This causes the scaling operator S_λ to commute with $\mathcal{D}$ for every $\lambda > 0$. Hence, existence of a fixed point f^* implies existence of a one parameter family of fixed points $\{S_\lambda f^*\}_{\lambda>0}$. In the case $E(f^*) \neq 0$, this family can be parameterized by the expectation $E(S_\lambda f^*) = E(f^*)/\lambda$. We first remove this symmetry and make the fixed point problem hyperbolic by introducing the family of maps $\{\mathcal{N}_\lambda\}_{\lambda>0}$ defined by

$$\mathcal{N}_\lambda(f) = S_\lambda\Big(c_\lambda(f) f * f + c_2 T\big(T(f*f) * T(f*f)\big)\Big)\,, \tag{2.1}$$

where $c_\lambda(f)$ is such that

$$E\big(\mathcal{N}_\lambda(f)\big) = 1\,. \tag{2.2}$$

The expression inside the outer brackets on the RHS of (2.1) differs from the map $\mathcal{D}$ only by the coefficient $c_\lambda(f)$. The following remark relates the maps $S_\lambda\mathcal{D}$ and $\mathcal{N}_\lambda$: If $c_\lambda(f_\lambda) = c_1$ for some fixed $\lambda > 0$ and some f_λ, then $S_\lambda\mathcal{D}(f_\lambda) = \mathcal{N}_\lambda(f_\lambda)$ by definition of $\mathcal{D}$ and $\mathcal{N}_\lambda$. This leads to the following criterion for the existence of a fixed point of $S_{\lambda^*}\mathcal{D}$.

Lemma 2.1. *If there exist a real $\lambda^* > 0$ and a fixed point f_{λ^*} of $\mathcal{N}_{\lambda^*}$ with $c_{\lambda^*}(f_{\lambda^*}) = c_1$, then f_{λ^*} is solution of the functional equation $f = S_{\lambda^*}\mathcal{D}(f)$.*

To prove existence of λ^* and f_{λ^*}, we will study the family $\{\mathcal{N}_\lambda\}$ in a neighborhood (λ^-, λ^+) of our best numerical value for λ^*. For each value of λ in this neighborhood, the contraction mapping principle will be used to prove existence of a fixed point f_λ of $\mathcal{N}_\lambda$. The maps $\mathcal{N}_\lambda$ are hyperbolic but not contracting in the neighborhood of their fixed point, due to an unstable direction that, roughly speaking, crosses transversally the manifold of functions with total mass equal to one. To cope with this problem, we will adopt later a standard strategy that consists of applying a variant of Newton's method. Once the existence of a fixed point $f_\lambda = \mathcal{N}_\lambda(f_\lambda)$ is established for all $\lambda \in [\lambda^-, \lambda^+]$, we will show that $c_{\lambda^-}(f_{\lambda^-}) < c_1 < c_{\lambda^+}(f_{\lambda^+})$. A continuity argument will finally yield the existence of a $\lambda^* \in (\lambda^-, \lambda^+)$ and a function f_{λ^*} satisfying the hypothesis of Lemma 2.1.

Before entering into more details, we introduce some notation and state a few results concerning the domains of definition and target spaces of $\mathcal{N}_\lambda$. Using the commutation and distributivity properties (1.21) and (1.22), one rewrites $\mathcal{N}_\lambda$ as

$$\mathcal{N}_\lambda(f) = c_\lambda(f)\mathcal{N}_\lambda^1(f) + c_2\mathcal{N}_\lambda^2(f)\,, \tag{2.3}$$

where

$$\mathcal{N}_\lambda^1(f) = S_\lambda(f*f)\,, \tag{2.4}$$

$$\mathcal{N}_\lambda^2(f) = T\big(T\mathcal{N}_\lambda^1(f) * T\mathcal{N}_\lambda^1(f)\big)\,. \tag{2.5}$$

From the condition (2.2), the coefficient $c_\lambda(f)$ is expressed in terms of the expectation of $\mathcal{N}_\lambda^1(f)$ and $\mathcal{N}_\lambda^2(f)$. Since $E(\mathcal{N}_\lambda^1(f)) = 2M(f)E(f)/\lambda$ and

$$\frac{1}{\lambda}E(\mathcal{N}_\lambda^2(f)) = E\big(T\big(T(f*f)*T(f*f)\big)\big) \equiv E_2(f) \tag{2.6}$$

one gets

$$c_\lambda(f) = \frac{\lambda - c_2 E_2(f)}{2M(f)E(f)}\,. \tag{2.7}$$

We will see that $E_2(f)$ is finite for $f \in \mathcal{B}_{\alpha\beta}$ with $\beta > 0$. Hence, $c_\lambda(f)$ is finite provided one excludes functions f in $\mathcal{H}$, i.e., functions such that $M(f)E(f) = 0$. We now state a result about the domains of definition and target spaces of the maps $\mathcal{N}_\lambda$.

Proposition 2.2. *For $\alpha \geq 0$, $\beta > 0$ and $\lambda > 0$, $\mathcal{N}_\lambda$ is well defined as a map from $\mathcal{B}_{\alpha\beta}\backslash\mathcal{H}$ to $\mathcal{B}_{\sigma\tau}$ for all $\sigma \leq 4\alpha/\lambda$ and $\tau \leq \lambda\beta$.*

Proof. The operators S_λ, T and the convolution $f \mapsto f*f$ are well defined on $L^1(\mathbb{R}_+)$. Next, we show that the convolution product maps $\mathcal{B}_{\zeta\eta} \times \mathcal{B}_{\zeta\eta}$ into $\mathcal{B}_{(4\zeta)\eta}$. Using Fubini's theorem, we get the following inequalities

$$\begin{aligned}
\|f*g\|_{(4\zeta)\eta} &\leq \int_0^\infty dx\; w_{(4\zeta)\eta}(x) \int_0^x dy\; |f(y)||g(x-y)| \\
&= \int_0^\infty dy|f(y)| \int_0^\infty dx\; w_{(4\zeta)\eta}(x+y)|g(x)| \\
&\leq \sup_{\substack{x>0\\y>0}} \Big(\frac{w_{(4\zeta)\eta}(x+y)}{w_{\zeta\eta}(x)w_{\zeta\eta}(y)}\Big)\|f\|_{\zeta\eta}\|g\|_{\zeta\eta} \\
&= \|f\|_{\zeta\eta}\|g\|_{\zeta\eta}\,.
\end{aligned} \tag{2.8}$$

The last equality follows from

$$\sup_{\substack{x>0\\y>0}} \frac{w_{(4\zeta)\eta}(x+y)}{w_{\zeta\eta}(x)w_{\zeta\eta}(y)} = \sup_{\substack{x>0\\y>0}} \exp\big(-\zeta h(x,y)\big)\,,$$

and

$$h(x,y) = \frac{x+y}{xy} - \frac{4}{x+y} \geq 0$$

for all $x, y > 0$. Next, S_λ is bounded as a map from $\mathcal{B}_{\zeta\eta}$ to $\mathcal{B}_{(\zeta/\lambda)(\lambda\eta)}$:

$$\begin{aligned}
\|S_\lambda f\|_{(\zeta/\lambda)(\lambda\eta)} &= \int_0^\infty w_{(\zeta/\lambda)(\lambda\eta)}(x/\lambda)|f(x)|dx \\
&\leq \sup_{x>0}\Big(\frac{w_{(\zeta/\lambda)(\lambda\eta)}(x/\lambda)}{w_{\zeta\eta}(x)}\Big)\|f\|_{\zeta\eta} \\
&= \|f\|_{\zeta\eta}\,.
\end{aligned} \tag{2.9}$$

Therefore $\mathcal{N}_\lambda^1$ maps $\mathcal{B}_{\alpha\beta}$ into $\mathcal{B}_{(4\alpha/\lambda)(\lambda\beta)}$, and hence into $\mathcal{B}_{\sigma\tau}$ for $\sigma \leq 4\alpha/\lambda$ and $\tau \leq \lambda\beta$. In order to check that $\mathcal{N}_\lambda^2$ maps $\mathcal{B}_{\alpha\beta}$ into $\mathcal{B}_{(4\alpha/\lambda)(\lambda\beta)}$, we first note that T is obviously bounded as a map from $\mathcal{B}_{\zeta\eta}$ to $\mathcal{B}_{\eta\zeta}$, with

$$\|Tf\|_{\eta\zeta} = \|f\|_{\zeta\eta}\,, \tag{2.10}$$

since $w_{\zeta\eta}(1/x) = w_{\eta\zeta}(x)$ for all $x > 0$. Using (2.8), (2.10) and the bound on $\mathcal{N}_\lambda^1$, one concludes that $\mathcal{N}_\lambda^2(f) \in \mathcal{B}_{(4\alpha/\lambda)(4\lambda\beta)} \subset \mathcal{B}_{(4\alpha/\lambda)(\lambda\beta)}$ for $f \in \mathcal{B}_{\alpha\beta}$. Finally, the bounds (2.8), (2.10), and the fact that the expectation of a function in $\mathcal{B}_{\sigma\tau}$ is finite for $\tau > 0$ imply that $c_\lambda(f)$ is finite for $f \in \mathcal{B}_{\alpha\beta}\backslash\mathcal{H}$ and $\beta > 0$.

■

Proposition 2.2 together with Remark 1.2 immediately imply that every fixed point $f \in \mathcal{B}_{\alpha\beta}\backslash\mathcal{H}$ of $\mathcal{N}_\lambda$ with $\alpha,\beta > 0$ and $\lambda \in (1,4)$ satisfies $f \in \mathcal{B}\backslash\mathcal{H}$. Furthermore, using the regularization properties of the convolution, one can show that every such fixed point is a smooth function. More precisely, we have the following proposition, whose proof can be found in the appendix.

Proposition 2.3. *Let $\alpha,\beta > 0$, $\lambda \in (1,4)$, and let $f \in \mathcal{B}_{\alpha\beta}\backslash\mathcal{H}$ be a fixed point of $\mathcal{N}_\lambda$. Then $f \in \mathcal{B}\backslash\mathcal{H}$, f is of class $\mathcal{C}^\infty(\mathbb{R}_+)$, and $f' \in \mathcal{B}$.*

The following theorem implies Theorem 1.4.

Theorem 2.4. *Let $\lambda^- \equiv 1.7562035$ and $\lambda^+ \equiv 1.7562048$. Then,*

(a) *For some $\alpha,\beta > 0$, there is a continuous family $\{f_\lambda\}_{\lambda\in[\lambda^-,\lambda^+]}$ of functions in $\mathcal{B}_{\alpha\beta}\backslash\mathcal{H}$ such that $\mathcal{N}_\lambda(f_\lambda) = f_\lambda$ for all $\lambda \in [\lambda^-,\lambda^+]$,*

(b) $c_{\lambda^-}(f_{\lambda^-}) < c_1 < c_{\lambda^+}(f_{\lambda^+})\,.$ (2.11)

Our main result follows from Theorem 2.4 and Proposition 2.3.

Proof of Theorem 1.4. Assume first that the map $\lambda \mapsto c_\lambda(f_\lambda)$ is continuous. Then Theorem 2.4 implies the existence of a $\lambda^* \in (\lambda^-,\lambda^+)$ for which $c_{\lambda^*}(f_{\lambda^*}) = c_1$ and $f_{\lambda^*} = \mathcal{N}_{\lambda^*} f_{\lambda^*}$, which using Lemma 2.1 implies that $S_{\lambda^*}\mathcal{D}(f_{\lambda^*}) = f_{\lambda^*}$, and using Proposition 2.3 that $f_{\lambda^*} \in (\mathcal{B}\backslash\mathcal{H})\cap\mathcal{C}^\infty(\mathbb{R}_+)$. It remains to be checked that the map $\lambda \mapsto c_\lambda(f_\lambda)$ is indeed continuous. One first observes that the linear functionals $f \mapsto M(f)$ and $f \mapsto E(f)$ are bounded as maps from $\mathcal{B}_{\sigma\tau}$ to $\mathbb{R}$, provided $\tau > 0$, and that

$$|M(f)| \leq \sup_{x>0}\Bigl(\frac{1}{w_{\sigma\tau}(x)}\Bigr)\|f\|_{\sigma\tau}\,, \tag{2.12}$$

$$|E(f)| \leq \sup_{x>0}\Bigl(\frac{x}{w_{\sigma\tau}(x)}\Bigr)\|f\|_{\sigma\tau}\,. \tag{2.13}$$

Hence, $f \mapsto E_2(f)$ is continuous as a map from $\mathcal{B}_{\alpha\beta}$ to $\mathbb{R}$ for every $\alpha,\beta > 0$, using in addition the bounds derived in the proof of Proposition 2.2. Therefore $f \mapsto c_\lambda(f)$

is continuous as a map from $\mathcal{B}_{\alpha\beta}\backslash\mathcal{H}$ to $\mathbb{R}$ for every $\alpha, \beta > 0$ and $\lambda \in \mathbb{R}$. Next, for each $f \in \mathcal{B}_{\alpha\beta}\backslash\mathcal{H}$ with $\alpha, \beta > 0$, the map $\lambda \mapsto c_\lambda(f)$ is continuous. The continuity of $\lambda \mapsto c_\lambda(f_\lambda)$ as a map from $[\lambda^-, \lambda^+]$ to $\mathbb{R}$ finally follows from the continuity of the family $\{f_\lambda\}_{\lambda \in [\lambda^-, \lambda^+]}$.

∎

The proof of Theorem 2.4 is in part computer–assisted. Once (a) is established, the verification of part (b) involves mainly an explicit calculation, that will be given in Section 5.3. The remainder of this section is devoted to the proof of part (a).

First, in order to simplify further our estimates, we introduce yet another family of operators, closely related to $\{\mathcal{N}_\lambda\}_{\lambda > 0}$. This family is defined by

$$\mathcal{N}_{\lambda,\kappa} = S_\kappa \mathcal{N}_\lambda ,\tag{2.14}$$

where $\lambda, \kappa > 0$.

Lemma 2.5. *Let* $\alpha \geq 0$ *and* $\beta, \lambda, \kappa > 0$. *Then* $\mathcal{N}_{\lambda,\kappa}$ *is well defined as a map from* $\mathcal{B}_{\alpha\beta}\backslash\mathcal{H}$ *to* $\mathcal{B}_{\sigma\tau}$ *for every* $\sigma \leq 4\alpha/\kappa\lambda$ *and* $\tau \leq \kappa\lambda\beta$, *and one has*

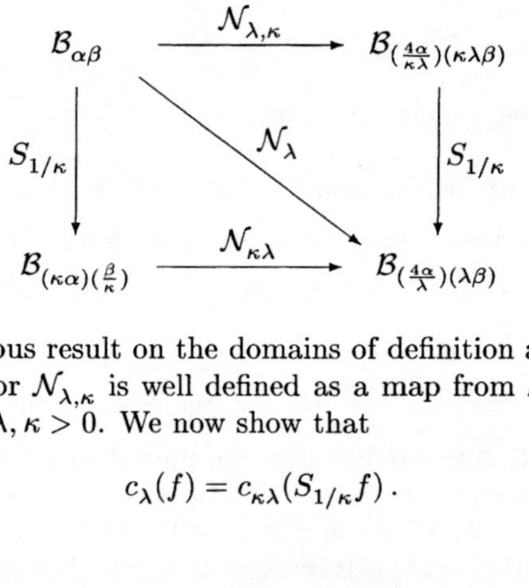

$$
\begin{array}{ccc}
\mathcal{B}_{\alpha\beta} & \xrightarrow{\ \mathcal{N}_{\lambda,\kappa}\ } & \mathcal{B}_{(\frac{4\alpha}{\kappa\lambda})(\kappa\lambda\beta)} \\[4pt]
\Big\downarrow{\scriptstyle S_{1/\kappa}} & {\scriptstyle \mathcal{N}_\lambda}\searrow & \Big\downarrow{\scriptstyle S_{1/\kappa}} \\[4pt]
\mathcal{B}_{(\kappa\alpha)(\frac{\beta}{\kappa})} & \xrightarrow[\ \mathcal{N}_{\kappa\lambda}\ ]{} & \mathcal{B}_{(\frac{4\alpha}{\lambda})(\lambda\beta)}
\end{array}
\tag{2.15}
$$

Proof. First, our previous result on the domains of definition and target spaces of $\mathcal{N}_\lambda$ implies that the operator $\mathcal{N}_{\lambda,\kappa}$ is well defined as a map from $\mathcal{B}_{\alpha\beta}\backslash\mathcal{H}$ to $\mathcal{B}_{(4\alpha/\kappa\lambda)(\kappa\lambda\beta)}$ whenever $\alpha \geq 0$ and $\beta, \lambda, \kappa > 0$. We now show that

$$c_\lambda(f) = c_{\kappa\lambda}(S_{1/\kappa}f).\tag{2.16}$$

Using (2.7), we see that

$$c_{\kappa\lambda}(S_{1/\kappa}f) = \frac{\kappa\lambda - c_2 E_2(S_{1/\kappa}f)}{2M(S_{1/\kappa}f)E(S_{1/\kappa}f)},$$

and the relations $M(S_{1/\kappa}f) = M(f)$, $E(S_{1/\kappa}f) = \kappa E(f)$ and $E_2(S_{1/\kappa}f) = \kappa E_2(f)$ lead to (2.16). Using (2.16), we compute

$$
\begin{aligned}
\mathcal{N}_{\kappa\lambda}(S_{1/\kappa}f) &= c_{\kappa\lambda}(S_{1/\kappa}f)\mathcal{N}^1_{\kappa\lambda}(S_{1/\kappa}f) + c_2\mathcal{N}^2_{\kappa\lambda}(S_{1/\kappa}f) \\
&= c_\lambda(f)\mathcal{N}^1_\lambda(f) + c_2\mathcal{N}^2_\lambda(f) \\
&= \mathcal{N}_\lambda(f),
\end{aligned}
$$

and conclude by observing that $\mathcal{N}_\lambda = S_{1/\kappa}\mathcal{N}_{\lambda,\kappa}$.

■

From Lemma 2.5 it follows that the fixed points of $\mathcal{N}_\lambda$ for $\lambda \in [\lambda^-, \lambda^+]$ are related to the fixed points of $\mathcal{N}_{\lambda^+,\kappa}$ for $\kappa \in [\lambda^-/\lambda^+, 1]$. Furthermore, the operators $\mathcal{N}_{\lambda,\kappa}$ are well defined as maps from $\mathcal{B}_{\alpha\beta}\backslash\mathcal{H}$ to $\mathcal{B}_{\alpha\beta}$ for λ and κ satisfying

$$\frac{1}{\lambda} \leq \kappa \leq \frac{4}{\lambda}. \tag{2.17}$$

This condition is easily seen to hold for $\kappa \in [\lambda^-/\lambda^+, 1]$ and $\lambda \in [\lambda^-, \lambda^+]$.

As mentioned earlier, the fixed points f_λ of the maps $\mathcal{N}_\lambda$ are not attractive. Numerically, the two largest eigenvalues of $D\mathcal{N}_\lambda(f_\lambda)$ are roughly 1.37 and 0.54 for $\lambda \approx \lambda^*$. In the context of computer–assisted proofs, the standard way of solving a hyperbolic fixed point problem is to turn it into a fixed point problem for a contraction by proceeding in the following way. We choose an invertible linear map M close to the inverse of $1 - D\mathcal{N}_{\lambda^*}(f_{\lambda^*})$ and define

$$\mathcal{M}_{\lambda,\kappa} = 1 + M(\mathcal{N}_{\lambda,\kappa} - 1). \tag{2.18}$$

In Section 7.2, we give a detailed description of M and establish its invertibility. Furthermore, we will see that M is bounded in $\mathcal{B}_{\alpha\beta}$ for all $\alpha, \beta \geq 0$. Hence, $\mathcal{M}_{\lambda,\kappa}$ is well defined as a map from $\mathcal{B}_{\alpha\beta}\backslash\mathcal{H}$ to $\mathcal{B}_{\alpha\beta}$ for all $\alpha \geq 0$, $\beta > 0$, and κ, λ satisfying (2.17).

The existence of the continuous family of fixed points $\{f_\lambda\}$ will follow from estimates on the contractions $\mathcal{M}_{\lambda^+,\kappa}$. These estimates are collected in the following proposition.

Proposition 2.6. *Let λ^+ and λ^- be defined as in Theorem 2.4. Then, for $\mu = 0.5$, $\nu = 0.9$ and $r = 9 \cdot 10^{-4}$, there is a function $f^0_{\lambda^+} \in \mathcal{B}_{\mu\nu}$ and two positive real numbers $q < 1$ and $\varepsilon < r(1-q)$ for which the following holds. For all $\kappa \in [\lambda^-/\lambda^+, 1]$, the operator $\mathcal{M}_{\lambda^+,\kappa}$ is well defined and continuously differentiable as a map from the closed ball $\overline{B_r(f^0_{\lambda^+})} \subset \mathcal{B}_{\mu\nu}\backslash\mathcal{H}$ to $\mathcal{B}_{\mu\nu}$ and satisfies for all $f \in \overline{B_r(f^0_{\lambda^+})}$*

$$\|\mathcal{M}_{\lambda^+,\kappa}(f^0_{\lambda^+}) - f^0_{\lambda^+}\|_{\mu\nu} \leq \varepsilon, \tag{2.19}$$

$$\|D\mathcal{M}_{\lambda^+,\kappa}(f)\| \leq q. \tag{2.20}$$

The fact that $\overline{B_r(f^0_{\lambda^+})}$ does not contain any function in $\mathcal{H}$ will follow from computing explicit bounds on the inverse of $M(f)E(f)$ for all $f \in \overline{B_r(f^0_{\lambda^+})}$. These bounds will be computed when evaluating the quantity $c_{\lambda^+}(f)$, cf. the remark preceding Section 6.1. We now show that Proposition 2.6 implies part (a) of Theorem 2.4.

Proof of Theorem 2.4 (a). By the contraction mapping principle, Proposition 2.6 implies the existence of a fixed point $f_{\lambda^+,\kappa} \in \mathcal{B}_{\mu\nu}\backslash\mathcal{H}$ of $\mathcal{M}_{\lambda^+,\kappa}$ for all $\kappa \in [\lambda^-/\lambda^+, 1]$.

From the invertibility of the operator M, the functions $f_{\lambda^+,\kappa}$ are also fixed points of the operators $\mathcal{N}_{\lambda^+,\kappa}$. Hence, the conjugation relation (2.15) ensures the existence of the family $\{f_\lambda\}$ of fixed points of $\mathcal{N}_\lambda$ for $\lambda \in [\lambda^-, \lambda^+]$. These fixed points are given by $f_\lambda = S_{1/\kappa} f_{\lambda^+,\kappa}$, $\kappa = \lambda/\lambda^+$. Since $S_{1/\kappa} f_{\lambda^+,\kappa} \in \mathcal{B}_{(\lambda-\mu/\lambda^+)\nu} \backslash \mathcal{H}$ for all $\kappa \in [\lambda^-/\lambda^+, 1]$, it follows that for all $\lambda \in [\lambda^-, \lambda^+]$, $f_\lambda \in \mathcal{B}_{\alpha\beta} \backslash \mathcal{H}$ for some $\alpha, \beta > 0$.

Finally, we prove the continuity of the family $\{f_\lambda\}$. From Proposition 2.3, it follows that $f_\lambda \in \mathcal{B} \cap \mathcal{C}^\infty(\mathbb{R}_+)$ with $f'_\lambda \in \mathcal{B}$. Since the functions $f_{\lambda^+,\kappa} = S_\kappa f_{\kappa\lambda^+}$ have the same properties, $\kappa \mapsto S_{1/\kappa} f_{\lambda^+,\kappa}$ is continuous as a map from $[\lambda^-/\lambda^+, 1]$ to $\mathcal{B}_{\alpha\beta}$ provided that the family $\{f_{\lambda^+,\kappa}\}_\kappa$ is continuous in $\mathcal{B}_{\mu\nu}$. In order to show that this is indeed the case, we check that $\{f_{\lambda^+,\kappa}\}_\kappa$ is continuous at $\kappa = \kappa_0$ for each $\kappa_0 \in [\lambda^-/\lambda^+, 1]$. Let us fix such a κ_0 and denote $f_0 \equiv f_{\lambda^+,\kappa_0}$. First, since the contraction mapping principle and Proposition 2.6 imply that f_0 belongs to the ball $B_{\tilde{r}}(f^0_{\lambda^+})$, where $\tilde{r} = \varepsilon/(1-q) < r$, then, for every $\tilde{\varepsilon} > 0$ satisfying $\tilde{\varepsilon} < r - \tilde{r}$, the ball $\overline{B_{\tilde{\varepsilon}}(f_0)}$ is contained in $B_r(f^0_{\lambda^+})$. Hence, by (2.20), the operators $\mathcal{M}_{\lambda^+,\kappa}$ are strict contractions there with rate q. Next, since $f_0 \in \mathcal{B} \cap \mathcal{C}^\infty(\mathbb{R}_+)$ with $f'_0 \in \mathcal{B}$, $\kappa \mapsto \mathcal{M}_{\lambda^+,\kappa}(f_0)$ is continuous as a map from $\mathbb{R}^*_+$ to $\mathcal{B}_{\alpha\beta}$ for all $\alpha, \beta > 0$. This implies that there is a $\delta = \delta(\tilde{\varepsilon})$ such that $\mathcal{M}_{\lambda^+,\kappa}$ maps the ball $\overline{B_{\tilde{\varepsilon}}(f_0)}$ into itself for each κ with $|\kappa - \kappa_0| < \delta$: for $f \in \overline{B_{\tilde{\varepsilon}}(f_0)}$, it follows from the continuity of the map $\kappa \mapsto \mathcal{M}_{\lambda^+,\kappa}(f_0)$ and $q < 1$ that

$$\begin{aligned}
\|\mathcal{M}_{\lambda^+,\kappa}(f) - f_0\|_{\mu\nu} &\leq \|\mathcal{M}_{\lambda^+,\kappa}(f) - \mathcal{M}_{\lambda^+,\kappa}(f_0)\|_{\mu\nu} + \|\mathcal{M}_{\lambda^+,\kappa}(f_0) - f_0\|_{\mu\nu} \\
&\leq q\tilde{\varepsilon} + \|\mathcal{M}_{\lambda^+,\kappa}(f_0) - \mathcal{M}_{\lambda^+,\kappa_0}(f_0)\|_{\mu\nu} \\
&< \tilde{\varepsilon}\,.
\end{aligned}$$

Therefore, the contraction mapping principle implies the existence of a fixed point of $\mathcal{M}_{\lambda^+,\kappa}$ in the ball $B_{\tilde{\varepsilon}}(f_0)$ whenever $|\kappa - \kappa_0| < \delta$. By uniqueness of the fixed points $f_{\lambda^+,\kappa}$, one concludes that $\|f_{\lambda^+,\kappa} - f_{\lambda^+,\kappa_0}\|_{\mu\nu} < \tilde{\varepsilon}$ for all κ satisfying $|\kappa - \kappa_0| < \delta$.

■

Proposition 2.6 reduces the proof of Theorem 1.4 to the verification of the estimates (2.11), (2.19) and (2.20). This verification is computer–assisted, and yields $q \approx 0.85$ and $\varepsilon \approx 1.15 \cdot 10^{-4}$. The function $f^0_{\lambda^+}$ has been numerically determined to be a very good approximation of the fixed point f_{λ^+}. It is given by the linear interpolation of 2^{17} positive numbers at well chosen points, and has been obtained by iterating a numerical version of the map $\mathcal{N}_{\lambda^+}$ (as described in Sections 4 and 5) and renormalizing the mass properly after each iteration in order to remove the unstable direction. Regarding the computation of the norm of the tangent map $D\mathcal{M}_{\lambda^+,\kappa}(f)$, we will take advantage of the fact that $D\mathcal{N}_\lambda(f)$ has very good contraction properties on certain subspaces of finite codimension provided f has some regularity. In particular, the nontrivial action of the operator M can be restricted to a finite dimensional subspace, and the computation of the norm of $D\mathcal{M}_{\lambda^+,\kappa}(f)$ essentially requires to explicitly evaluate $D\mathcal{M}_{\lambda^+,\kappa}(f)$ on finitely many basis vectors.

The remainder of this paper is devoted to the proof of Proposition 2.6 and the verification of inequality (2.11). In Section 3, we review the basic approach of computer–assisted proofs, and extend it to function spaces of L^1–type. In Sections 4 and 5, we give a detailed account of the rigorous implementation on a computer of the maps $\mathcal{N}_\lambda$ and of the computation of bound (2.19) and inequality (2.11). Section 6 is devoted to the tangent maps $D\mathcal{N}_\lambda$ and their contraction properties, whereas Section 7 deals with $D\mathcal{M}_{\lambda+,\kappa}$ and the computation of the bound (2.20). `Section.8` is available as a supplement to this paper, and contains the source code of the program (`proof.f`) and two input data files (`fpoint.lp` and `fpoint.lm`). The program has been written in Fortran 77[†] and consists of a (short) main program and several subroutines ordered in a "bottom–up" hierarchy, accordingly to the organization of the paper. Except for Section 3, references to the program are collected in remarks at the end of each section. For a description of the input data files, see Sections 5.2 and 5.3.

3. Constructive Analysis in $\mathcal{B}_{\alpha\beta}$

Computer–assisted proofs rely on the ability, first, to discretize the problem under study in terms of objects that are representable on a computer, and, second, to have a rigorous control on the errors arising from the discretization. We note that in this respect, arithmetic operations are special, since controlling them rigorously requires an explicit knowledge of how rounding is performed by the computer. Nevertheless, we emphasize that the main difficulties related to discretization are usually concerned with the specific transformations involved in the functional equation under study, the control of numerical rounding being typically of no particular relevance.

To address discretization issues, one introduces the notions of bounds and standard sets. Denoting, for any set Σ, by $\mathcal{P}(\Sigma)$ the set of all subsets of Σ, we start by defining what we call a bound in the context of computer–assisted proofs.

Definition 3.1. *Let φ be a map from $D_\varphi \subseteq \Sigma$ to Σ'. Denote by f the set map obtained by lifting φ in the canonical way, that is, $D_f = \mathcal{P}(D_\varphi)$ and $f(S) = \{s' \in \Sigma' \mid s' = \varphi(s)$ for some $s \in S\}$ for every $S \in D_f$. We say that a set map $g : \mathcal{P}(\Sigma) \supseteq D_g \to \mathcal{P}(\Sigma')$ is a bound on φ if $D_f \supseteq D_g$ and if $f(S) \subseteq g(S)$ for all $S \in D_g$.*

We remark that g being a bound on φ means $\varphi(s) \in g(S)$ whenever $s \in S \in D_g$. Bounds of this type have the following two properties. First, they make it possible to estimate complicated maps in terms of simpler ones, since the composition of two bounds, if well defined, provides a bound on the composition. Second, they can be implemented on a computer. Indeed, given a set Σ, we start by specifying a finite collection of sets $\mathrm{std}(\Sigma) \subset \mathcal{P}(\Sigma)$ that can be represented on the computer with a given data type. The elements of $\mathrm{std}(\Sigma)$ are referred to as the *standard sets* of Σ. Next,

---

[†] Mostly standard Fortran 77. Some extensions that can be found on most popular compilers (SUN(TM), Microsoft(TM)PowerFortran,...) are used for convenience. In particular, we make use of the double precision complex data type `complex*16`.

given a map φ: $\Sigma \supseteq D_\varphi \to \Sigma'$, we construct a bound on φ within the class of maps g: $\mathrm{std}(\Sigma) \supseteq D_g \to \mathrm{std}(\Sigma')$. Finally, it is in general possible to characterize the images of g in $\mathrm{std}(\Sigma')$ constructively and implement this map on the computer. We note that one can usually choose $\mathrm{std}(\Sigma)$ and $\mathrm{std}(\Sigma')$ specifically adapted to the map φ in order to improve the bounds g that can be constructed.

Unless specified otherwise, the standard sets for a Cartesian product $\Sigma \times \Sigma'$ will be defined by setting $\mathrm{std}(\Sigma \times \Sigma') = \mathrm{std}(\Sigma) \times \mathrm{std}(\Sigma')$.

3.1. Operations Involving Real and Complex Numbers

In our application of the above mentioned procedure to the case of real numbers, we have followed the approach of [KSW] which is based on the 64 bit IEEE standard for floating point arithmetics. This standard specifies two things: a format for floating point numbers (IEEE numbers) and rules concerning rounding after the operations $+, -, *, /$ and $\sqrt{\ }$. We will not discuss the detail of the implementation, but refer the interested reader to the corresponding section of [KSW]. We first choose a subset $\mathcal{S}$ of IEEE numbers, the "safe range", for which no underflows nor overflows can occur. The standard sets of $\mathbb{R}$ and $\mathbb{R}_+^*$ are defined as follows.

Definition 3.2. *We define* $\mathrm{std}(\mathbb{R})$ *as the collection of all (closed) intervals* $[a, b]$ *with* $a \le b$ *elements of* $\mathcal{S}$. *We define* $\mathrm{std}(\mathbb{R}_+^*)$ *as the subset of* $\mathrm{std}(\mathbb{R})$ *made of intervals* $[a, b]$ *with* $a > 0$.

To represent an interval in $\mathrm{std}(\mathbb{R})$ on the computer, we use for convenience the data type for complex numbers available in Fortran. Given $a \le b \in \mathcal{S}$, the procedure `sbound` returns the interval $[a, b]$, whereas, given the interval $[a, b] \in \mathrm{std}(\mathbb{R})$, `rl` and `ru`, respectively, returns a and b. We add two more functions, `siconst` and `srconst`, which, given $r \in \mathcal{S}$ an integral constant, and, respectively, $r \in \mathcal{S}$ an IEEE number, return the (unique) singleton in $\mathrm{std}(\mathbb{R})$ containing r.

By using the IEEE specifications related to the rounding occurring after the operations $+, -, *, /$ and $\sqrt{\ }$, one first writes two functions, `rup` and `rdown`, which, given r_c the rounded result of any of these operations, compute an upper bound and a lower bound, respectively, on the exact result r. If these bounds do not belong to $\mathcal{S}$, a flag is raised and the program stops. In a straightforward manner, one next constructs bounds in $\mathrm{std}(\mathbb{R})$ (in the sense of Definition 3.1) on the maps $x \mapsto -x$ (`sneg`), $|x|$ (`sabs`), $1/x$ (`sinv`), x^2 (`spower2`), $\sqrt{x}$ (`ssqrt`), and $(x, y) \mapsto x + y$ (`ssum`), $x - y$ (`sdiff`), $x * y$ (`sprod`) and x/y (`squot`). We will also need a bound on the function $x \mapsto \exp(x)$. The precision with which this function is evaluated is not specified by the IEEE standard. Hence, we make use of the bounds constructed so far and compose them in the following way. First, we use $\exp(nx) = \exp(x)^n$ to restrict the Taylor expansion of $\exp(x)$ to cases where $|x| < 0.03$. Next, we compute the first three terms in the expansion and bound the tail by a geometrical series. This bound is implemented in the function sexp. We note that it is only involved in the computation of the weight $w_{\alpha\beta}$ and is not required to be of great accuracy. Finally, we will need to evaluate for x close to zero and $n = 0, \ldots, 3$,

$$\mathrm{Log}_n(x) \equiv -(-x)^{-n} \sum_{k=n+1}^{\infty} \frac{(-x)^k}{k}. \tag{3.1}$$

Note that the second factor is just the tail of the Taylor expansion of $\log(1+x)$. In particular, $\mathrm{Log}_0(x) = \log(1+x)$. One easily checks that the inequalities

$$-\sum_{k=1}^{m} \frac{(-x)^k}{k+n} - \left|\frac{x^{m+1}}{m+n+1}\right| \leq \mathrm{Log}_n(x) \leq -\sum_{k=1}^{m} \frac{(-x)^k}{k+n} + \left|\frac{x^{m+1}}{m+n+1}\right|$$

are valid for all $m \geq 1$. With $m = 4$, a sufficiently accurate bound is constructed in selogne from the previous inequalities.

We end this section with the discussion of a bound on the discrete convolution. For $r = (r_0, \ldots, r_{n-1})$ and $s = (s_0, \ldots, s_{n-1}) \in \mathbb{R}^n$, the discrete convolution $r * s$ is the element of $\mathbb{R}^{2n-1}$ given by

$$(r * s)_k = \sum_{i+j=k} r_i s_j, \quad k = 0, \ldots, 2n-2. \tag{3.2}$$

Computing $r * s$ according to (3.2) involves $\mathcal{O}(n^2)$ operations, and becomes impractical for large n. The standard strategy is to go into Fourier space where the convolution becomes the pointwise product of vectors. The gain in computational time is due to the fact that the discrete Fourier transform can be implemented with an $\mathcal{O}(n \log_2 n)$ algorithm, known as the Fast Fourier Transform (FFT). More precisely, the discrete Fourier transform is a map from $\mathbb{C}^n$ to $\mathbb{C}^n$ defined by

$$\big(\mathcal{F}(z)\big)_k = \sum_{j=0}^{n-1} \exp\Big(i\frac{2\pi kj}{n}\Big) z_j, \quad k = 0, \ldots, n-1, \tag{3.3}$$

for $z = (z_0, \ldots, z_{n-1}) \in \mathbb{C}^n$. The inverse Fourier transform $\mathcal{F}^{-1}$ is given by

$$\big(\mathcal{F}^{-1}(z)\big)_k = \frac{1}{n}\sum_{j=0}^{n-1} \exp\Big(-i\frac{2\pi kj}{n}\Big) z_j, \quad k = 0, \ldots, n-1. \tag{3.4}$$

For $r, s \in \mathbb{R}^n$ as above, one has the well known relation

$$(r * s)_k = (\mathcal{F}^{-1}(\mathcal{F}(\tilde{r}) \cdot \mathcal{F}(\tilde{s})))_k, \quad k = 0, \ldots, 2n-2, \tag{3.5}$$

where $\tilde{r} = (r_0, \ldots, r_{n-1}, 0, \ldots, 0) \in \mathbb{R}^{2n}$ and $\tilde{s} = (s_0, \ldots, s_{n-1}, 0, \ldots, 0) \in \mathbb{R}^{2n}$. An efficient implementation of (3.3) and (3.4) follows from the observation that if n is even, then $\mathcal{F}(z)$ can be decomposed into the sum of the Fourier transform of two vectors in $\mathbb{C}^{n/2}$. Hence, for n a power of 2, one can repeat this decomposition $\log_2 n$ times until only the Fourier transform of a single complex number remains to be computed. $\mathcal{F}(z)$ is finally obtained by summing the intermediate Fourier transforms, which needs $\mathcal{O}(n)$ operations.

In order to implement a bound on the discrete Fourier transform, we first need to choose the standard sets of $\mathbb{C}$. For our purpose, it is sufficient to define them in terms of $\mathrm{std}(\mathbb{R})$ as follows.

Definition 3.3. *We define* $\mathrm{std}(\mathbb{C})$ *to be the collection of all sets* $R + i * I$ *of the form*

$$R + i * I = \{x + i \cdot y \in \mathbb{C} \mid x \in R,\, y \in I\} \tag{3.6}$$

with R *and* I *elements of* $\mathrm{std}(\mathbb{R})$.

The only operations in $\mathbb{C}$ involved in the FFT algorithm are the addition and product, bounds on which are readily implemented from our bounds acting on $\mathrm{std}(\mathbb{R})$. One also needs bounds on the trigonometric factors appearing in (3.3) and (3.4). From the periodicity properties of the functions sin and cos, one first notes that it is sufficient to construct a bound on the maps $(l, m) \mapsto \sin(l\pi/m), \cos(l\pi/m)$ where $m \geq 4$ is a power of 2 and l ranges in $\{0, \ldots, m/4\}$. The case $l = 0$ is trivial. For $l = 1$ and $m = 4, 8, \ldots$, one evaluates sin and cos recursively: For $m = 4$ one has $\cos(\pi/4) = \sin(\pi/4) = 1/\sqrt{2}$, and for $m > 4$ a power of 2 one uses the half angle formulas

$$\cos(x/2) = \sqrt{\frac{1 + \cos(x)}{2}}, \quad \sin(x/2) = \frac{1}{2}\frac{\sin(x)}{\cos(x/2)}.$$

Finally, for $l > 1$, one applies the double angle formulas.

Remark. Bounds on the Fourier transform and inverse Fourier transform are implemented in the procedure `fft` according to the FFT algorithm. We do not enter into the details of this algorithm, and refer the reader to [PFTV], from which the code has been adapted to interval analysis using the bounds described above. Adapting again to interval analysis a code from [PFTV], a bound on the discrete convolution $r \mapsto r * r$ is implemented in the procedure `fastconvolution1`, while the general case $(r, s) \mapsto r * s$ is implemented in `fastconvolution2`. Those bounds are restricted to vectors whose dimension is a power of 2. In the sequel, we actually compute the discrete convolution of vectors of the form $r = (0, r_1, \ldots, r_n, 0)$, n a power of 2. The first two elements and last two elements of such convolution are trivially zero and are updated directly in the procedures `fastconvolution1` and `fastconvolution2`. For convenience later on, see Section 4.4, we also add at the beginning and the end of the result one element zero.

3.2. Standard Sets of $\mathcal{B}_{\alpha\beta}$

We now describe the standard sets of the Banach spaces $\mathcal{B}_{\alpha\beta}$. As mentioned above, the choice of these sets should be adapted to the problem in order to optimize the bounds that one needs to construct. Although functions in $\mathcal{B}_{\alpha\beta}$ are in general irregular, the fixed points of the maps $\mathcal{N}_\lambda$ are smooth. Furthermore, these maps are continuous and preserve the regularity. Therefore, we will take for our standard sets of $\mathcal{B}_{\alpha\beta}$ balls centered at regular functions. To represent a regular function on the computer, we will rely on the approximation scheme of spline interpolation.

A spline function of order n is a function in $\mathcal{C}^{n-1}$ which is piecewise polynomial, each of the polynomials being of degree n. For our purpose, it is sufficient to consider splines of order one as the centers of our standard sets, i.e., continuous piecewise affine functions. This choice is a compromise between the quality of the approximation and the simplicity of the bounds that we will have to construct. Note that increasing the order of the interpolation does not lead in general to better approximations. Indeed, for a function $f \in \mathcal{C}^\infty([a,b])$ and a typical partition of $[a,b]$ with mesh size $\varepsilon > 0$, the associated interpolation of f by a spline g of order $n-1$ satisfies for a norm of L^1–type

$$\|f - g\| \approx \varepsilon^n \|f^{(n)}\|.$$

Hence, depending upon the behavior of $f^{(n)}$, it can become better to consider finer partitions of $[a,b]$ rather than to increase the order of the interpolation.

We now introduce a few objects that will be used to define the standard sets of $\mathcal{B}_{\alpha\beta}$.

Definition 3.4. *For $n \geq 2$, we denote by $\mathcal{P}_n$ the set of all partitions p of $\mathbb{R}_+^*$ of the form*

$$p = \{0 < x_0 < x_1 < \ldots < x_n < \infty\}. \tag{3.7}$$

Furthermore, we denote by $\mathcal{P}_n^u$ the subset of $\mathcal{P}_n$ made of uniform partitions, i.e., partitions $p = \{x_i\}_{i=0}^n \in \mathcal{P}_n$ satisfying $x_i - x_{i-1} = \varepsilon$, $i = 1, \ldots, n$, for some $\varepsilon > 0$.

The uniform partitions have been introduced in order to simplify the implementation of a bound on the convolution operator. For $p = \{x_i\}_{i=0}^n \in \mathcal{P}_n$ and $\lambda > 0$, we adopt the convention to denote by λp the partition $\{\lambda x_i\}_{i=0}^n$. Next, we describe more precisely the piecewise affine functions we will work with.

Definition 3.5. *We define $\mathcal{A}$ to be the set of all functions $\rho \in \mathcal{C}^0(\mathbb{R}_+)$ for which there is an $n \geq 2$ and a partition $p = \{x_i\}_{i=0}^n \in \mathcal{P}_n$ such that ρ is affine on $[x_{i-1}, x_i]$ for $i = 1, \ldots, n$ and $\rho(x) = 0$ for $x \notin (x_0, x_n)$. Furthermore, $\mathcal{A}^u$ denotes the subset of $\mathcal{A}$ consisting of those functions for which p can be chosen uniform.*

We note that $\mathcal{A}, \mathcal{A}^u \subset \mathcal{B}_{\alpha\beta}$ for all $\alpha, \beta \geq 0$. Given a partition $p = \{x_i\}_{i=0}^n \in \mathcal{P}_n$ and a set of values $v = \{v_i\}_{i=0}^n \in \mathbb{R}^{n+1}$ satisfying $v_0 = v_n = 0$, we denote by $\mathcal{T}_1(p, v)$

the linear interpolation of (p,v) in $\mathcal{A}$, i.e.,

$$\mathcal{T}_1(p,v)(x) = \begin{cases} v_i + \frac{v_{i+1}-v_i}{x_{i+1}-x_i}(x-x_i) \text{ for } x \in [x_i, x_{i+1}] \text{ and } i \in \{0,\dots,n-1\}, \\ 0 \qquad \text{otherwise.} \end{cases} \tag{3.8}$$

Conversely, associated with every function $\rho \in \mathcal{A}$, there is a pair (p,v) in $\mathcal{P}_n \times \mathbb{R}^{n+1}$, for some $n \geq 2$, satisfying $\mathcal{T}_1(p,v) = \rho$. If $\rho \not\equiv 0$ and if one imposes a minimality condition on n, then the associated pair (p,v) is unique.

Definition 3.6. *For $\rho \not\equiv 0$ a function in $\mathcal{A}$, let*

$$n(\rho) \equiv \min\{n \geq 2 \mid \exists\ (p,v) \in \mathcal{P}_n \times \mathbb{R}^{n+1} \text{ such that } \mathcal{T}_1(p,v) = \rho\},$$

and define $\pi(\rho)$ to be the (unique) element of $\mathcal{P}_{n(\rho)} \times \mathbb{R}^{n(\rho)+1}$ satisfying $\mathcal{T}_1(\pi(\rho)) = \rho$.

Note that by definition of $\mathcal{A}$, one has always $\pi(\rho) = (\cdot, \{\rho_i\}_{i=0}^{n(\rho)})$ with $\rho_0 = \rho_{n(\rho)} = 0$. In order to define the standard sets of $\mathcal{A}$ and $\mathcal{A}^u$, we need to choose the standard sets of $\mathcal{P}_n$ and $\mathcal{P}_n^u$.

Definition 3.7. *For $n \geq 2$, we define $\mathrm{std}(\mathcal{P}_n)$ to be the collection of all sets $(X_0,\dots,X_n)$ of the form*

$$(X_0,\dots,X_n) = \{\{x_i\}_{i=0}^n \in \mathcal{P}_n \mid x_0 \in X_0,\dots,x_n \in X_n\}, \tag{3.9}$$

with $X_0,\dots,X_n$ any increasing sequence of $n+1$ pairwise disjoint elements of $\mathrm{std}(\mathbb{R}_+^*)$. Similarly, we define $\mathrm{std}(\mathcal{P}_n^u)$ as the collection of all sets (A,E) of the form

$$(A,E) = \{p \in \mathcal{P}_n^u \mid p = \{a + i\varepsilon\}_{i=0}^n, a \in A \text{ and } \varepsilon \in E\}, \tag{3.10}$$

with $A, E \in \mathrm{std}(\mathbb{R}_+^*)$.

Note that $\mathrm{std}(\mathcal{P}_n^u)$ is not a subset of $\mathrm{std}(\mathcal{P}_n)$. Indeed, the sets (A,E) contain only uniform partitions, whereas there are always non–uniform partitions in each set $(X_0,\dots,X_n)$ which is not a singleton. The standard sets of $\mathcal{A}$ and $\mathcal{A}^u$ are defined in terms of $\mathrm{std}(\mathcal{P}_n)$ and $\mathrm{std}(\mathcal{P}_n^u)$ as follows.

Definition 3.8. *Let $N = 2^{20}$. We define $\mathrm{std}(\mathcal{A})$, respectively $\mathrm{std}(\mathcal{A}^u)$, to be the collection of all sets (P,V) of the form*

$$(P,V) = \{\rho \in \mathcal{A} \mid \rho = \mathcal{T}_1(p,v), p \in P \text{ and } v \in V\}, \tag{3.11}$$

with $P \in \mathrm{std}(\mathcal{P}_n)$, respectively $P \in \mathrm{std}(\mathcal{P}_n^u)$, $V \in \mathrm{std}(\mathbb{R}^{n+1})$ and $2 \leq n \leq N$.

Finally, we introduce the standard sets of $\mathcal{B}_{\alpha\beta}$. They will be of two types, denoted by $\mathrm{std}(\mathcal{B}_{\alpha\beta})$ and $\mathrm{std}(\mathcal{B}_{\alpha\beta})^u$.

Definition 3.9. *Let $\alpha \geq 0$ and $\beta \geq 0$. We define* $\mathrm{std}(\mathcal{B}_{\alpha\beta})$*, respectively* $\mathrm{std}(\mathcal{B}_{\alpha\beta})^u$*, to be the collection of all sets (P, V, G) of the form*

$$(P,V,G) = \{f \in \mathcal{B}_{\alpha\beta} \mid f = \rho + g,\ \rho \in (P,V),\ g \in \mathcal{B}_{\alpha\beta} \text{ and } \|g\|_{\alpha\beta} \leq G\}, \tag{3.12}$$

with $(P, V) \in \mathrm{std}(\mathcal{A})$, respectively $(P, V) \in \mathrm{std}(\mathcal{A}^u)$, and $G \in \mathcal{S}$, $G \geq 0$.

Hence, a set (P, V, G) is the union of all balls of radius G that are centered at piecewise affine functions belonging to (P, V).

Remark. In our program, the data type with which a set (P, V, G) is represented, with $P \in \mathrm{std}(\mathcal{P}_n)$ and $V \in \mathrm{std}(\mathbb{R}^{n+1})$, is a $2 \times (n+2)$ matrix, say `f`, with entries of complex data type. (Recall that we use the Fortran data type for complex numbers to represent the elements of $\mathrm{std}(\mathbb{R})$ and $\mathrm{std}(\mathbb{R}^*_+)$.) The entries `f(1,0)` up to `f(1,`n`)` contain V, and `f(1,`$n+1$`)` contains the interval $[0, G]$. The entries `f(0,0)` up to `f(0,`n`)` contain the partition P. If $P = (A, E) \in \mathrm{std}(\mathcal{P}^u_n)$, then `f(0,0)`$= A$ and `f(0,`$n+1$`)`$=$ E. If $P \notin \mathrm{std}(\mathcal{P}^u_n)$, then `f(0,`$n+1$`)`$=$ $[0, 0]$. Given an integer $n \geq 2$ and $a, \varepsilon \in \mathcal{S}$, $a, \varepsilon > 0$, the procedure `fzero` returns a standard set $(P, V, G) \in \mathrm{std}(\mathcal{B}_{\alpha\beta})^u$ where $G = 0$, $V = ([0, 0], \ldots, [0, 0])$ and P contains the partition $\{a + i\varepsilon\}_{i=0}^n$. Finally, given $(P, V, G) \in \mathrm{std}(\mathcal{B}_{\alpha\beta})$ and an integer i, the procedure `get_f_on_i` returns two elements of $\mathrm{std}(\mathbb{R}^*_+)$ and two elements of $\mathrm{std}(\mathbb{R})$ containing respectively x_{i-1}, x_i, ρ_{i-1} and ρ_i for all $\rho \in (P, V)$, $\pi(\rho) = (\{x_j\}, \{\rho_j\})_{j=0}^n$.

We end this section with a few comments about the strategy that we will adopt when constructing bounds on the various maps that enter the definition of $\mathcal{N}_\lambda$. Some of these maps are linear and preserve $\mathcal{A}$. Let $\mathcal{L}$ be such a map. Then, for $f = \rho + g$ with $\rho \in \mathcal{A}$, the piecewise affine part ρ and the general term g can be treated separately, and since the piecewise affine parts carry the relevant information, it is natural to describe the affine part of $\mathcal{L}(f)$ by $\mathcal{L}(\rho)$ and its general term by $\mathcal{L}(g)$. Moreover, the choice of the standard set image in $\mathrm{std}(\mathcal{A})$ containing $\mathcal{L}(\rho)$ is straightforward. For instance, the product of a function $f \in \mathcal{B}_{\alpha\beta}$ by a scalar $\lambda \in \mathbb{R}$ is bounded using $\pi(\lambda\rho) = (p, \lambda v)$, where $(p, v) = \pi(\rho)$, and $\|\lambda g\|_{\alpha\beta} = |\lambda|\,\|g\|_{\alpha\beta}$. This bound is implemented in the procedure `fmult`. In general, however, the maps that will be considered do not preserve $\mathcal{A}$. Let $\mathcal{U} : \mathcal{B}_{\alpha\beta} \to \mathcal{B}_{\zeta\eta}$ be such a transformation. For $f = \rho + g$ with $\rho \in \mathcal{A}$ and $g \in \mathcal{B}_{\alpha\beta}$, we write

$$\mathcal{U}(\rho + g) = \mathcal{U}(\rho) + \mathcal{W}(\rho, g).$$

Since $\mathcal{U}(\rho) \notin \mathcal{A}$, we will consider the linear interpolation $\tilde{\rho} \in \mathcal{A}$ of $\mathcal{U}(\rho)$ at well chosen points. This choice will usually be a compromise between the quality of the approximation and the simplicity of the implementation. One then has

$$\mathcal{U}(\rho + g) = \tilde{\rho} + \mathcal{W}(\rho, g) + \big(\mathcal{U}(\rho) - \tilde{\rho}\big),$$

and the last two terms on the RHS correspond to the general term $\tilde{g}$ of $\mathcal{U}(\rho + g)$. They will be bounded using first

$$\|\tilde{g}\|_{\zeta\eta} \leq \|\mathcal{W}(\rho, g)\|_{\zeta\eta} + \|\mathcal{U}(\rho) - \tilde{\rho}\|_{\zeta\eta}.$$

Next, explicit formulas involving ρ, g and the values of $\mathcal{U}(\rho)$ at the chosen interpolation points, together with the use of interval analysis, will lead to a rigorous upper bound on the previous expression, and hence to the representable $G \in \mathcal{S}$ entering Definition 3.9. We note that the elements of std($\mathbb{R}$) defining the piecewise affine function $\tilde{\rho}$ consist in general of intervals of non zero length. Nevertheless, since the bound G has been computed for all reals in those intervals, one can "close" each of them by picking arbitrarily one of the representable numbers it contains. This will prevent the standard sets containing the piecewise affine part from "opening up" substantially when bounds are composed, in particular when evaluating convolution products.

4. Operations Involving Functions

In this section, we construct bounds (in the sense of Section 3) on the various maps that enter the definition of the transformations $\mathcal{N}_\lambda$. Most of the bounds given here follow from direct calculations and are easy to prove. We have grouped some of these calculations in the appendix.

In the following, we will usually consider $f \in \mathcal{B}_{\alpha\beta}$ of the form $f = \rho + g$, where $\rho \in \mathcal{A}$ will always stand for the piecewise affine part of f and $g \in \mathcal{B}_{\alpha\beta}$ for the general term of f. Furthermore, when not explicitly mentioned otherwise, $n(\rho)$ and $\pi(\rho) = (p, v)$ will be denoted by n and $(\{x_i\}_{i=0}^n, \{\rho_i\}_{i=0}^n)$, respectively. Finally, we denote the interval $[x_{i-1}, x_i]$ by I_i, $i = 1, \ldots, n$.

4.1. Elementary Operations

We start with the map $f \mapsto \|f\|_{\alpha\beta}$, a bound on which is constructed from std($\mathcal{B}_{\alpha\beta}$) to std($\mathbb{R}_+$) using the triangle inequality and, for the piecewise affine part, the estimate

$$\begin{aligned} \|\rho\|_{\alpha\beta} &= \sum_{i=1}^{n} \int_{I_i} w_{\alpha\beta}(x)\, |\rho(x)|\, dx \\ &\le \sum_{i=1}^{n} \sup_{x \in I_i} \left(w_{\alpha\beta}(x) \right) (x_i - x_{i-1}) \frac{|\rho_{i-1}| + |\rho_i|}{2}. \end{aligned} \tag{4.1}$$

The convexity of $w_{\alpha\beta}$ leads to

$$\sup_{x \in I_i} w_{\alpha\beta}(x) = \max\{w_{\alpha\beta}(x_{i-1}), w_{\alpha\beta}(x_i)\}. \tag{4.2}$$

We next consider the mass $M(f)$ and the expectation $E(f)$ of a function $f \in \mathcal{B}_{\alpha\beta}$, for which it will be sufficient to construct bounds from std$(\mathcal{B}_{\alpha\beta})^u$ to std($\mathbb{R}$). By linearity, we can first treat separately the affine part ρ, and by using $\rho_0 = \rho_n = 0$, a direct calculation yields

$$M(\rho) = \varepsilon \sum_{i=1}^{n-1} \rho_i, \tag{4.3}$$

$$E(\rho) = \varepsilon \sum_{i=1}^{n-1} \rho_i x_i, \tag{4.4}$$

where $\varepsilon = x_1 - x_0$. Using $M(\rho) - |M(g)| \le M(f) \le M(\rho) + |M(g)|$ and the corresponding inequality for $E(f)$, we get the desired bounds by estimating the mass and expectation of the general term g with (2.12) and (2.13). The supremum of $1/w_{\alpha\beta}$ appearing in (2.12) is taken at $x_c = \sqrt{\alpha/\beta}$, which leads to

$$\sup_{x>0}\left(\frac{1}{w_{\alpha\beta}(x)}\right) = \exp(-2\sqrt{\alpha\beta}). \tag{4.5}$$

Similarly, one computes

$$\sup_{x>0}\left(\frac{x}{w_{\alpha\beta}(x)}\right) = \frac{1+\sqrt{1+4\alpha\beta}}{2\beta}\exp\left(-\sqrt{1+4\alpha\beta}\right). \tag{4.6}$$

We end this section with the discussion of a bound on the addition of two functions $f_1, f_2 \in \mathcal{B}_{\alpha\beta}$. Due to the linearity of this map and the fact that the addition of two functions in $\mathcal{A}$ is again in $\mathcal{A}$, it is natural to choose for the general term of $f_1 + f_2$ the addition of the general terms of those two functions, whose norm is bounded by using the triangle inequality. It then remains to construct a bound on the map $+ : \mathcal{A} \times \mathcal{A} \to \mathcal{A}$. Let $\rho_1, \rho_2 \in \mathcal{A}$ such that $\pi(\rho_1) = (p_1, v_1) \in \mathcal{P}_n \times \mathbb{R}^{n+1}$ and $\pi(\rho_2) = (p_2, v_2) \in \mathcal{P}_m \times \mathbb{R}^{m+1}$. If ρ_1 and ρ_2 have no common nodes, then $(p, w) \equiv \pi(\rho_1 + \rho_2) \in \mathcal{P}_{n+m+1} \times \mathbb{R}^{n+m+2}$, with p being the refined partition made of the ordered union of p_1 and p_2. The last is valid only if ρ_1 and ρ_2 have no common nodes and we shall construct a bound whose domain is restricted to such cases. Hence, denoting $p = \{y_i\}_{i=0}^{n+m+1}$ and $w = \{w_i\}_{i=0}^{n+m+1}$, one defines for each $i = 0, \ldots, n+m+1$,

$$w_i \equiv (\rho_1 + \rho_2)(y_i) = \begin{cases} (v_1)_j + \rho_2((p_1)_j) & \text{if } \exists\, j \text{ such that } y_i = (p_1)_j, \\ (v_2)_j + \rho_1((p_2)_j) & \text{if } \exists\, j \text{ such that } y_i = (p_2)_j. \end{cases}$$

To implement this bound with interval analysis, we must check first that the nodes in $\mathrm{std}(\mathbb{R}_+^*)$ of the standard sets $P_1 \in \mathrm{std}(\mathcal{P}_n)$ and $P_2 \in \mathrm{std}(\mathcal{P}_m)$ containing the partition p_1 and p_2 are pairwise disjoint intervals. This implies that every function in (P_1, V_1) is linear on each node of P_2, and vice versa. This in turn implies that a bound on the evaluation $\rho_2((p_1)_j)$ and $\rho_1((p_2)_j)$ is readily obtained from (3.8) using interval analysis. We note finally that when the standard set containing ρ_1 (ρ_2) is in $\mathrm{std}(\mathcal{A}^u)$, i.e., with P_1 (P_2) of the form (A, E), we first proceed to the evaluation of the nodes in terms of A and E.

Finally, one constructs a bound on the difference of two functions by composing the previous bound with a bound on the unary minus $\mathcal{B}_{\alpha\beta} \ni f \mapsto -f$ obtained from $||-g||_{\alpha\beta} = ||g||_{\alpha\beta}$ and $\pi(-\rho) = (p, -v)$, where $(p, v) = \pi(\rho)$.

Remark. A bound on the weight $w_{\alpha\beta}$ is computed in the procedure `sw`. The inequality (4.1) is implemented in `snorm_pl`, whereas (4.2) and (4.6) are implemented in `ssup_of_w` and `ssup_of_x_over_w`, respectively. For several intervals $I \subset \mathbb{R}_+^*$, we will need to evaluate later the quantities $\sup_{x\in I} 1/w_{\alpha\beta}(x)$, $\int_I w$, and $|I|^{-1}\int_I w$, where $|I|$ denotes the length of I. By using the convexity of w, a bound on the first quantity is computed in `ssup_of_winverse`, whereas the other two quantities are bounded in `sint_of_w`. The other bounds described in this section are implemented in the procedures `snorm`, `smass`, `sexpectation`, `fadd` and `fdiff`.

4.2. The Scaling Operator

It will be sufficient for our purpose to construct a bound on

$$\begin{aligned} S_\lambda : \mathcal{B}_{\alpha\beta} &\to \mathcal{B}_{\frac{\alpha}{4}\gamma} \\ f(x) &\mapsto h(x) = \lambda f(\lambda x), \end{aligned} \tag{4.7}$$

acting from $\mathrm{std}(\mathcal{B}_{\alpha\beta})^u$ to $\mathrm{std}(\mathcal{B}_{(\alpha/4)\gamma})^u$. We recall that the scaling operators are bounded under constraints which translate in this particular case into $\lambda \le 4$ (if $\alpha > 0$) and $\gamma \le \lambda\beta$. It will be checked by the program that these inequalities are satisfied for the values of λ, α, β and γ we will use. Since S_λ is linear and preserves $\mathcal{A}^u$, we can treat separately the piecewise affine part and the general term. A bound on $S_\lambda : \mathcal{A}^u \to \mathcal{A}^u$ is obtained from the relation $\pi(S_\lambda\rho) = (p/\lambda, \lambda v)$, where $(p, v) = \pi(\rho)$. For the general term we estimate using (4.5),

$$\begin{aligned} ||S_\lambda g||_{\frac{\alpha}{4}\gamma} &\le \sup_{x>0} \frac{w_{\frac{\alpha}{4}\gamma}(x/\lambda)}{w_{\alpha\beta}(x)} ||g||_{\alpha\beta} \\ &= \exp\Big(-2\sqrt{\alpha(1-\lambda/4)(\beta-\gamma/\lambda)}\Big)\, ||g||_{\alpha\beta}, \end{aligned} \tag{4.8}$$

the last equality being valid under the conditions on λ, α, β and γ mentioned above. Note that the scaling operator (4.7) is a strict contraction for $\gamma/\beta < \lambda < 4$. This will be used to improve the bound on $\mathcal{N}_\lambda^1(f) = S_\lambda f * S_\lambda f$ that we shall construct later. In (4.7), taking a larger target space, i.e., $\mathcal{B}_{(\alpha/\sigma)\gamma}$ with $\sigma > 4$, would lead to a better contraction. Nevertheless, $\sigma = 4$ is the largest value for which $\mathcal{N}_\lambda^1$ maps $\mathcal{B}_{\alpha\beta}$ into $\mathcal{B}_{\alpha\gamma}$, cf. (2.8).

Remark. A bound on $S_\lambda : \mathcal{A}^u \to \mathcal{A}^u$ is implemented in the procedure `fscale_pl`, and the bound (4.8) in `fscale_gen`. Those two procedures are called in `fscale` to build the desired bound on the operator (4.7).

4.3. The Operator T

We now construct a bound on the operator

$$\begin{aligned} T : \mathcal{B}_{\alpha\beta} &\to \mathcal{B}_{\beta\alpha} \\ f(x) &\mapsto h(x) = \frac{1}{x^2} f(\frac{1}{x}), \end{aligned} \tag{4.9}$$

acting from $\mathrm{std}(\mathcal{B}_{\alpha\beta})^u$ to $\mathrm{std}(\mathcal{B}_{\beta\alpha})$. For $f = \rho + g$ with $\rho \in \mathcal{A}^u$ and $g \in \mathcal{B}_{\alpha\beta}$, one has $Tf = T\rho + Tg$. Since $T\rho$ is not piecewise linear, we must first choose a function $\tilde{\rho} \in \mathcal{A}$ which approximates $T\rho$. Denoting again $\pi(\rho) = (\{x_i\}_{i=0}^n, \{\rho_i\}_{i=0}^n)$, we consider for $\tilde{\rho}$ the linear interpolation of $T\rho$ at the nodes

$$\tilde{x}_i = 1/x_{n-i}, \quad i = 0, \ldots, n. \tag{4.10}$$

Therefore, we define $\tilde{\rho}$ to be

$$\tilde{\rho} = \mathcal{T}_1(\tilde{p}, \tilde{v}), \tag{4.11}$$

where $\tilde{p} = \{\tilde{x}_i\}_{i=0}^n$, and $\tilde{v} = \{\tilde{\rho}_i\}_{i=0}^n$ with

$$\tilde{\rho}_i \equiv (T\rho)(\tilde{x}_i) = x_{n-i}^2 \rho_{n-i}, \quad i = 0, \ldots, n. \tag{4.12}$$

Next, the general term $\tilde{g}$ of Tf is given by $\tilde{g} = T\rho - \tilde{\rho} + Tg$, and we use (2.10) to estimate

$$\|\tilde{g}\|_{\beta\alpha} \leq \|T\rho - \tilde{\rho}\|_{\beta\alpha} + \|g\|_{\alpha\beta}. \tag{4.13}$$

In order to bound the first term on the RHS of (4.13), we use again (2.10) together with the linearity of T and the fact that it is an involution. This leads to

$$\begin{aligned} \|T\rho - \tilde{\rho}\|_{\beta\alpha} &= \|\rho - T\tilde{\rho}\|_{\alpha\beta} \\ &\leq \sum_{i=1}^{n} \sup_{x\in I_i} w_{\alpha\beta}(x) \int_{I_i} \left|(\rho - T\tilde{\rho})(x)\right| dx. \end{aligned} \tag{4.14}$$

Finally, an explicit bound on the integral appearing in the previous expression follows from a direct calculation and is given in the next lemma.

Lemma 4.1. *Let $\rho \in \mathcal{A}^u$, and $\tilde{\rho}$ be defined as in (4.11). With $\pi(\rho) = (\{x_i\}_{i=0}^n, \{\rho_i\}_{i=0}^n)$, $\varepsilon = x_1 - x_0$ and $I_i = [x_{i-1}, x_i]$, one has*

$$\begin{aligned} \int_{I_i} |(\rho - T\tilde{\rho})(x)|\, dx \leq \frac{\varepsilon}{4}\Big(\log\Big(1 + \frac{\varepsilon}{x_{i-1}}\Big)|\rho_i - \rho_{i-1}| \\ + \varepsilon\left|\frac{\rho_i}{x_{i-1}} - \frac{\rho_{i-1}}{x_i}\right| + \varepsilon(x_i - \varepsilon/2)\left|\frac{\rho_i}{x_{i-1}^2} - \frac{\rho_{i-1}}{x_i^2}\right|\Big). \end{aligned} \tag{4.15}$$

Remark. A bound on $T : \mathcal{B}_{\alpha\beta} \to \mathcal{B}_{\beta\alpha}$ is implemented as described here in the procedure `ft`.

4.4. The Convolution

For $\alpha, \beta > 0$ and $\gamma \in [\alpha, 4\alpha]$, we consider in this section the operator

$$\begin{aligned} C : \mathcal{B}_{\alpha\beta} \times \mathcal{B}_{\alpha\beta} &\to \mathcal{B}_{\gamma\beta} \\ (f, h) &\mapsto f * h. \end{aligned} \tag{4.16}$$

As mentioned before, we specifically introduced standard sets of piecewise affine functions defined on uniform partitions in order to simplify the construction of a bound on the convolution. Hence, our bound will act from $\mathrm{std}(\mathcal{B}_{\alpha\beta})^u \times \mathrm{std}(\mathcal{B}_{\alpha\beta})^u$ to $\mathrm{std}(\mathcal{B}_{\gamma\beta})^u$. To simplify further the explicit expressions which we shall derive below, we restrict its domain to pairs (F_1, F_2) for which the standard sets (A_1, E_1) and (A_2, E_2) containing the partitions associated with the affine functions in F_1 and F_2 satisfy $E_1 = E_2$ and both E_1 and E_2 are singletons. This ensures that all affine functions in F_1 and F_2 are defined on (uniform) partitions with identical mesh size.

Let $f = \rho + g_f$ and $h = \sigma + g_h$, with $\rho, \sigma \in \mathcal{A}^u$ and $g_f, g_h \in \mathcal{B}_{\alpha\beta}$. Then, one has

$$f * h = \rho * \sigma + \rho * g_h + g_f * h. \tag{4.17}$$

The relevant information is carried by the term $\rho * \sigma$. Since it does not belong to $\mathcal{A}$, we will proceed as in the previous section and approximate it by a function $\tilde{\rho} \in \mathcal{A}^u$. The general term of $f * h$ will be given by

$$\tilde{g} = (\rho * \sigma - \tilde{\rho}) + \rho * g_h + g_f * h. \tag{4.18}$$

One can estimate the last two terms on the RHS of the previous expression using the bound (2.8). However, estimating the norm of the first term requires an explicit expression for $(\rho * \sigma)(x)$, $x > 0$. We now derive this expression, which will be used also to specify $\tilde{\rho}$. We first state an intermediate result whose proof can be found in the appendix.

Lemma 4.2. *If $\rho, \sigma \in \mathcal{A}^u$ have uniform partitions with identical mesh size ε, then $(\rho * \sigma)'' \in \mathcal{A}^u$. Furthermore, assume $n(\rho) = n(\sigma) \equiv n$, denote $\pi(\rho) = (\{x_i\}, \{\rho_i\})_{i=0}^n$, $\pi(\sigma) = (\{y_i\}, \{\sigma_i\})_{i=0}^n$, and define $\{s_k\}_{k=0}^{2n}$ to be the discrete convolution of $\{\rho_i\}_{i=0}^n$ and $\{\sigma_i\}_{i=0}^n$, i.e.,*

$$s_k = \sum_{i+j=k} \rho_i \sigma_j. \tag{4.19}$$

Then $n((\rho * \sigma)'') = 2n$, and $(\rho * \sigma)'' = \mathcal{T}_1(\{z_k\}_{k=0}^{2n}, \{v_k\}_{k=0}^{2n})$ where

$$z_k = x_0 + y_0 + k\varepsilon, \tag{4.20}$$

$$v_k = \frac{1}{\varepsilon}(s_{k+1} - 2s_k + s_{k-1}), \tag{4.21}$$

with the convention $s_{-1} = s_{2n+1} = 0$.

We now specify the nature of $\rho * \sigma$.

Lemma 4.3. *Let $\rho, \sigma \in \mathcal{A}^u$ as in Lemma 4.2 and define p to be the partition $\{z_k\}_{k=0}^{2n}$ where z_k is given by (4.20). Then $\rho * \sigma$ is in $\mathcal{C}^2(\mathbb{R}^+)$, has support in $[z_0, z_{2n}]$ and is there identical to its natural cubic spline approximation φ at the nodes of p, i.e., with the boundary conditions $\varphi''(z_0) = \varphi''(z_{2n}) = 0$.*

Proof. First, it follows from $\text{supp}(\rho) \subseteq (x_0, x_n)$ and $\text{supp}(\sigma) \subseteq (y_0, y_n)$ that the support of $\rho * \sigma$ is in $(x_0 + y_0, x_n + y_n)$. Then, the regularity properties of the convolution imply $\rho * \sigma \in \mathcal{C}^2(\mathbb{R}^+)$. To see that $\rho * \sigma$ is equal to its natural cubic spline approximation at the nodes $a \equiv z_0 < z_1 < \ldots < z_{2n} \equiv b$, one shows that it minimizes the quantity

$$\int_a^b \varphi''(x)^2 dx$$

over all $\varphi \in \mathcal{C}^2[a, b]$ satisfying $\varphi(z_k) = (\rho * \sigma)(z_k)$, $k = 0, \ldots, 2n$. Let φ be such a function. Setting $\varphi_0 \equiv \rho * \sigma$, one has

$$\begin{aligned}\int_a^b \varphi''(x)^2 dx &= \int_a^b (\varphi_0''(x) + (\varphi - \varphi_0)''(x))^2 dx \\ &= \int_a^b \varphi_0''(x)^2 dx + 2\int_a^b \varphi_0''(x)(\varphi - \varphi_0)''(x) dx + \int_a^b (\varphi - \varphi_0)''(x)^2 dx. \quad (4.22)\end{aligned}$$

We will see that the second term on the RHS is zero, yielding

$$\int_a^b \varphi''(x)^2 dx \geq \int_a^b \varphi_0''(x)^2 dx.$$

The conclusion then follows from a well known result in spline theory, see for instance [N], which ensures that such a minimization problem has a unique solution given by the natural cubic spline interpolation of the data points entering the constraints of the minimization problem. It remains to see that the second term on the RHS of (4.22) is zero, i.e., that φ_0'' and $(\varphi - \varphi_0)''$ are orthogonal in $L^2[a, b]$. From Lemma 4.2, we have $\varphi_0'' \in \mathcal{A}_p$, where $\mathcal{A}_p$ denotes the subspace of $L^2[a, b]$ consisting of all functions $\tau \in \mathcal{A}$ with $\pi(\tau) = (p_\tau, \cdot)$ and p_τ a subpartition of p. Next, we observe that every $\psi \in \mathcal{C}^2[a, b]$ with $\psi(z_k) = 0$, $k = 0, \ldots, 2n$, satisfies $\psi'' \in \mathcal{A}_p^\perp$: a basis of $\mathcal{A}_p$ is given by $\{\tau_k\}_{k=1}^{2n-1}$, τ_k being the "hat" function centered at z_k, i.e., with χ_I the characteristic function of the interval I,

$$\tau_k(x) = \chi_{[z_{k-1}, z_k]}(x)(x - z_{k-1}) + \chi_{(z_k, z_{k+1}]}(x)(z_{k+1} - x),$$

and a simple calculation using integration by parts leads to

$$\int_a^b \tau_k(x)\psi''(x) dx = 0,$$

$k = 1, \ldots, 2n - 1$. We conclude the proof by noting that the conditions of the minimization problem are $(\varphi - \varphi_0)(z_k) = 0$, $k = 0, \ldots, 2n$.

■

As a consequence, it follows that $\rho * \sigma$ is given on each interval $[z_k, z_{k+1}]$, $k = 0, \ldots, 2n-1$, by the cubic polynomial

$$(\rho * \sigma)(z_k + \theta) = C_0(k) + C_1(k)\theta + C_2(k)\theta^2 + C_3(k)\theta^3 \,, \tag{4.23}$$

$\theta \in [0, \varepsilon]$, where the coefficients $C_i(k)$ take the form

$$\begin{aligned} C_0(k) &= \frac{\varepsilon}{6}(s_{k+1} + 4s_k + s_{k-1}), \\ C_1(k) &= \frac{1}{2}(s_{k+1} - s_{k-1}), \\ C_2(k) &= \frac{1}{2\varepsilon}(s_{k+1} - 2s_k + s_{k-1}), \\ C_3(k) &= \frac{1}{6\varepsilon^2}(s_{k+2} - 3s_{k+1} + 3s_k - s_{k-1}), \end{aligned} \tag{4.24}$$

using again the convention $s_{-1} = s_{2n+1} = 0$. Indeed, $C_2(k)$ is just $(\rho * \sigma)''(z_k)/2$ and has been directly computed in Lemma 4.2. Using $(\rho * \sigma)(z_0) = (\rho * \sigma)(z_{2n}) = 0$, the remaining coefficients are obtained from $C_2(k)$ and the formula for natural cubic spline interpolation, see for instance [ANW].

A natural choice for the affine part $\tilde{\rho}$ of $\rho * \sigma$ would be to consider the linear interpolation of the values of $\rho * \sigma$ at the points z_k, $k = 0, \ldots, 2n$. However, that would amount to double the number of parameters and would lead eventually to memory problems when estimating the precision of the approximate fixed point $f^0_{\lambda+}$. Hence, we choose here to define $\tilde{\rho}$ in terms of the same number of parameters as ρ and σ, and we consider

$$\tilde{\rho} = \mathcal{T}_1(\{z_{2l}\}_{l=0}^n, \{C_0(2l)\}_{l=0}^n), \tag{4.25}$$

Note that the nodes z_{2l} generate a uniform partition, so that $\tilde{\rho} \in \mathcal{A}^u$.

To conclude this section, we come back to the general term $\tilde{g}$ of $f * h$ given by (4.18). We first use the triangle inequality and (2.8) to get

$$||\tilde{g}||_{\gamma\beta} \le ||\rho * \sigma - \tilde{\rho}||_{\gamma\beta} + ||\rho||_{\alpha\beta}||g_h||_{\alpha\beta} + ||g_f||_{\alpha\beta}||h||_{\alpha\beta}. \tag{4.26}$$

The bound on the map $f \mapsto ||f||_{\alpha\beta}$ described earlier allows us to estimate the last two terms on the RHS of (4.26). A bound on the first term is obtained by a direct calculation using (4.25) and the explicit expression (4.23). The result is formulated in the next lemma.

Lemma 4.4. *Let ρ, σ and $\tilde{\rho}$ defined as above. Then*

$$||\rho * \sigma - \tilde{\rho}||_{\gamma\beta} \le \varepsilon^3 \sum_{l=0}^{n-1} \sup_{x \in I_l} w_{\gamma\beta}(x) \Big(\frac{4}{3}|C_2(2l+1)| + \frac{3\varepsilon}{4}\big(|C_3(2l)| + |C_3(2l+1)|\big)\Big). \tag{4.27}$$

where $I_l = [z_{2l}, z_{2l+2}]$.

Remarks.

- By definition of $\mathcal{A}^u$, the first and last two elements of the discrete convolution (4.19) are trivially zero, so that only the convolution of $\{\rho_i\}_{i=1}^{n-1}$ and $\{\sigma_i\}_{i=1}^{n-1}$ needs to be computed. Furthermore, recall that in order to simplify the implementation of the bound on the discrete convolution, we have restricted its domain to the standard sets of $\mathrm{std}(\mathbb{R}^n)$ for which n is a power of 2. Hence, our bound on the convolution (4.16) is defined only on elements of $\mathrm{std}(\mathcal{B}_{\alpha\beta})^u$ with partitions in $\mathrm{std}(\mathcal{P}_n^u)$ such that $n-1$ is a power of 2.
- Bounds on the quantities $\varepsilon^{i-1}C_i(k)$, $i=0,\ldots,3$ respectively, are computed in the procedure `cubic_spline_coeff` and saved in the vectors `st0`, `st1`, `st2` and `st3`. Note that the interpolation (4.25) and the bound (4.27) provide a bound on the convolution from $\mathrm{std}(\mathcal{A}^u)\times\mathrm{std}(\mathcal{A}^u)$ to $\mathrm{std}(\mathcal{B}_{\gamma\beta})$. It is implemented in the procedure `fcubic_to_pwlinear`. Finally, `fconvolute2` computes the desired bound on (4.16), making first use of `fastconvolution2` to get the discrete convolution (4.19), whereas `fconvolute1` is adapted to the special case $f=h$. Those two subroutines have a call to `sexp_of_tconv`, which has been introduced to compute an accurate bound on the expectation of $\mathcal{N}_\lambda^2(f)$, cf. (2.6), and will be explained in Section 5.1.

4.5. The Identity

Another operator we need to consider is the identity. Indeed, we recall that ultimately we want to compose the bounds constructed so far in order to get bounds on the maps of interest. However, the bounds constructed so far do not have always matching range and domain, and cannot in general be composed as such. In particular, the bound on the operator T applies in $\mathrm{std}(\mathcal{B}_{\alpha\beta})$ whereas the convolution is defined only on $\mathrm{std}(\mathcal{B}_{\alpha\beta})^u\times\mathrm{std}(\mathcal{B}_{\alpha\beta})^u$. Furthermore, the bound on the convolution is defined for pairs whose affine parts satisfy constraints on the mesh of their partitions. Hence, we need a bound on the identity map $I:\mathcal{B}_{\alpha\beta}\to\mathcal{B}_{\alpha\beta}$ defined from $\mathrm{std}(\mathcal{B}_{\alpha\beta})$ to $\mathrm{std}(\mathcal{B}_{\alpha\beta})^u$ such that the affine part of all functions in every standard set image is ensured to possess a given partition.

Let $p=(x_0,\ldots,x_n)\in\mathcal{P}_n^u$ a fixed but arbitrary uniform partition, and $f=\rho+g$ with $\rho\in\mathcal{A}$ and $g\in\mathcal{B}_{\alpha\beta}$. For the new affine part $\tilde{\rho}$ of f with partition p, we would like to consider the linear spline interpolation of ρ at the nodes of p. However, in order for $\tilde{\rho}$ to be in $\mathcal{A}^u$, one must ensure $\tilde{\rho}$ to be continuous, so that we define

$$\tilde{\rho}=\mathcal{T}_1\big(p,\{0,\rho(x_1),\ldots,\rho(x_{n-1}),0\}\big). \tag{4.28}$$

Then, from

$$f=\tilde{\rho}+\big(\rho-\tilde{\rho}\big)+g,$$

the new general term $\tilde{g}$ reads $(\rho-\tilde{\rho})+g$ and its norm is simply bounded by

$$\|\rho-\tilde{\rho}+g\|_{\alpha\beta}\le\|\rho-\tilde{\rho}\|_{\alpha\beta}+\|g\|_{\alpha\beta}. \tag{4.29}$$

The first term on the RHS is bounded using the bounds constructed previously on the norm in $\mathcal{B}_{\alpha\beta}$ and the difference of two functions.

For every uniform partition p, the previous construction leads to a specific bound on the identity map. This bound can be optimized from case to case by adapting p to the function ρ so that $\|\rho - \tilde{\rho}\|_{\alpha\beta}$ is minimal. Again, our approach is to seek for a compromise between accuracy and simplicity of the implementation. First, we choose not to increase the number of parameters from ρ to $\tilde{\rho}$, so that $n \leq n(\rho)$. Hence, the only free parameters for p are the first node x_0 and last node x_n. Given a $\tau > 0$, the interval (x_0, x_n) is chosen to be the smallest interval such that $|\rho(x)| < \tau$ for $x \notin (x_0, x_n)$. This interval might be fairly different from $\operatorname{supp}(\rho)$, leading to a mesh ε smaller than $|\operatorname{supp}(\rho)|/n$ and hence a better approximation of ρ on regions where the information is more relevant. The cutoff τ may vary from place to place in the proof and has been determined empirically.

Remark. Given an integer $n \geq 2$, and two positive representable numbers x_0 and s, the procedure `fidentity` constructs a standard set in $\operatorname{std}(\mathcal{P}_n^u)$ containing the uniform partition p with $\operatorname{supp}(p) = (x_0, x_0 + s)$, and computes a bound on the identity map as described above. The representable numbers x_0 and s which describe the support adapted to a given function are determined in the procedure `rsupport`.

5. The Maps $\mathcal{N}_\lambda$

The goal of this section is to explain how bound (2.19) of Proposition 2.6 is computed and how inequality (2.11) of Theorem 2.4 is checked. A major step is to compute $\mathcal{N}_{\lambda^+}$ and $\mathcal{N}_{\lambda^-}$ on various functions of $\mathcal{B}_{\mu\nu}$, with μ, ν, λ^- and λ^+ as in Proposition 2.6. We will see in Section 5.2 that these maps must be estimated from $\mathcal{B}_{\mu\nu}$ to $\mathcal{B}_{\mu(\lambda^+\nu/\lambda^-)}$, a space slightly smaller than $\mathcal{B}_{\mu\nu}$, since $\lambda^+/\lambda^- \approx 1 + 7 \cdot 10^{-7}$. In the sequel we denote $\delta = \lambda^-/\lambda^+$, and begin in the next section by describing the construction of a bound on the maps $\mathcal{N}_\lambda : \mathcal{B}_{\mu\nu} \to \mathcal{B}_{\mu(\nu/\delta)}$.

5.1. A Bound on $\mathcal{N}_\lambda$

We recall that for $\lambda > 0$, $\mathcal{N}_\lambda : \mathcal{B}_{\mu\nu}/\mathcal{H} \to \mathcal{B}_{\mu(\nu/\delta)}$ is well defined provided $1/\delta \leq \lambda \leq 4$, cf. Proposition 2.2, and is given by

$$\mathcal{N}_\lambda(f) = c_\lambda(f)\mathcal{N}_\lambda^1(f) + c_2\mathcal{N}_\lambda^2(f), \tag{5.1}$$

where

$$\mathcal{N}_\lambda^1(f) = S_\lambda(f * f), \tag{5.2}$$

$$\mathcal{N}_\lambda^2(f) = T(T\mathcal{N}_\lambda^1(f) * T\mathcal{N}_\lambda^1(f)), \tag{5.3}$$

$$c_\lambda(f) = \frac{\lambda}{2}\frac{1 - c_2 E(\mathcal{N}_\lambda^2(f))}{M(f)E(f)}. \tag{5.4}$$

The expression (5.4) is more convenient for our present purpose than (2.7). In principle, one readily gets a bound on $\mathcal{N}_\lambda$ by composing the bounds constructed in the previous section. However, one can without too much effort improve this bound in two ways.

First, the distributivity and commutativity properties of the operators involved in (5.1) give us the freedom to choose the order in which the bounds are composed. The order can affect the estimates, since in general these properties are not shared by the bounds. Regarding $\mathcal{N}^1_\lambda$, the fact that S_λ preserves $\mathcal{A}^u$ yields slightly better estimates by using

$$\mathcal{N}^1_\lambda(f) = S_\lambda f * S_\lambda f, \tag{5.5}$$

instead of (5.2). Furthermore, in order to get as much contraction as possible from the scaling operator, cf. Section 4.2, one chooses the sequence of spaces

$$\mathcal{N}^1_\lambda : \mathcal{B}_{\mu\nu} \xrightarrow{S_\lambda} \mathcal{B}_{\frac{\mu}{4}(\nu/\delta)} \xrightarrow{*} \mathcal{B}_{\mu(\nu/\delta)}. \tag{5.6}$$

A bound on (5.6) follows by composing the bounds of Section 4. We now turn to $\mathcal{N}^2_\lambda$ and, as above, let S_λ act first, considering (5.3) with $\mathcal{N}^1_\lambda$ as in (5.5). Regarding the choice of spaces, we note that one could exploit the operators T and the outer convolution to estimate $\mathcal{N}^2_\lambda$ in the smaller space $\mathcal{B}_{\mu(\nu/\delta)}$, as needed. That would permit us to consider ν instead of ν/δ in (5.6) for which S_λ is a better contraction. Nevertheless, δ is so close to one that it does not lead to any significant improvement, and for convenience one simply constructs a bound on $\mathcal{N}^2_\lambda$ by composing the previous bound on $\mathcal{N}^1_\lambda$ and a bound on the map

$$\mathcal{B}_{\mu(\nu/\delta)} \xrightarrow{T} \mathcal{B}_{(\nu/\delta)\mu} \xrightarrow{*} \mathcal{B}_{(\nu/\delta)\mu} \xrightarrow{T} \mathcal{B}_{\mu(\nu/\delta)}. \tag{5.7}$$

The target space of the convolution above is chosen in order to minimize the norm of the general term arising from the convolution of the piecewise affine part.

The second improvement concerns the computation of the coefficient $c_\lambda(f)$. An estimation of the quantity $E(\mathcal{N}^2_\lambda(f))$ entering (5.4) is of poor quality if obtained by composing the bound on $\mathcal{N}^2_\lambda$ described above and the bound on the expectation as given in the previous section. Exploiting the structure of $\mathcal{N}^2_\lambda$ and the fact that it maps into a smaller space, due to the outer convolution, leads to a substantial improvement. Defining

$$\begin{aligned} \mathcal{E} : \mathcal{B}_{\alpha\beta} \times \mathcal{B}_{\alpha\beta} &\to \mathbb{R} \\ (f,h) &\mapsto E(T(f * h)), \end{aligned} \tag{5.8}$$

one has $E(\mathcal{N}^2_\lambda(f)) = \mathcal{E}(T\mathcal{N}^1_\lambda(f), T\mathcal{N}^1_\lambda(f))$. In order to construct a bound on $\mathcal{E}$ acting from $\mathrm{std}(\mathcal{B}_{\alpha\beta})^u \times \mathrm{std}(\mathcal{B}_{\alpha\beta})^u$ to $\mathrm{std}(\mathbb{R})$, we first observe that for $g \in \mathcal{B}_{\zeta\eta}$,

$$|E(Tg)| \le \int_0^\infty \frac{1}{x}|g(x)|\,dx \le \sup_{x>0} \frac{x}{w_{\eta\zeta}(x)} \|g\|_{\zeta\eta}. \tag{5.9}$$

Next, for $f = \rho + g_f$ and $h = \sigma + g_h$, with $\rho, \sigma \in \mathcal{A}^u$ and $g_f, g_h \in \mathcal{B}_{\alpha\beta}$, one has

$$\mathcal{E}(f,h) = E\big(T(\rho * \sigma)\big) + E\big(T(\rho * g_h + g_f * h)\big), \tag{5.10}$$

and, since $\rho * g_h + g_f * h \in \mathcal{B}_{(4\alpha)\beta}$, one obtains from (5.9) the following estimate on the second term in the RHS of (5.10),

$$|E\big(T(\rho * g_h + g_f * h)\big)| \le \sup_{x>0} \frac{x}{w_{\beta(4\alpha)}(x)} (\|\rho\|_{\alpha\beta}\|g_h\|_{\alpha\beta} + \|g_f\|_{\alpha\beta}\|h\|_{\alpha\beta}). \tag{5.11}$$

Finally, the first term on the RHS of (5.10) can be computed explicitly. We use the same notation as in Section 4.4. Then, $\rho * \sigma$ is given on each interval $[z_k, z_{k+1}]$, $k = 0, \ldots, 2n-1$, by the cubic polynomial

$$(\rho * \sigma)(z_k + \theta) = C_0(k) + C_1(k)\theta + C_2(k)\theta^2 + C_3(k)\theta^3,$$

where the coefficients $C_i(k)$ are given by (4.24). Hence,

$$\begin{aligned} E\big(T(\rho * \sigma)\big) &= \int_0^\infty \frac{1}{x}(\rho * \sigma)(x)\, dx \\ &= \sum_{k=0}^{2n-1} \int_0^1 \frac{1}{\theta + z_k/\varepsilon}\big(C_0 + \varepsilon\theta C_1(k) + \varepsilon^2\theta^2 C_2(k) + \varepsilon^3\theta^3 C_3(k)\big)\, d\theta, \end{aligned} \tag{5.12}$$

and using

$$\int_0^1 \frac{x^n}{x + 1/a} dx = a^{-n} \int_0^a \frac{x^n}{x+1} dx = \mathrm{Log}_n(a),$$

where Log_n is defined in (3.1), one can integrate each term in (5.12) and gets finally

$$E\big(T(\rho * \sigma)\big) = \sum_{k=0}^{2n-1} \sum_{m=0}^{3} \varepsilon^m C_m(k)\ \mathrm{Log}_m\Big(\frac{\varepsilon}{z_k}\Big). \tag{5.13}$$

Remark. A bound on (5.8) is implemented in `sexp_of_tconv`. Since the quantities entering (5.13) are computed during the estimation of the convolution, this subroutine is called in `fconvolute1` and `fconvolute2`. A bound on $\mathcal{N}_\lambda$ is implemented in `fN`. This subroutine also returns a standard set containing the coefficient $c_\lambda(f)$ that will be used to check (2.11), treating separately the special case where the value of $E(f)$ is known exactly, cf. Section 5.3.

5.2. Existence of the Family of Fixed Points: First Estimate

We now explain how the quantity ε entering inequality (2.19) of Proposition 2.6 is computed. Recall that it consists in an upper bound on

$$\|\mathcal{M}_{\lambda^+,\kappa}(f^0_{\lambda^+}) - f^0_{\lambda^+}\|_{\mu\nu}, \tag{5.14}$$

uniform in $\kappa \in [\delta, 1]$, $\delta = \lambda^-/\lambda^+$, where $\lambda^- < \lambda^+$, μ and ν are given in Proposition 2.6, and $f^0_{\lambda^+}$ is an approximate fixed point in $\mathcal{A}^u$. From the definition of $\mathcal{N}_{\lambda^+,\kappa}$ and $\mathcal{M}_{\lambda^+,\kappa}$, cf. (2.14) and (2.18), it follows that

$$\|\mathcal{M}_{\lambda^+,\kappa}(f^0_{\lambda^+}) - f^0_{\lambda^+}\|_{\mu\nu} \leq \|M\| \, \|\mathcal{N}_{\lambda^+,\kappa}(f^0_{\lambda^+}) - f^0_{\lambda^+}\|_{\mu\nu}, \tag{5.15}$$

and from (2.9) one obtains

$$\begin{aligned} \|\mathcal{N}_{\lambda^+,\kappa}(f^0_{\lambda^+}) - f^0_{\lambda^+}\|_{\mu\nu} &\leq \|S_\kappa(\mathcal{N}_{\lambda^+}(f^0_{\lambda^+}) - f^0_{\lambda^+})\|_{\mu\nu} + \|S_\kappa f^0_{\lambda^+} - f^0_{\lambda^+}\|_{\mu\nu} \\ &\leq \|\mathcal{N}_{\lambda^+}(f^0_{\lambda^+}) - f^0_{\lambda^+}\|_{\mu(\nu/\delta)} + \|(S_\kappa - 1)f^0_{\lambda^+}\|_{\mu\nu}. \end{aligned} \tag{5.16}$$

The last inequality is valid since $f^0_{\lambda^+}$ and $\mathcal{N}_{\lambda^+}(f^0_{\lambda^+})$ belong to $\mathcal{B}_{\mu(\nu/\delta)}$. Indeed, $f^0_{\lambda^+}$ has compact support in $(0,\infty)$, and $\mathcal{N}_{\lambda^+}$ preserves this property. Therefore, $\mathcal{N}_{\lambda^+}(f^0_{\lambda^+}) \in \mathcal{B}_{\alpha\beta}$ and $f^0_{\lambda^+} \in \mathcal{B}_{\alpha\beta}$ for all $\alpha, \beta \geq 0$. By composing the bounds constructed in the previous sections, one gets an estimate for the first term on the RHS of (5.16). At this point, the only dependence on the parameter κ lies in the second term of (5.16), which one bounds uniformly using the following result.

Lemma 5.1. *Let $0 < \kappa \leq 1$ and $f \in W^1_1(\mathbb{R}^+, w_{\alpha\beta}(x)dx)$. If $f' \in \mathcal{B}_{\alpha\gamma}$ for some $\gamma > \beta/\kappa$, then*

$$\|(S_\kappa - 1)f\|_{\alpha\beta} \leq (1-\kappa)\big(\|f\|_{\alpha\beta} + \|xf'(x)\|_{\alpha(\beta/\kappa)}\big). \tag{5.17}$$

By definition, the function $f^0_{\lambda^+} \in \mathcal{A}^u$ satisfies the hypothesis of Lemma 5.1 for all $\alpha, \beta \geq 0$. Furthermore, the bound (5.17) is decreasing in κ. Hence, one has for all $\kappa \in [\delta, 1]$,

$$\|(S_\kappa - 1)f^0_{\lambda^+}\|_{\mu\nu} \leq (1-\delta)\big(\|f^0_{\lambda^+}\|_{\mu\nu} + \|x(f^0_{\lambda^+})'(x)\|_{\mu(\nu/\delta)}\big). \tag{5.18}$$

Collecting the inequalities (5.15), (5.16) and (5.18) yields the desired bound ε on (5.14), uniform in $\kappa \in [\delta, 1]$. The only missing information is the norm of the operator M appearing in (5.15), which will be given in Section 7.2.

We end this section with the

Proof of Lemma 5.1. For $f \in W^1_1(\mathbb{R}^+, dx)$, one can rewrite

$$S_\kappa f(x) - f(x) = (\kappa - 1)f(x) + \kappa \int_x^{\kappa x} f'(y)dy.$$

Hence,

$$\|(S_\kappa - 1)f\|_{\alpha\beta} \leq (1-\kappa)\|f\|_{\alpha\beta} + \kappa \int_0^\infty dx\, w_{\alpha\beta}(x) \int_{\kappa x}^x |f'(y)|dy. \tag{5.19}$$

Furthermore,

$$\int_0^\infty dx\, w_{\alpha\beta}(x) \int_{\kappa x}^{x} |f'(y)|dy = \int_0^\infty dy\,|f'(y)| \int_y^{y/\kappa} w_{\alpha\beta}(x)dx$$
$$\leq \frac{1-\kappa}{\kappa} \int_0^\infty y|f'(y)| \max\{w_{\alpha\beta}(y), w_{\alpha\beta}(y/\kappa)\}dy$$
$$\leq \frac{1-\kappa}{\kappa} \sup_{y\geq 0} \frac{\max\{w_{\alpha\beta}(y), w_{\alpha\beta}(y/\kappa)\}}{w_{\alpha(\beta/\kappa)}(y)} \|yf'(y)\|_{\alpha(\beta/\kappa)}.$$

Finally, one checks that the supremum in the previous expression is bounded by one for $\kappa \leq 1$, which leads to (5.17).

■

Remark. The quantity ε is computed in the procedure `compute_residual`. This procedure also returns a bound on the first term in the RHS of (5.16), a quantity that will be used in Section 5.3. The bound (5.17) is implemented in the procedure `snorm_of_Skappam1`, where the second term on the RHS of (5.17) is estimated using, for $\rho \in \mathcal{A}$ with $\pi(\rho) = (\{x_i\}, \{\rho_i\})_{i=0}^n$,

$$\|x\rho'(x)\|_{\alpha(\beta/\kappa)} \leq \frac{1}{2} \sum_{i=1}^{n} \sup_{x\in I_i} w_{\alpha(\beta/\kappa)}(x)|\rho_i - \rho_{i-1}|(x_i + x_{i-1}). \tag{5.20}$$

Finally, the (positive) representable numbers that describe the approximate fixed point $f^0_{\lambda^+} \in \mathcal{A}^u$ are contained in the file `fpoint.lp`. The first two numbers in this file are the boundary points of $\text{supp}(f^0_{\lambda^+})$. They determine the partition $p \in \mathcal{P}^n_u$, $n = 2^{17} + 1$, satisfying $\pi(f^0_{\lambda^+}) = (p, \cdot)$. The last 2^{17} numbers are the (nonzero) entries of the vector v, where $\pi(f^0_{\lambda^+}) = (\cdot, v)$. Given a nonnegative $G \in \mathcal{S}$, the subroutine `read_fp` reads the file `fpoint.lp` and constructs a standard set (P, V, G) with $(P, V) \in \text{std}(\mathcal{A}^u)$ containing $f^0_{\lambda^+}$.

5.3. Existence of the Fixed Point f_{λ^*}

Recall that once the existence of the continuous family $\{f_\lambda\}$ of fixed points of $\mathcal{N}_\lambda$ is established for $\lambda \in [\lambda^-, \lambda^+]$, our main result, namely the existence of a $\lambda^* \in [\lambda^-, \lambda^+]$ and a function f^* satisfying $S_{\lambda^*}\mathcal{D}(f^*) = f^*$, follows from

$$c_{\lambda^-}(f_{\lambda^-}) < c_1 < c_{\lambda^+}(f_{\lambda^+}),$$

where c_1 is given by (1.10) and $c_\lambda(f)$ by (5.4). Checking this inequality amounts to computing for each of the three quantities involved a standard set in std(IR).

Let us start with $c_{\lambda^+}(f_{\lambda^+})$. Suppose that one has a standard set, say in $\mathrm{std}(\mathcal{B}_{\mu\nu})$, containing the fixed point f_{λ^+}. Then one readily gets a standard set in $\mathrm{std}(\mathbb{R})$ containing $c_{\lambda^+}(f_{\lambda^+})$ by composing our bounds to compute

$$c_{\lambda^+}(f_{\lambda^+}) = \frac{\lambda^+}{2}\frac{1 - c_2 E(\mathcal{N}^2_{\lambda^+}(f_{\lambda^+}))}{M(f_{\lambda^+})}. \tag{5.21}$$

The previous expression follows from (5.4) and $E(f_{\lambda^+}) = 1$, a property satisfied by definition of the maps $\mathcal{N}_\lambda$, cf. (2.2). In order to check that $c_1 < c_{\lambda^+}(f_{\lambda^+})$, the size of the standard set obtained from (5.21) must be small enough, which ultimately requires to localize well enough the fixed point f_{λ^+}. In particular, Proposition 2.6 implies only that $f_{\lambda^+} \in B_{\tilde{r}}(f^0_{\lambda^+})$ with $\tilde{r} = \varepsilon/(1-q)$. This cannot be used to construct a suitable standard set containing f_{λ^+}, since the ball $B_{\tilde{r}}(f^0_{\lambda^+})$ also contains the fixed point $f_{\lambda^+,\delta} = S_\delta f_{\lambda^-}$ for which $c_{\lambda^+}(S_\delta f_{\lambda^-}) = c_{\lambda^-}(f_{\lambda^-}) < c_1$. In order to get a suitable standard set, we first use that the approximate fixed point $f^0_{\lambda^+}$ has been numerically determined as a very good approximation of f_{λ^+}, and exploit our bounds to compute

$$\begin{aligned}\|\mathcal{M}_{\lambda^+}(f^0_{\lambda^+}) - f^0_{\lambda^+}\|_{\mu\nu} &\le \|M\|\,\|\mathcal{N}_{\lambda^+}(f^0_{\lambda^+}) - f^0_{\lambda^+}\|_{\mu\nu} \\ &\le \varepsilon',\end{aligned} \tag{5.22}$$

with $\varepsilon' \approx 4.97 \cdot 10^{-7}$ (to be compared with $\varepsilon \approx 1.15 \cdot 10^{-4}$). Next, since by Proposition 2.6, $\mathcal{M}_{\lambda^+} = \mathcal{M}_{\lambda^+,1}$ is a contraction on the ball $B_r(f^0_{\lambda^+}) \in \mathcal{B}_{\mu\nu}$ with rate $q < 1$, one infers from (5.22) and the contraction mapping principle that

$$\|f_{\lambda^+} - f^0_{\lambda^+}\|_{\mu\nu} \le \frac{\varepsilon'}{1-q}.$$

Finally, one constructs in $\mathrm{std}(\mathcal{B}_{\mu\nu})$ the standard set whose affine part is given by the singleton $\{f^0_{\lambda^+}\}$ and whose general term has norm $\varepsilon'/(1-q)$. This set contains f_{λ^+} and allows to check that $c_1 < c_{\lambda^+}(f_{\lambda^+})$.

We now consider $c_{\lambda^-}(f_{\lambda^-})$, setting again $\delta = \lambda^-/\lambda^+$. For convenience, we work with the fixed point $f_{\lambda^+,\delta}$ of $\mathcal{M}_{\lambda^+,\delta}$ whose existence is guaranteed in $B_r(f^0_{\lambda^+}) \in \mathcal{B}_{\mu\nu}$ by Proposition 2.6. Lemma 2.5 implies that $f_{\lambda^-} = S_{1/\delta} f_{\lambda^+,\delta}$ and identity (2.16) leads to

$$c_{\lambda^-}(f_{\lambda^-}) = c_{\lambda^+}(f_{\lambda^+,\delta}) = \delta\frac{\lambda^+}{2}\frac{1 - c_2 E(\mathcal{N}^2_{\lambda^+}(f_{\lambda^+,\delta}))}{M(f_{\lambda^+,\delta})}, \tag{5.23}$$

where $E(f_{\lambda^+,\delta}) = 1/\delta$ has been used. In order to check that $c_{\lambda^-}(f_{\lambda^-}) < c_1$ using the previous relation, we must localize $f_{\lambda^+,\delta}$ closely enough. For this purpose, we have determined a very good approximation $f^0_{\lambda^-}$ to the fixed point f_{λ^-}. As $f^0_{\lambda^+}$, it is given by the linear interpolation of 2^{17} positive values at well chosen points. First, we check using our bounds that $f^0_{\lambda^-}$ satisfies

$$\|S_\delta f^0_{\lambda^-} - f^0_{\lambda^+}\|_{\mu\nu} < r, \tag{5.24}$$

with r as in Proposition 2.6, and,

$$\begin{aligned}
\|\mathcal{M}_{\lambda^+,\delta}(S_\delta f^0_{\lambda^-}) - S_\delta f^0_{\lambda^-}\|_{\mu\nu} &\leq \|M\|\,\|\mathcal{N}_{\lambda^+,\delta}(S_\delta f^0_{\lambda^-}) - S_\delta f^0_{\lambda^-}\|_{\mu\nu} \\
&= \|M\|\,\|S_\delta \mathcal{N}_{\delta\lambda^+}(f^0_{\lambda^-}) - S_\delta f^0_{\lambda^-}\|_{\mu\nu} \\
&\leq \|M\|\,\|\mathcal{N}_{\lambda^-}(f^0_{\lambda^-}) - f^0_{\lambda^-}\|_{\mu(\nu/\delta)} \\
&\leq \varepsilon'',
\end{aligned} \tag{5.25}$$

with $\varepsilon'' \approx 4.97 \cdot 10^{-7}$. Inequality (5.24) ensures that $S_\delta f^0_{\lambda^-} \in B_r(f^0_{\lambda^+})$. Hence, Proposition 2.6 and inequality (5.25) imply by the contraction mapping principle that

$$\|f_{\lambda^+,\delta} - S_\delta f^0_{\lambda^-}\|_{\mu\nu} \leq \frac{\varepsilon''}{1-q}.$$

As above, this leads to the construction of a suitable standard set in $\mathrm{std}(\mathcal{B}_{\mu\nu})$ containing the fixed point $f_{\lambda^+,\delta}$.

To conclude, we emphasize that the accuracy of the bounds on (5.21) and (5.23) is crucial, since it determines how close to λ^* one can take λ^- and λ^+, and since, on the other hand, the size of the interval $[\lambda^-, \lambda^+]$ must be small enough in order to prove the existence of the family $\{f_\lambda\}_{\lambda\in[\lambda^-,\lambda^+]}$.

Remark. The bounds ε' and ε'' are computed in the subroutine `compute_residual` introduced earlier. The computations of $c_{\lambda^+}(f_{\lambda^+})$ and $c_{\lambda^-}(f_{\lambda^-})/\delta$ are carried out in the subroutine `fN`, in which a bound on the maps $\mathcal{N}_\lambda$ is implemented as explained in Section 5.1. The remainder of the procedure described in this section is worked out at the end of the main program. The (positive) representable numbers that describe $f^0_{\lambda^-}$ are contained in the file `fpoint.lm`. This file is organized in the same way as `fpoint.lp`, and a standard set containing $f^0_{\lambda^-}$ is constructed by the subroutine `read_fp` described in Section 5.2.

6. Contractivity Properties of $D\mathcal{N}_\lambda$

As mentioned in Section 2, the tangent map $D\mathcal{N}_\lambda(f)$ is a contraction on certain subspaces of $\mathcal{B}_{\alpha\beta}$ with finite codimension. The main goal of this section is to describe these subspaces and compute the contraction factors. Those will be used in Section 7 to estimate the norm of the tangent map of $\mathcal{M}_{\lambda^+,\kappa}$. We first introduce some notations and check that $\mathcal{N}_\lambda$ is C^1 on its domain of definition. The (Fréchet) derivative of $\mathcal{N}_\lambda$ at $f \in \mathcal{B}_{\alpha\beta}$ is explicitly given by

$$D\mathcal{N}_\lambda(f)h = S_\lambda\Big(2c_\lambda(f)\, f * h + 4c_2 T\big(T(f*f) * T(f*h)\big) + \delta_\lambda(f,h)\, f * f\Big), \tag{6.1}$$

where the variation $\delta_\lambda(f,h)$ of $c_\lambda(f)$ is such that

$$E\big(D\mathcal{N}_\lambda(f)h\big) = 0. \tag{6.2}$$

Indeed, since all functions in the range of $\mathcal{N}_\lambda$ have the same expectation, the tangent space contains only functions with expectation zero. Defining $\mathcal{N}_\lambda^1$ and $\mathcal{N}_\lambda^2$ as in (2.4) and (2.5), we rewrite (6.1) as

$$D\mathcal{N}_\lambda(f)h = c_\lambda(f)D\mathcal{N}_\lambda^1(f)h + c_2 D\mathcal{N}_\lambda^2(f)h + \delta_\lambda(f,h)\mathcal{N}_\lambda^1(f), \tag{6.3}$$

where

$$D\mathcal{N}_\lambda^1(f)h = 2S_\lambda(f*h), \tag{6.4}$$
$$D\mathcal{N}_\lambda^2(f)h = 2T\big(T\mathcal{N}_\lambda^1(f)*TD\mathcal{N}_\lambda^1(f)h\big). \tag{6.5}$$

From the condition (6.2), δ_λ is expressed in terms of the expectation of the three terms on the RHS of (6.3). Using the relations (1.23), one gets

$$\delta_\lambda(f,h) = -c_\lambda(f)\Big(\frac{M(h)}{M(f)} + \frac{E(h)}{E(f)}\Big) - \lambda c_2 \frac{E\big(D\mathcal{N}_\lambda^2(f)h\big)}{2M(f)E(f)}. \tag{6.6}$$

Now, for $\alpha \geq 0$ and $\beta, \lambda > 0$, one easily checks that the estimates of Proposition 2.2 imply that whenever $\mathcal{N}_\lambda : \mathcal{B}_{\alpha\beta}/\mathcal{H} \to \mathcal{B}_{\sigma\tau}$ is well defined, i.e., $\sigma \leq 4\alpha/\lambda$, $\tau \leq \lambda\beta$, it is continuously differentiable on $\mathcal{B}_{\alpha\beta}/\mathcal{H}$.

We will need later to estimate $D\mathcal{N}_\lambda(f)$ for $\mathcal{N}_\lambda : \mathcal{B}_{\alpha\beta} \to \mathcal{B}_{\alpha\gamma}$ with γ slightly larger than β. In this case, the conditions for $D\mathcal{N}_\lambda(f)$ to be bounded become $\gamma/\beta \leq \lambda \leq 4$. We will see below that under the stronger conditions $\gamma/\beta < \lambda < 4$, the tangent map $D\mathcal{N}_\lambda(f)$ is actually compact provided f is sufficiently regular, i.e., there is a sequence of subspaces of finite codimension on which $D\mathcal{N}_\lambda(f)$ converges to zero. These subspaces are defined as follows.

Definition 6.1. *For $p = \{x_0, \ldots, x_n\}$ a partition in $\mathcal{P}_n$ and $a, b > 0$, we define the following subspaces of $\mathcal{B}_{\alpha\beta}$,*

$$\mathcal{L}_\alpha^a = \{h \in \mathcal{B}_{\alpha\eta}, \eta = 0 \mid \operatorname{supp}(h) \subseteq (0,a)\},$$
$$\mathcal{C}_p = \{h \in L^1(\mathbb{R}_+) \mid \operatorname{supp}(h) \subseteq (x_0,x_n), \int_{x_{i-1}}^{x_i} h(x)\,dx = 0 \text{ for } i = 1,\ldots,n\},$$
$$\mathcal{R}_\beta^b = \{h \in \mathcal{B}_{\zeta\beta}, \zeta = 0 \mid \operatorname{supp}(h) \subseteq (b,\infty)\}.$$

Furthermore, we denote by $\mathcal{B}_{\alpha\beta}^p$ the following subspace of $\mathcal{B}_{\alpha\beta}$,

$$\mathcal{B}_{\alpha\beta}^p = \mathcal{L}_\alpha^{x_0} \oplus \mathcal{C}_p \oplus \mathcal{R}_\beta^{x_n}. \tag{6.7}$$

For a small enough and b large enough, it turns out that when restricted to $\mathcal{L}_\alpha^a$, $\mathcal{R}_\beta^b$, and $\mathcal{C}_p$ respectively, the tangent map of $\mathcal{N}_\lambda$ at a function f in $W_1^1(\mathbb{R}_+, w_{\alpha\beta}(x)dx)$ has norms of value $\mathcal{O}(e^{-1/a}\|f\|_{\alpha\beta})$, $\mathcal{O}(e^{-b}\|f\|_{\alpha\beta})$, and $\mathcal{O}(|p|\,\|f'\|_{\alpha\beta})$, where $|p|$ denotes the mesh size of the partition p. More generally, if $\mathcal{C}_p$ consists of functions whose $n-1$

first moments vanish on every interval of the partition p, the norm of the tangent map restricted to $\mathcal{C}_p$ is $\mathcal{O}(|p|^n \| f^{(n)} \|_{\alpha\beta})$, provided that the base function f is regular enough. For our purpose, it is sufficient to consider $n = 1$.

Before deriving explicitly the contraction factors, we remark that we will need to evaluate later $D\mathcal{N}_\lambda(f)$ on the complement of $\mathcal{B}^p_{\alpha\beta}$ in $\mathcal{B}_{\alpha\beta}$. It is easily seen that for a given partition p, every $h \in \mathcal{B}_{\alpha\beta}$ can be uniquely decomposed into a sum $h = g + \tau$ where $g \in \mathcal{B}^p_{\alpha\beta}$ and where τ is constant on each interval of the partition p and satisfies $\text{supp}(\tau) = \text{supp}(p)$. More precisely, with $p = \{x_0, \ldots, x_n\}$ and χ_I the characteristic function of the interval I, one has

$$\mathcal{B}_{\alpha\beta} = \mathcal{B}^p_{\alpha\beta} \oplus \mathcal{V}^p, \tag{6.8}$$

where $\mathcal{V}^p$ is the n-dimensional vector space defined by

$$\mathcal{V}^p = \{\tau \,|\, \tau = \sum_{i=1}^{n} \lambda_i \chi_{[x_{i-1}, x_i]}, \lambda_i \in \mathbb{R}\}. \tag{6.9}$$

Section 6.4 is devoted to the construction of a bound on the map $D\mathcal{N}_\lambda(f) : \mathcal{V}^p \to \mathcal{B}_{\alpha\gamma}$.

Remark. In (6.3), there are factors that depend only on the base function f. These factors, namely $\mathcal{N}^1_\lambda(f)$, $T\mathcal{N}^1_\lambda(f)$ and their norms, together with $c_\lambda(f)$, $E(f)$ and $M(f)$, are computed once and for all in the subroutine `compute_constant_terms` using the bounds of Section 4. (This subroutine makes use of `snorm_of_der_pl`, a function commented in the final remark of Section 6.1.) According to Proposition 2.6, $\lambda = \lambda^+$ and f is represented by the standard set in $\text{std}(\mathcal{B}_{\mu\nu})^u$ whose affine part is the singleton $\{f^0_{\lambda^+}\}$ and whose general term g satisfies $\|g\|_{\mu\nu} \le 9 \cdot 10^{-4}$. Finally, for given standard sets containing $M(h), E(h)$ and $E(D\mathcal{N}^2_\lambda(f)h)$, a bound on $\delta_\lambda(f, h)$ is computed in the procedure `sdelta1` using (6.6).

6.1. Oscillatory Functions

We derive now an upper bound on the norm of the operator $D\mathcal{N}_\lambda(f) : \mathcal{C}_p \to \mathcal{B}_{\alpha\gamma}$, with $\mathcal{C}_p$ as in Definition 6.1 and with $f = \rho + g \in \mathcal{B}_{\alpha\beta}/\mathcal{H}$, $\rho \in \mathcal{A}^u$. For the first two terms in (6.3), and for $\|g\|_{\alpha\beta}$ small, the contraction factor will come from the convolution in $D\mathcal{N}^1_\lambda$. Hence, we first use the bounds obtained in Proposition 2.2 and get in full generality

$$\|D\mathcal{N}_\lambda(f)h\|_{\alpha\gamma} \le \big(|c_\lambda(f)| + 2c_2 \|\mathcal{N}^1_\lambda(f)\|_{\alpha\gamma}\big)\, \|D\mathcal{N}^1_\lambda(f)h\|_{\alpha\gamma} + |\delta_\lambda(f,h)|\, \|\mathcal{N}^1_\lambda(f)\|_{\alpha\gamma}. \tag{6.10}$$

In the previous expression, only the quantities that depend on h remain to be estimated.

Let us begin with $\delta_\lambda(f, h)$. For $h \in \mathcal{C}_p$, one has $M(h) = 0$ and the first term in (6.6) vanishes. Next, $E(h)$ is expressed in term of the largest interval in the partition p.

Denoting $p = \{x_0, \ldots, x_n\}$ and $I_i = [x_{i-1}, x_i]$, $i = 1, \ldots, n$, the identity $\int_{I_i} h(x)dx = 0$ implies

$$\Big|\int_{I_i} xh(x)\,dx\Big| = \Big|\int_{I_i} (x - \frac{x_i + x_{i-1}}{2})h(x)\,dx\Big| \le \frac{1}{2}(x_i - x_{i-1})\int_{I_i} |h(x)|\,dx,$$

which in turn yields

$$|E(h)| \le \frac{1}{2}\max_{i=1,\ldots,n}\{x_i - x_{i-1}\}\sup_{x>0}\Big(\frac{1}{w_{\alpha\beta}(x)}\Big)\|h\|_{\alpha\beta}. \tag{6.11}$$

Finally, since $D\mathcal{N}_\lambda^2(f)h \in \mathcal{B}_{\alpha(4\gamma)}$, it follows from (2.13) that

$$|E(D\mathcal{N}_\lambda^2(f)h)| \le 2\sup_{x>0}\Big(\frac{x}{w_{\alpha(4\gamma)}(x)}\Big)\|\mathcal{N}_\lambda^1(f)\|_{\alpha\gamma}\|D\mathcal{N}_\lambda^1(f)h\|_{\alpha\gamma}. \tag{6.12}$$

Inserting (6.11) and (6.12) into (6.6) leads to an estimate for the second term on the RHS of (6.10). In order to bound the RHS of (6.12) and the first term on the RHS of (6.10), it then remains to estimate $\|D\mathcal{N}_\lambda^1(f)h\|_{\alpha\gamma}$.

In order to treat $D\mathcal{N}_\lambda^1(f)$, one has the possibility to exploit, as in the previous section, the distributivity of the scaling operator S_λ with respect to the convolution. It turns out that the order is not crucial and we consider for simplicity

$$D\mathcal{N}_\lambda^1 : \mathcal{B}_{\alpha\beta} \times \mathcal{C}_p \xrightarrow{\ *\ } \mathcal{B}_{(4\alpha)\beta} \xrightarrow{2S_\lambda} \mathcal{B}_{\alpha\gamma}. \tag{6.13}$$

We begin with the convolution and use the following result.

Lemma 6.2. *Let $f \in W_1^1(\mathbb{R}_+, w_{\alpha\beta}(x)\,dx)$ and $h \in \mathcal{C}_p$ with $p = \{x_0, \ldots, x_n\}$. Then,*

$$\|f * h\|_{(4\alpha)\beta} \le \frac{1}{2}c_{\alpha\beta}(p)\|f'\|_{\alpha\beta}\|h\|_{\alpha\beta}, \tag{6.14}$$

where, denoting $I_i = [x_{i-1}, x_i]$,

$$c_{\alpha\beta}(p) = \max_{i=1,\ldots,n}\Big\{\sup_{x\in I_i}\frac{1}{w_{\alpha\beta}(x)}\int_{I_i} w_{\alpha\beta}(x)\,dx\Big\}. \tag{6.15}$$

Since for $\rho \in \mathcal{A}$ one has by definition $\rho \in W_1^1(\mathbb{R}_+, w_{\alpha\beta}(x)dx)$, the previous lemma together with Proposition 2.2 imply, with $f = \rho + g$, $\rho \in \mathcal{A}^u$, $g \in \mathcal{B}_{\alpha\beta}$, and $h \in \mathcal{C}_p$,

$$\begin{aligned}\|f * h\|_{(4\alpha)\beta} &\le \|\rho * h\|_{(4\alpha)\beta} + \|g\|_{\alpha\beta}\|h\|_{\alpha\beta}\\ &\le \Big(\frac{1}{2}c_{\alpha\beta}(p)\|\rho'\|_{\alpha\beta} + \|g\|_{\alpha\beta}\Big)\|h\|_{\alpha\beta}.\end{aligned} \tag{6.16}$$

Next, since $\gamma/\beta \le \lambda \le 4$, inequality (4.8) applies (with α replaced by 4α), and we finally obtain

$$\|D\mathcal{N}^1_\lambda(f)h\|_{\alpha\gamma} \le 2e^{-A}\Big(\frac{1}{2}c_{\alpha\beta}(p)\|\rho'\|_{\alpha\beta} + \|g\|_{\alpha\beta}\Big)\|h\|_{\alpha\beta}, \tag{6.17}$$

where $A = 2\sqrt{\alpha(4-\lambda)(\beta-\gamma/\lambda)}$.

A few comments are in order. In (6.17), the contraction factor is not only given by $c_{\alpha\beta}(p)$ but also by how close in $\mathcal{B}_{\alpha\beta}$ the base function f is to a regular function together with the norm of that function in $W^1_1(\mathbb{R}_+, w_{\alpha\beta}(x)dx)$. The fact that the fixed point whose existence we want to prove is smooth plays an important role here. To make a connection with Proposition 2.6, the quantity $\|g\|_{\alpha\beta}$ in (6.17) is the radius of the ball on which the tangent maps $D\mathcal{M}_{\lambda+,\kappa}$ need to be contractions. All the other terms can be made as small as we wish by letting the size of the largest interval in p go to zero, cf. (6.11) and (6.15). Note that $c_{\alpha\beta}(p)$ depends sensitively on α and β, and optimizing this factor requires to consider a partition p with smaller intervals where the weight $w_{\alpha\beta}$ varies strongly. We will encounter later other optimization criteria for p. We shall denote by p_r the partition p which we will eventually choose, cf. Section 7.1.

We end this section with the

Proof of Lemma 6.2. Define the function h_1 by

$$h_1(x) = \int_{x_0}^{x} h(\xi)\, d\xi,$$

for $x \in (x_0, x_n)$, and $h_1(x) = 0$ otherwise. Note that $h_1' = h$ and, by definition of h, $h_1(x_i) = 0$ for $i = 0, \ldots, n$. Hence, integration by parts leads to

$$\|f * h\|_{(4\alpha)\beta} = \|f' * h_1\|_{(4\alpha)\beta} \le \|f'\|_{\alpha\beta}\|h_1\|_{\alpha\beta}.$$

It remains to estimate the norm of h_1 in term of h. For $i = 1, \ldots, n$ and $x \in [x_{i-1}, x_i]$, one has

$$|h_1(x)| = \frac{1}{2}\Big(\Big|\int_{x_{i-1}}^{x} h(\xi)\, d\xi\Big| + \Big|\int_{x}^{x_i} h(\xi)\, d\xi\Big|\Big) \le \frac{1}{2}\int_{I_i} |h(\xi)|\, d\xi,$$

which in turn yields

$$\begin{aligned}
\|h_1\|_{\alpha\beta} &= \sum_{i=1}^{n} \int_{I_i} w_{\alpha\beta}(x)\, |h_1(x)|\, dx, \\
&\le \frac{1}{2}\sum_{i=1}^{n} \int_{I_i} w_{\alpha\beta}(x) \int_{I_i} |h(\xi)|\, d\xi\, dx \\
&\le \frac{1}{2}\max_{i=1,\ldots,n}\Big\{\sup_{x\in I_i}\frac{1}{w_{\alpha\beta}(x)}\int_{I_i} w_{\alpha\beta}(x)\, dx\Big\}\|h\|_{\alpha\beta}.
\end{aligned}$$

∎

Remark. The quantity $c_{\alpha\beta}(p_r)$ is computed in the subroutine `swsupint`. (See the final remark of Section 7.1 for a description of the parameters related to the partition p_r.) The estimates (6.11), (6.12) and (6.17) are implemented in `fDN_center` to compute (6.10), with a call to the subroutine `sdelta1` to get $\delta_\lambda(f,h)$. The quantity $\|\rho'\|_{\alpha\beta}$ entering (6.17) is bounded in `snorm_of_der_pl` by

$$\|\rho'\|_{\alpha\beta} \leq \sum_{j=1}^{m} \sup_{x\in I_j} w_{\alpha\beta}(x)|\rho_j - \rho_{j-1}|, \tag{6.18}$$

where $\pi(\rho) = (\{y_j\},\{\rho_j\})_{j=0}^m$ and $I_j = [y_{j-1}, y_j]$.

6.2. Functions with Support Near the Origin

In this section, we consider $D\mathcal{N}_\lambda(f)$ acting on functions $h \in \mathcal{L}_\alpha^a$ for a small enough. As in the previous section, but for different reasons, the contractivity properties of $D\mathcal{N}_\lambda(f)$ are entirely due to the term $D\mathcal{N}_\lambda^1(f)$. Indeed, since the functions f which will be considered have in general a support given by $\mathbb{R}_+$, the support of $D\mathcal{N}_\lambda^1(f)\,h$ for $h \in \mathcal{L}_\alpha^a$ is also equal to $\mathbb{R}_+$ due to the convolution. Hence, the size of $D\mathcal{N}_\lambda^2(f)h$ is essentially given by the size of $D\mathcal{N}_\lambda^1(f)\,h$, and we proceed as before starting with the bound (6.10) on $\|D\mathcal{N}_\lambda(f)h\|_{\alpha\gamma}$.

The last term in the expression (6.6) for $\delta_\lambda(f,h)$ is again bounded using (6.12). The h-dependent coefficients of the first two terms in (6.6) are given by $M(h)$ and $E(h)$, which are bounded using

$$|M(h)| \leq \frac{1}{w_{\alpha\beta}(a)}\|h\|_{\alpha\beta}, \tag{6.19}$$

$$|E(h)| \leq \frac{a}{w_{\alpha\beta}(a)}\|h\|_{\alpha\beta}, \tag{6.20}$$

provided $a \leq \sqrt{\alpha/\beta}$ for the first inequality, and $a \leq \bigl(1+\sqrt{1+4\alpha\beta}\bigr)/2\beta$ for the second inequality, cf. the discussion of (4.5) and (4.6).

It remains to bound $D\mathcal{N}_\lambda^1(f)h$ in $\mathcal{B}_{\alpha\gamma}$. We consider

$$D\mathcal{N}_\lambda^1 : \mathcal{B}_{\alpha\beta} \times \mathcal{L}_\alpha^a \xrightarrow{\;*\;} \mathcal{B}_{(\eta\alpha)\beta} \xrightarrow{\;2S_\lambda\;} \mathcal{B}_{\alpha\gamma}, \tag{6.21}$$

with $\eta \in [\lambda, 4]$ a parameter to be chosen later. For the convolution, we use the

Lemma 6.3. *Let $f \in \mathcal{B}_{\alpha\beta}$ and $h \in \mathcal{L}_\alpha^a$. Then, for $1 \leq \eta \leq 4$,*

$$\|f * h\|_{(\eta\alpha)\beta} \leq \exp\Bigl(-\frac{\alpha\sqrt{\eta}(2-\sqrt{\eta})}{a}\Bigr)\|f\|_{\alpha\beta}\|h\|_{\alpha\beta}. \tag{6.22}$$

Proof. Exploiting $\mathrm{supp}(h) \subseteq (0, a)$, we proceed as in Proposition 2.2 and get

$$\|f * h\|_{(\eta\alpha)\beta} \leq \sup_{\substack{x>0 \\ a>y>0}} \exp\bigl(-\alpha g(x,y)\bigr)\|f\|_{\alpha\beta}\|h\|_{\alpha\beta},$$

where

$$g(x,y) = \frac{x+y}{xy} - \frac{\eta}{x+y}.$$

Since $g(x,y) \geq 0$ for $\eta \leq 4$, one has $\sup \exp(-\alpha g) = \exp(-\alpha \inf g)$ and, using $\eta \geq 1$, we compute

$$\inf_{\substack{x>0 \\ a>y>0}} g(x,y) = \inf_{a>y>0} \frac{\sqrt{\eta}(2-\sqrt{\eta})}{y} = \frac{\sqrt{\eta}(2-\sqrt{\eta})}{a}.$$

■

We now turn to the scaling operator. Since $\mathrm{supp}(f * h) = \mathbb{R}_+$ for functions f that will be considered, the following general bound is optimal,

$$\|S_\lambda g\|_{\alpha\gamma} \leq \exp\bigl(-2\sqrt{\alpha(1-\lambda/\eta)(\beta-\gamma/\lambda)}\bigr)\|g\|_{(\eta\alpha)\beta}, \tag{6.23}$$

which is valid provided $\gamma/\beta \leq \lambda \leq \eta$. From (6.23) and (6.22), we get a bound on $\|D\mathcal{N}_\lambda^1(f)h\|_{\alpha\gamma}$. We now optimize the parameter η. Since ultimately we will get the needed contraction factor by choosing a small enough, and since (6.23) does not depend on a, we consider (6.22) only. For $\lambda \geq 1$, the maximum of $\sqrt{\eta}(2-\sqrt{\eta})$ on $[\lambda, 4]$ is taken at $\eta = \lambda$, and one gets finally

$$\|D\mathcal{N}_\lambda^1(f)h\|_{\alpha\gamma} \leq 2\exp\Bigl(-\frac{\alpha\sqrt{\lambda}(2-\sqrt{\lambda})}{a}\Bigr)\|f\|_{\alpha\beta}\|h\|_{\alpha\beta}. \tag{6.24}$$

Recall that (6.24) is valid provided $\gamma/\beta \leq \lambda \leq 4$. Furthermore, it leads for a small enough to a strict contraction only if $\lambda < 4$: this is the first compactness condition.

Before ending this section, let us comment on the optimization of the contraction factor. Instead of (6.21), one can consider $D\mathcal{N}_\lambda^1(f)h = 2(S_\lambda f * S_\lambda h)$ with $S_\lambda : \mathcal{B}_{\alpha\beta} \to \mathcal{B}_{(\alpha/\eta)\gamma}$ and $\eta \in [\lambda, 4]$ a parameter to be optimized. Since $S_\lambda h \in \mathcal{L}^{a/\lambda}_{(\alpha/\eta)\gamma}$ is of order $\mathcal{O}(e^{-1/a})$ if $\eta > \lambda$, one gets a second a-dependent contraction factor from the convolution. However, optimizing η leads to the same bound as (6.24), and we use (6.21) for convenience of implementation.

Remark. The bounds (6.19), (6.20) and (6.24) are implemented in the procedure `fDN_left` to compute (6.10). The conditions on a under which (6.19) and (6.20) are valid are first checked, namely $a \leq \sqrt{\alpha/\beta}$ and $a \leq \bigl(1+\sqrt{1+4\alpha\beta}\bigr)/2\beta$. An explicit check of $\gamma/\beta \leq \lambda \leq 4$ is also necessary. Up to now, this inequality was implicitly verified when bounds were computed, as in (6.17) for instance.

6.3. Functions with Support Near Infinity

We now consider functions $h \in \mathcal{R}_\beta^b$ with b large enough. Here, the situation differs from the previous cases in the sense that the term $D\mathcal{N}_\lambda^2(f)h$ is small independently of the size of $D\mathcal{N}_\lambda^1(f)h$. Indeed, the property of h to have support away from the origin is preserved by $D\mathcal{N}_\lambda^1(f)$. After applying the transformation T, one obtains a function whose support is near the origin, and the result from the previous section related to the convolution yields a second exponentially small factor. Hence, we simply start with the triangle inequality to get from (6.3)

$$\|D\mathcal{N}_\lambda(f)h\|_{\alpha\gamma} \le |c_\lambda(f)| \|D\mathcal{N}_\lambda^1(f)h\|_{\alpha\gamma} + c_2\|D\mathcal{N}_\lambda^2(f)h\|_{\alpha\gamma} + |\delta_\lambda(f,h)|\,\|\mathcal{N}_\lambda^1(f)\|_{\alpha\gamma}. \tag{6.25}$$

Let us begin with the first term. The main contraction factor is here entirely due to the scaling operator acting on $\mathcal{R}_\beta^b$. Furthermore, for $f \in \mathcal{B}_{\alpha\beta}$, the map $h \mapsto f * h$ preserves $\mathcal{R}_\beta^b$. Hence, one has the choice of the order in which the scaling and the convolution are composed. By letting the scaling act first, one gains a (b-independent) contraction factor when applying this operator to the function f. Recall that $S_\lambda : \mathcal{B}_{\alpha\beta} \to \mathcal{B}_{(\alpha/4)\gamma}$ is a strict contraction for $\gamma/\beta < \lambda < 4$. One can improve this factor by considering $\mathcal{B}_{0\gamma}$ for the target space of S_λ. Hence, we consider finally

$$D\mathcal{N}_\lambda^1/2 : \mathcal{B}_{\alpha\beta} \times \mathcal{R}_\beta^b \xrightarrow{S_\lambda} \mathcal{B}_{0\gamma} \times \mathcal{R}_\gamma^{b/\lambda} \xrightarrow{*} \mathcal{B}_{\alpha\gamma}. \tag{6.26}$$

Provided $\gamma/\beta \le \lambda$, the scaling operator in (6.26) is bounded, and, since $S_\lambda h$ has again support away from the origin, the convolution above is well defined even for $\alpha > 0$. For $f \in \mathcal{B}_{\alpha\beta}$, one estimates as usual

$$\begin{aligned}\|S_\lambda f\|_{0\gamma} &\le \sup_{x>0} \frac{w_{0\gamma}(x/\lambda)}{w_{\alpha\beta}(x)} \|f\|_{\alpha\beta} \\ &= \exp\Bigl(-2\sqrt{\alpha(\beta-\gamma/\lambda)}\Bigr)\|f\|_{\alpha\beta},\end{aligned} \tag{6.27}$$

and for $h \in \mathcal{R}_\beta^b$, one uses the knowledge about the support of h to get

$$\begin{aligned}\|S_\lambda h\|_{0\gamma} &\le \sup_{x>b} \frac{w_{0\beta}(x/\lambda)}{w_{\alpha\gamma}(x)} \|h\|_{\alpha\beta} \\ &= \exp(-\alpha/b - b(\beta-\gamma/\lambda))\|h\|_{\alpha\beta},\end{aligned} \tag{6.28}$$

the last equality being valid if $b \ge \sqrt{\alpha/(\beta-\gamma/\lambda)}$. Next, we consider the convolution in (6.26). For $f \in \mathcal{B}_{0\gamma}$ and $h \in \mathcal{R}_\gamma^{b/\lambda}$, we proceed as in Proposition 2.2 and get

$$\begin{aligned}\|f * h\|_{\alpha\gamma} &\le \sup_{\substack{x>0\\ y>b/\lambda}} \Bigl(\frac{w_{\alpha\gamma}(x+y)}{w_{0\gamma}(x)w_{0\gamma}(y)}\Bigr)\|f\|_{0\gamma}\|h\|_{0\gamma} \\ &= \sup_{\substack{x>0\\ y>b/\lambda}} \exp\Bigl(\frac{\alpha}{x+y}\Bigr)\|f\|_{0\gamma}\|h\|_{0\gamma} \\ &= \exp(\alpha\lambda/b)\|f\|_{0\gamma}\|h\|_{0\gamma}.\end{aligned} \tag{6.29}$$

Finally, (6.27), (6.28) and (6.29) lead to

$$\|D\mathcal{N}_\lambda^1(f)h\|_{\alpha\gamma} \le 2e^{-A}\exp\big(-b(\beta-\gamma/\lambda)\big)\|h\|_{\alpha\beta}\|f\|_{\alpha\beta}, \tag{6.30}$$

where $A = 2\sqrt{\alpha(\beta-\gamma/\lambda)} - \alpha(\lambda-1)/b$. Although the convolution deteriorates the b-independent factor given by the scaling, (6.26) is still a good choice due to the large values of b that will be considered. Proceeding in this way is not crucial, but allows to take smaller values for b, thereby saving about 10 percent of the computation time devoted to the evaluation of $D\mathcal{N}_\lambda(f)$ on $\mathcal{V}^p$, the space of piecewise constant functions. We conclude by observing that (6.30) yields a bound which is exponentially small in b only if $\gamma/\beta < \lambda$: this is the second compactness condition.

Next, we consider the second term in (6.25). One has

$$\|D\mathcal{N}_\lambda^2(f)h\|_{\alpha\gamma} \le 2\|T\mathcal{N}_\lambda^1(f) * TD\mathcal{N}_\lambda^1(f)h\|_{\gamma\alpha}.$$

From $D\mathcal{N}_\lambda^1(f)h \in \mathcal{R}_\gamma^{b/\lambda}$ it follows that $TD\mathcal{N}_\lambda^1(f)h \in \mathcal{L}_\gamma^{\lambda/b}$, and applying Lemma 6.3 with $\eta = 1$ leads to

$$\|D\mathcal{N}_\lambda^2(f)h\|_{\alpha\gamma} \le 2\exp\Big(-\frac{\gamma b}{\lambda}\Big)\|\mathcal{N}_\lambda^1(f)\|_{\alpha\gamma}\|D\mathcal{N}_\lambda^1(f)h\|_{\alpha\gamma}. \tag{6.31}$$

It remains to estimate $\delta_\lambda(f,h)$. The expectation of $D\mathcal{N}_\lambda^2(f)h$ is simply bounded by

$$|E(D\mathcal{N}_\lambda^2(f)h)| \le \sup_{x>0}\Big(\frac{x}{w_{\alpha\gamma}(x)}\Big)\|D\mathcal{N}_\lambda^2(f)h\|_{\alpha\gamma}. \tag{6.32}$$

Note that in the previous cases, we used the properties of the convolution near the origin to bound this quantity according to (6.12). Here, these properties have been used already in the bound (6.31) to extract a second exponentially small factor in b. Therefore, inserting (6.31) into (6.32) leads to a better estimate than (6.12). Finally, for $h \in \mathcal{R}_\beta^b$ and b large, one has the following bounds on $M(h)$ and $E(h)$

$$|M(h)| \le \frac{1}{w_{\alpha\beta}(b)}\|h\|_{\alpha\beta}, \qquad |E(h)| \le \frac{b}{w_{\alpha\beta}(b)}\|h\|_{\alpha\beta}, \tag{6.33}$$

provided $b \ge \sqrt{\alpha/\beta}$ for the first inequality, and $b \ge (1+\sqrt{1+4\alpha\beta})/2\beta$ for the second inequality.

Remark. The bounds (6.30),(6.31), (6.32) and (6.33) are implemented in `fDN_right` to estimate (6.25). The validity conditions of (6.28) and (6.33), namely $b \ge \sqrt{\alpha/(\beta-\gamma/\lambda)}$ ($\ge \sqrt{\alpha/\beta}$) and $b \ge (1+\sqrt{1+4\alpha\beta})/2\beta$, are explicitly checked.

6.4. Piecewise Constant Functions

Finally, we consider the case of functions h in $\mathcal{V}^p$. On this space, the tangent map $D\mathcal{N}_\lambda(f)$ is not a contraction and the relevant information is contained in the images $D\mathcal{N}_\lambda(f)h$ of the basis vectors h of $\mathcal{V}^p$. Therefore, in order to keep track of this information, we need to construct a bound on the tangent map in the sense of Section 3. For $p = \{x_0, \ldots, x_n\}$ and $I_i = (x_i, x_{i-1})$, a basis of $\mathcal{V}^p$ is given by $\{\chi_{I_i}\}_{i=1}^n$. Hence, we introduce the following set $\mathcal{X}$ of characteristic functions,

$$\mathcal{X} = \{c\chi_{[a,a+\delta]} | \, c \in \mathbb{R}, a > 0, \delta > 0\},$$

and we construct a bound on $D\mathcal{N}_\lambda : \mathcal{B}_{\alpha\beta} \times \mathcal{X} \to \mathcal{B}_{\alpha\gamma}$ acting from $\mathrm{std}(\mathcal{B}_{\alpha\beta})^u \times \mathrm{std}(\mathcal{X})$ to $\mathrm{std}(\mathcal{B}_{\alpha\gamma})$, where we define $\mathrm{std}(\mathcal{X})$ to be the collection of all sets of the form

$$(A, B, C) = \{h \in \mathcal{X} \mid h = c\chi_{[a,a+\delta]} \text{ with } a \in A, \delta \in B, c \in C\} \tag{6.34}$$

for $C \in \mathrm{std}(\mathbb{R})$ and $A, B \in \mathrm{std}(\mathbb{R}_+^*)$.

Note that, once a bound on $D\mathcal{N}_\lambda^1 : \mathcal{B}_{\alpha\beta} \times \mathcal{X} \to \mathcal{B}_{\alpha\gamma}$ has been obtained, composing it with the bounds of Section 4 readily yields bounds on the first two terms of $D\mathcal{N}_\lambda(f)h$, cf. (6.3) and (6.5). To compute the coefficient $\delta_\lambda(f, h)$ in the third term of (6.3), the only missing quantities are the mass and the expectation of $h \in \mathcal{X}$. Those are obtained from the equalities

$$M(\chi_{[a,a+\delta]}) = \delta \, , \qquad E(\chi_{[a,a+\delta]}) = \delta(a + \delta/2). \tag{6.35}$$

It remains to construct a bound on $D\mathcal{N}_\lambda^1$. We consider

$$D\mathcal{N}_\lambda^1/2 : \mathcal{B}_{\alpha\beta} \times \mathcal{X} \xrightarrow{S_\lambda} \mathcal{B}_{\frac{\alpha}{4}\gamma} \times \mathcal{X} \xrightarrow{*} \mathcal{B}_{\alpha\gamma}. \tag{6.36}$$

The reason for this choice is as follows. Some of the functions h will have support close to the origin or far away from the origin. In such cases, we know from the previous sections that the scaling in (6.36) is a very good contraction. Hence, considering (6.36) will automatically yield an extra contraction factor and improve the bound on the convolution between $S_\lambda h$ and the general term of $S_\lambda f$.

A bound on $S_\lambda : \mathcal{X} \to \mathcal{X}$ is easily obtained from

$$S_\lambda \chi_{[a,a+\delta]} = \lambda \chi_{[a/\lambda,(a+\delta)/\lambda]}.$$

Next, we construct a bound on the convolution defined from $\mathrm{std}(\mathcal{B}_{\zeta\eta})^u \times \mathrm{std}(\mathcal{X})$ to $\mathrm{std}(\mathcal{B}_{\gamma\eta})^u$, with $\gamma \in [\zeta, 4\zeta]$. Let $f = \rho + g$, $\rho \in \mathcal{A}^u$ and $g \in \mathcal{B}_{\zeta\eta}$. One has $f*h = \rho*h + g*h$, and the second term of this equality will be a part of the general term $\tilde{g}$ of $f * h$ and will be treated as usual. The first term contains the relevant information and needs to be computed explicitly. In the sequel, we consider for simplicity $h = \chi_{[a,a+\delta]}$. Let

$\pi(\rho) = (\{x_i\}, \{\rho_i\})_{i=0}^n$ and denote by ε the mesh of the uniform partition associated with ρ. If $\varepsilon \geq \delta$, the function $\rho * h$ takes a simpler form than in the case $\varepsilon < \delta$, and we restrict the domain of our bound to such cases in order to simplify the implementation. Define

$$y_k = a + x_0 + k\varepsilon, \qquad k = 0, \ldots, n+1, \tag{6.37}$$

and $I_k = [y_k, y_{k+1}]$, $k = 0, \ldots, n$. It is clear from the properties of the convolution that $\rho * h$ is continuous and has a support equal to $(y_0, y_n + \delta)$. Next, a short computation shows that provided $\varepsilon \geq \delta$, $\rho * h$ is given on the interval I_k by

$$(\rho * h)(y_k + \theta) = \begin{cases} \delta(\rho_k - \delta\rho'_{k-1}/2) + \theta\delta\rho'_{k-1} + \theta^2(\rho'_k - \rho'_{k-1})/2\,, & 0 \leq \theta \leq \delta, \\ \delta(\rho_k - \delta\rho'_k/2) + \theta\delta\rho'_k\,, & \delta \leq \theta \leq \varepsilon, \end{cases} \tag{6.38}$$

with the convention that $\rho_{-1} = \rho_{n+1} = 0$, and where

$$\rho'_k = \frac{\rho_{k+1} - \rho_k}{\varepsilon}.$$

Indeed, one has

$$(\rho * h)(y_k + \theta) = \int_a^{a+\delta} \rho(y_k + \theta - x)\, dx = \int_0^\delta \rho(x_k + \theta - \xi)\, d\xi. \tag{6.39}$$

Two cases arise: if $\theta \geq \delta$, the function ρ in the above integral is given by

$$\rho(x) = \rho_k + (x - x_k)\rho'_k. \tag{6.40}$$

Inserting (6.40) into (6.39) and integrating lead to the second part of (6.38). For $\theta < \delta$, we rewrite (6.39) as

$$(\rho * h)(y_k + \theta) = \int_0^\theta \rho(x_k + \theta - \xi)\, d\xi + \int_\theta^\delta \rho(x_k + \theta - \xi)\, d\xi. \tag{6.41}$$

In the first term, ρ is again given by (6.40), whereas in the second term one has

$$\rho(x) = \rho_k + (x - x_k)\rho'_{k-1}. \tag{6.42}$$

Inserting (6.40) and (6.42) into (6.41) yields the first part of (6.38). Next, we define the affine part $\tilde{\rho}$ of $f * h$ to be the linear interpolation of $\rho * h$ at the nodes $\{y_k\}$. More precisely, we consider

$$\tilde{\rho} = \mathcal{T}_1(\{y_i\}_{i=0}^{n+1}, \{\tilde{\rho}_i\}_{i=0}^{n+1}), \tag{6.43}$$

where

$$\tilde{\rho}_k = \delta(\rho_k - \delta\rho'_{k-1}/2), \qquad k = 0, \ldots, n+1. \tag{6.44}$$

Note that $\tilde{\rho} \in \mathcal{A}^u$. Finally, the general term of $f * h$ is given by $\tilde{g} = \rho * h - \tilde{\rho} + g * h$ and one gets

$$\|\tilde{g}\|_{\gamma\eta} \leq \|\rho * h - \tilde{\rho}\|_{\gamma\eta} + \|g\|_{\zeta\eta}\|h\|_{\zeta\eta}. \tag{6.45}$$

For $h = \chi_{[a,a+\delta]}$, one simply uses that

$$\|h\|_{\zeta\eta} \leq \delta \sup_{x\in[a,a+\delta]} w_{\zeta\eta}(x).$$

To bound the first term on the RHS of (6.45), we first note that on the interval I_k, $k = 0, \ldots, n$,

$$\begin{aligned}\tilde{\rho}(y) &= \tilde{\rho}_k + (y - y_k)\frac{\tilde{\rho}_{k+1} - \tilde{\rho}_k}{\varepsilon} \\ &= \delta(\rho_k - \delta\rho'_{k-1}/2) + \delta(y - y_k)(\rho'_k - \delta(\rho'_k - \rho'_{k-1})/2\varepsilon).\end{aligned}$$

From this formula and the expression (6.38) for $\rho * h$, one computes for $\theta \in [0, \delta]$,

$$|(\rho * h - \tilde{\rho})(y_k + \theta)| = \theta\Big(\delta - \frac{\theta}{2} - \frac{\delta^2}{2\varepsilon}\Big)|\rho'_k - \rho'_{k-1}|, \tag{6.46}$$

and for $\theta \in [\delta, \varepsilon]$,

$$|(\rho * h - \tilde{\rho})(y_k + \theta)| = \frac{\delta^2}{2}\Big(1 - \frac{\theta}{\varepsilon}\Big)|\rho'_k - \rho'_{k-1}|. \tag{6.47}$$

Therefore, integrating (6.46) and (6.47) leads to

$$\begin{aligned}\|\rho * h - \tilde{\rho}\|_{\gamma\eta} &\leq \sum_{k=0}^{n} \sup_{x\in I_k} w_{\gamma\eta}(x) \int_{I_k} |(\rho * \chi - \tilde{\rho})(y)|\, dy \\ &= \delta^2\Big(\frac{1}{4} - \frac{\delta}{6\varepsilon}\Big)\sum_{k=0}^{n} \sup_{x\in I_k} w_{\zeta\eta}(x)|\rho_{k+1} - 2\rho_k + \rho_{k-1}|\,,\end{aligned} \tag{6.48}$$

with the convention $\rho_{-1} = \rho_{n+1} = 0$.

Remark. A set $(A, B, C) \in \mathrm{std}(\mathcal{X})$ is represented on the computer by a vector, say `fb`, with `fb(1)`$=A$, `fb(2)`$=B$, and `fb(3)`$=C$. The scaling $S_\lambda : \mathcal{X} \to \mathcal{X}$ is implemented in `fscale_chi`. A bound on the convolution in (6.36) is implemented in the procedure `fconv_chi` from (6.43), (6.45) and (6.48), where we first check the condition $\varepsilon \geq \delta$. We note that for the purpose of the proof of Proposition 2.6, the base function f is always represented by the same standard set in $\mathrm{std}(\mathcal{B}_{\alpha\beta})^u$. Hence, the only quantity in (6.44) that may change from basis vector to basis vector is δ. By choice of the partition p_r, see Section 7.1, most of the basis vectors have equal δ, and the computation of the $\tilde{\rho}_k$'s is carried out only once for such basis vectors. Finally, the bounds on the scaling and on the convolution in (6.36) and the bounds from Section 4 are composed in the subroutine `fDN_chi` to implement a bound on $D\mathcal{N}_\lambda(f) : \mathcal{X} \to \mathcal{B}_{\alpha\gamma}$, $f \in \mathcal{B}_{\alpha\beta}$.

7. The Tangent Maps $D\mathcal{M}_{\lambda,\kappa}$

In this section we explain how a uniform upper bound on the contraction rate of the operators $\mathcal{M}_{\lambda^+,\kappa}$ in a neighborhood of the fixed point f^* is obtained for all $\kappa \in [\lambda^-/\lambda^+, 1]$. This will complete the proof of Proposition 2.6. We recall that the operators $\mathcal{M}_{\lambda^+,\kappa}$ are given in terms of the original maps $\mathcal{N}_{\lambda^+,\kappa} = S_\kappa \mathcal{N}_{\lambda^+}$ by

$$\mathcal{M}_{\lambda^+,\kappa} = 1 + M(\mathcal{N}_{\lambda^+,\kappa} - 1), \tag{7.1}$$

where M is some fixed invertible linear map close to the inverse of $1 - D\mathcal{N}_{\lambda^*}(f_{\lambda^*})$. Since $\mathcal{N}_{\lambda^+,\kappa}$ is already a good contraction on the subspaces $\mathcal{B}^p_{\alpha\beta}$ for certain partitions p, we need M to be different from the identity only on the finite dimensional subspace $\mathcal{V}^p$, cf. (6.8). In Section 7.1, we introduce some notation and express the norm of a linear map in $\mathcal{B}_{\alpha\beta}$ in terms of its norms when restricted to $\mathcal{B}^p_{\alpha\beta}$ and $\mathcal{V}^p$. The description of M is given in Section 7.2. The last section is devoted to the final estimate needed to prove Proposition 2.6.

7.1. Decomposition of the Operator Norm

Let $p = \{x_0, \ldots, x_n\}$ be a partition in $\mathcal{P}_n$. In order to express the projector on $\mathcal{V}^p$, we introduce two maps associated with p: the finite rank operator $\mathcal{I}_p : \mathcal{B}_{\alpha\beta} \to \mathbb{R}^n$ defined by

$$\mathcal{I}_p f = \left\{ \frac{1}{|I_i|} \int_{I_i} f(x)dx \right\}_{i=1,\ldots,n}, \tag{7.2}$$

and $\mathcal{J}_p : \mathbb{R}^n \to \mathcal{V}^p$ defined by

$$\mathcal{J}_p \{f_i\}_{i=1}^n = \sum_{i=1}^n f_i \chi_{I_i}, \tag{7.3}$$

where $I_i = (x_{i-1}, x_i]$ and $|I|$ is the Lebesgue measure of $I \subset \mathbb{R}$. With this notation, the projector $\mathcal{Q}_p$ on $\mathcal{V}^p$ may be written as

$$\mathcal{Q}_p = \mathcal{J}_p \mathcal{I}_p. \tag{7.4}$$

Let A be a bounded linear map in $\mathcal{B}_{\alpha\beta}$. One has

$$\|Af\|_{\alpha\beta} \le \|f\|_p \max\{\|A|_{\mathcal{V}^p}\|, \|A|_{\mathcal{B}^p_{\alpha\beta}}\|\}, \tag{7.5}$$

where $\|\cdot\|_p$ is the norm in $\mathcal{B}_{\alpha\beta}$ given by

$$\|f\|_p = \|\mathcal{Q}_p f\|_{\alpha\beta} + \|(1 - \mathcal{Q}_p) f\|_{\alpha\beta}. \tag{7.6}$$

The norms $\|\cdot\|_p$ and $\|\cdot\|_{\alpha\beta}$ are equivalent, with

$$\|f\|_{\alpha\beta} \le \|f\|_p \le K_p^{\alpha\beta}\|f\|_{\alpha\beta} \tag{7.7}$$

for some constant $K_p^{\alpha\beta}$. From the definition of $\mathcal{B}_{\alpha\beta}^p$ and its subspaces $\mathcal{L}_\alpha^a$, $\mathcal{C}_p$, $\mathcal{R}_\beta^b$, it follows that

$$\|A|_{\mathcal{B}_{\alpha\beta}^p}\| = \max\{\|A|_{\mathcal{L}_\alpha^{x_0}}\|, \|A|_{\mathcal{C}_p}\|, \|A|_{\mathcal{R}_\beta^{x_n}}\|\}. \tag{7.8}$$

Furthermore, one has

$$\|A|_{\mathcal{V}^p}\| = \max_{i=1,\dots,n} \|A\eta_i\|_{\alpha\beta}, \tag{7.9}$$

with η_i the characteristic function of I_i normalized in $\mathcal{B}_{\alpha\beta}$, i.e.,

$$\eta_i = \left(\int_{I_i} w_{\alpha\beta}(x)\,dx\right)^{-1} \chi_{I_i}. \tag{7.10}$$

Inserting (7.7), (7.8) and (7.9) into (7.5), one gets

$$\|A\| \le K_p^{\alpha\beta} \max\{\{\|A\eta_i\|_{\alpha\beta}\}_{i=1}^n, \|A|_{\mathcal{L}_\alpha^{x_0}}\|, \|A|_{\mathcal{C}_p}\|, \|A|_{\mathcal{R}_\beta^{x_n}}\|\}. \tag{7.11}$$

For $A = D\mathcal{M}_{\lambda^+,\kappa}(f)$, evaluating the quantities in the RHS of this expression will yield the desired bound on the norm of the tangent map of $\mathcal{M}_{\lambda^+,\kappa}$. The bounds obtained in the previous section will allow us to estimate each of the last three quantities in one step, by evaluating in turn $\|D\mathcal{M}_{\lambda^+,\kappa}(f)h\|_{\alpha\beta}$ for all h in the unit balls of $\mathcal{L}_\alpha^{x_0}$, $\mathcal{C}_p$ and $\mathcal{R}_\beta^{x_n}$. In contrast, the contractivity of $\mathcal{M}_{\lambda^+,\kappa}$ on $\mathcal{V}^p$ follows from the specific choice of the operator M, and an explicit computation of the n quantities $\|A\eta_i\|_{\alpha\beta}$ is required. This accounts for most of the computation time of the proof.

This leads us to the problem of optimizing the partition p in (7.11) with respect to $A = D\mathcal{M}_{\lambda^+,\kappa}(f)$. Roughly speaking, the size of the intervals in $p = (x_0, \dots, x_n)$ is determined by the contraction rate of A on $\mathcal{C}_p$ that we need to obtain. Hence, the number of intervals n is fixed by x_0 and x_n. In order to minimize n, we want to maximize x_0 and minimize x_n. These two parameters determine the contraction rate of A on $\mathcal{L}_\alpha^{x_0}$ and $\mathcal{R}_\beta^{x_n}$. Increasing α and β improves the contraction and allows to consider larger x_0 and smaller x_n. However, large values of α and β deteriorate the estimate (2.19) of Proposition 2.6, i.e., the precision of the approximate fixed point. Good values for α and β have been found empirically to be $\alpha = 0.5$ and $\beta = 0.9$, for which $x_0 = 0.065$, $x_n = 11.83$ and a (non–uniform) partition of 5050 intervals give the desired bound (2.20). In the sequel, we will refer to this partition as p_r and denote $n_r = 5050$.

We end this section with the computation of the equivalence constant $K_p^{\alpha\beta}$. First, we estimate $\|\mathcal{Q}_p f\|_{\alpha\beta}$. From

$$|(\mathcal{I}_p f)_i| = \left|\frac{1}{|I_i|}\int_{I_i} f(x)\,dx\right| \le \frac{1}{|I_i|}\sup_{x\in I_i}\left(\frac{1}{w_{\alpha\beta}(x)}\right)\int_{I_i} w_{\alpha\beta}(x)\,|f(x)|\,dx, \tag{7.12}$$

it follows

$$\|\mathcal{Q}_p f\|_{\alpha\beta} = \sum_{i=1}^{n} |(\mathcal{I}_p f)_i| \int_{I_i} w_{\alpha\beta}(x)\, dx$$
$$\leq \max_{i=1,\dots,n} \Big(\frac{1}{|I_i|} \sup_{x\in I_i} \frac{1}{w_{\alpha\beta}(x)} \int_{I_i} w_{\alpha\beta}(x)\, dx \Big) \|f\|_{\alpha\beta}.$$

Hence, the following inequality

$$\|f\|_p = \|\mathcal{Q}_p f\|_{\alpha\beta} + \|(1-\mathcal{Q}_p)f\|_{\alpha\beta} \leq \|f\|_{\alpha\beta} + 2\|\mathcal{Q}_p f\|_{\alpha\beta},$$

implies

$$K_p^{\alpha\beta} \leq 1 + 2 \max_{i=1,\dots,n} \Big(\frac{1}{|I_i|} \sup_{x\in I_i} \frac{1}{w_{\alpha\beta}(x)} \int_{I_i} w_{\alpha\beta}(x)\, dx \Big). \tag{7.13}$$

Note that the previous upper bound tends to 3 from above when n increases and when the size of each interval goes to zero. Also, the weight contributes to this bound by its largest variation on the intervals $\{I_i\}$. We have already encountered a similar situation, cf. (6.15), and we chose to consider a non–uniform partition with a higher density of nodes where the weight varies strongly. For the partition p_r introduced above, one has $K_{p_r}^{\mu\nu} < 3.15$.

Remark. An upper bound on the equivalence constant $K_{p_r}^{\mu\nu}$ is computed in the procedure `compute_equiv_const`, using `swsupint` to estimate the second term in the RHS of (7.13). The first and last points in the partition p_r are $x_0 = 0.065$ and $x_{n_r} = 11.83$, respectively. The first 100 (`npr1`) intervals are uniform with mesh $\varepsilon_{r_1} = (x_{n_r} - x_0)10^{-4}$ (`sepspr1`), whereas the remaining 4950 (`npr2`) intervals are uniform with mesh $\varepsilon_{r_2} = 2\,\varepsilon_{r_1}$ (`sepspr2`).

7.2. The Operator M

As mentioned earlier, M should be a good approximation to the inverse of $1 - D\mathcal{N}_{\lambda^*}(f_{\lambda^*})$, and needs to be different from the identity on the finite dimensional space $\mathcal{V}^{p_r}$ only. Hence, for a certain partition $p \in \mathcal{P}_m$ to be chosen later, we write

$$M = \big(1 - \mathcal{Q}_p D\mathcal{N}_{\lambda^+}(f_{\lambda^+}^0)\mathcal{Q}_p\big)^{-1}, \tag{7.14}$$

where $f_{\lambda^+}^0$ is the explicit approximate fixed point of $\mathcal{N}_{\lambda^+}$ entering the statement of Proposition 2.6. The previous expression involves the $m \times m$ matrix

$$A = \mathcal{I}_p D\mathcal{N}_{\lambda^+}(f_{\lambda^+}^0)\mathcal{J}_p, \tag{7.15}$$

and can be rewritten as

$$M = (1 - \mathcal{J}_p A \mathcal{I}_p)^{-1} = 1 + \mathcal{J}_p A (1-A)^{-1} \mathcal{I}_p. \tag{7.16}$$

Since we look only for an approximation, the operations involved in the computation of the matrices A and $A(1-A)^{-1}$ need not to be exact. Hence, the use of interval analysis is not required here and we will rely on numerics only. The result of this operation will be denoted by B, i.e.,

$$B \approx A(1-A)^{-1}. \tag{7.17}$$

With the notation

$$C_p = \mathcal{J}_p C \mathcal{I}_p,$$

for C an $m \times m$ real matrix, M is finally defined by

$$M = 1 + B_p. \tag{7.18}$$

We note that the numerical invertibility of $1 - A$ does not imply the invertibility of M. Since this property is required in order for the fixed points of $\mathcal{N}_{\lambda+,\kappa}$ and of $1 + M(\mathcal{N}_{\lambda+,\kappa} - 1)$ to be in correspondence, we must check that M is indeed invertible. We exhibit a matrix C for which $(1+B)C$ is invertible. This implies that the matrix $1 + B$ is invertible, which in turn ensures the invertibility of M. For C, we consider the matrix $1 - A$ that has been previously numerically determined. Then, we check rigorously with interval analysis that the matrix X given by

$$X = (1+B)C - 1, \tag{7.19}$$

satisfies

$$\|X\| < 1, \tag{7.20}$$

for some norm on $\mathbb{R}^m$. From this inequality, it then follows that $1 + X$ is invertible. The norm on $\mathbb{R}^m$ we use in the program is $\|x\| = \max_{i=1,\ldots,m} |x_i|$, that is, for C a real matrix with coefficients $\{c_{ij}\}$,

$$\|C\| = \max_{i=1,\ldots,m} \sum_{j=1}^{n} |c_{ij}|. \tag{7.21}$$

We now discuss the choice of the partition p used in the definition of M. This partition will be denoted by p_s. Since the decomposition $\mathcal{B}^{p_r}_{\alpha\beta} \oplus \mathcal{V}^{p_r}$ has been introduced in order to isolate the subspace $\mathcal{B}^{p_r}_{\alpha\beta}$ on which $\mathcal{N}_{\lambda+,\kappa}$ is a contraction and since the non trivial action of M should turn $\mathcal{M}_{\lambda+,\kappa}$ into a contraction on $\mathcal{V}^{p_r}$, it is natural to require

$$\mathcal{B}^{p_r}_{\alpha\beta} \subseteq \mathrm{Ker}(B_{p_s}).$$

This is in particular true if p_s is a subpartition of p_r, i.e.,

$$p_s \subseteq p_r. \tag{7.22}$$

There is no need for p_s to be equal to p_r. In particular, p_s could have fewer nodes than p_r, which would improve performance with respect to memory and computation time.

By trial and error, we have determined a small partition which satisfies (7.22) and leads to a contraction on $\mathcal{V}^{p_r}$. This (uniform) partition contains $m_s = 500$ intervals. Hence, B is a 500×500 matrix with entries in $\mathcal{S}$, the set of (safe) representable real numbers.

For technical reason, the matrix A is not computed according to (7.15) with $p = p_s$. This would amount to computing the matrix elements $a_{ij} = (\mathcal{I}_{p_s} D\mathcal{N}_{\lambda^+}(f^0_{\lambda^+})\mathcal{J}_{p_s}\hat{x}_j)_i$, where $\{\hat{x}_j\}$ is the canonical basis of $\mathbb{R}^{m_s}$. To avoid the writing of special procedures, we want to use our bound on $D\mathcal{N}_{\lambda^+}$ acting on $\mathcal{X}$ even though interval analysis is not required. However, the intervals in p_s are too large for the $\mathcal{J}_{p_s}\hat{x}_j$ to be in the domain of this bound. (Recall the restriction on the domain of the convolution between a characteristic and a piecewise linear function in Section 6.4.) Hence, we first divide each interval in p_s into d subintervals. This leads to a partition $p_t \in \mathcal{P}_{dm_s}$ whose intervals are now small enough for $d = 10$. With $\{\hat{y}_k\}$ denoting the canonical basis of $\mathbb{R}^{dm_s}$, one has $\mathcal{J}_{p_s}\hat{x}_j = \sum_{l=1}^{d} \mathcal{J}_{p_t}\hat{y}_{d(j-1)+l}$. Next, in order to save some computation time, we exploit the continuity of $D\mathcal{N}_{\lambda^+}(f^0_{\lambda^+})$ to compute an approximated matrix A given by

$$a_{ij} = (\mathcal{I}_{p_s} D\mathcal{N}_{\lambda^+}(f^0_{\lambda^+})\tilde{\mathcal{J}}_{p_t}\hat{y}_{k(j)})_i, \tag{7.23}$$

where $\tilde{\mathcal{J}}_{p_t} = d\mathcal{J}_{p_t}$ and $k(j) = d(j - 1/2)$.

We recall that in (7.23), the function $D\mathcal{N}_{\lambda^+}(f^0_{\lambda^+})\tilde{\mathcal{J}}_{p_t}\hat{y}_{k(j)}$ is given by our bound as a sum $\rho + g$, with $\rho \in \mathcal{A}$ and g a general term. For the purpose of computing A, g is discarded and it remains to discuss the map $\mathcal{I}_{p_s} : \mathcal{A} \to \mathbb{R}^{m_s}$. We will need later to evaluate this map rigorously and we now describe how to bound it. Let $p \in \mathcal{P}_m$ and $\pi(\rho) = (p_\rho, \cdot)$. Define $\tilde{p} \equiv p \cup p_\rho = \{y_j\}_{j=0}^{N}$ and $\tilde{\rho}_j = \rho(y_j)$. Then, writing $\tilde{I}_j = (y_{j-1}, y_j)$, one has for $i = 1, \ldots, m$,

$$\left(\mathcal{I}_p\rho\right)_i = \frac{1}{|I_i|}\sum_{\substack{j \\ \tilde{I}_j \subseteq I_i}} \int_{\tilde{I}_j} \tilde{\rho}(x)\, dx = \frac{1}{|I_i|}\sum_{\substack{j \\ \tilde{I}_j \subseteq I_i}} |\tilde{I}_j| \frac{\tilde{\rho}_j + \tilde{\rho}_{j-1}}{2}. \tag{7.24}$$

We restrict the domain of this bound to those ρ's for which the support of the partition p contains the support of p_ρ. By proceeding so, we ensure that no information is lost when projecting on $\mathcal{V}^p$.

We end this section by deriving an expression for the operator norm of M in $\mathcal{B}_{\alpha\beta}$. Recall that this quantity was needed in Section 5, cf. (5.15), (5.22) and (5.25). We start with the trivial estimate

$$\|M\| \leq 1 + \|B_{p_s}\|, \tag{7.25}$$

and express the norm of the finite rank operator B_{p_s} in terms of the partition p_s and the matrix elements of B. Let $p = \{x_0, \ldots, x_n\} \in \mathcal{P}_n$, $I_i = (x_{i-1}, x_i)$ and let C be an

$n \times n$ matrix with real entries $\{c_{ij}\}$. For $f \in \mathcal{B}_{\alpha\beta}$, one estimates

$$\begin{aligned}\|\mathcal{J}_p C \mathcal{I}_p f\|_{\alpha\beta} &= \sum_{i=1}^{n} \Big|\sum_{j=1}^{n} c_{ij} (\mathcal{I}_p f)_j\Big| \int_{I_i} w_{\alpha\beta}(x)\, dx \\ &\le \sum_{j=1}^{n} |(\mathcal{I}_p f)_j| \sum_{i=1}^{n} |c_{ij}| \int_{I_i} w_{\alpha\beta}(x)\, dx,\end{aligned}$$

and, using our previous bound (7.12) on $|(\mathcal{I}_p f)_j|$, one gets

$$\|C_p\| \le N_p^{\alpha\beta}(C), \tag{7.26}$$

where

$$N_p^{\alpha\beta}(C) = \max_{j=1,\ldots,n} \Big(\frac{1}{|I_j|} \sup_{x \in I_j} \Big(\frac{1}{w_{\alpha\beta}(x)} \Big) \sum_{i=1}^{n} |c_{ij}| \int_{I_i} w_{\alpha\beta}(x)\, dx \Big). \tag{7.27}$$

Remark. A bound on the map $\mathcal{I}_p : \mathcal{A} \to \mathbb{R}^n$ is implemented in the procedure `projection`, checking first the condition on its domain of definition and using `fadd` to compute $\tilde{p}$ and $\{\tilde{p}_j\}$. In the procedure `compute_matrix`, the matrix B (`bm`) is computed using (7.17) and (7.23). A call to `show_invertibility` verifies that $1 + B$ is invertible. For the numerical inversion of $1 - A$, we use the standard algorithm of Gauss elimination, implemented in the subroutine `gaussj`. The operator norms of M and B_{p_s} are computed in `compute_matrix_norm`. Finally, the partition p_s satisfies $\mathrm{supp}(p_s) = \mathrm{supp}(p_r)$ and is uniform with mesh $\varepsilon_s = 10\, \varepsilon_{r_2}$ (`sepsps`). Hence, it contains 500 (`mps`) intervals.

7.3. Existence of the Family of Fixed Points: Second Estimate

In this section, we derive a uniform bound on the norm of the tangent maps $D\mathcal{M}_{\lambda^+,\kappa}(f)$ for κ in $[\lambda^-/\lambda^+, 1]$, and $f \in B_r(f^0_{\lambda^+}) \subset \mathcal{B}_{\mu\nu}$ with $\mu = 0.5$, $\nu = 0.9$ and $r = 9 \cdot 10^{-4}$. In the sequel, we set $\delta = \lambda^-/\lambda^+$. By definition, one has

$$D\mathcal{M}_{\lambda^+,\kappa}(f) = 1 + M\big(S_\kappa D\mathcal{N}_{\lambda^+}(f) - 1\big). \tag{7.28}$$

For $f \in \mathcal{B}_{\alpha\beta}/\mathcal{H}$, $D\mathcal{N}_\lambda(f)$ is bounded from $\mathcal{B}_{\alpha\beta}$ to $\mathcal{B}_{\alpha\gamma}$ provided $\gamma/\beta \le \lambda \le 4$. Hence, with $\beta = \nu$ and $\gamma = \nu/\kappa$, $S_\kappa D\mathcal{N}_{\lambda^+}(f)$ is bounded as a map from $\mathcal{B}_{\mu\nu}$ to $\mathcal{B}_{\mu\nu}$ provided $1/\kappa \le \lambda^+ \le 4$. One concludes that for all $\kappa \in [\delta, 1]$, $D\mathcal{M}_{\lambda^+,\kappa}(f)$ is bounded as a map from $\mathcal{B}_{\mu\nu}$ to $\mathcal{B}_{\mu\nu}$ provided $1/\delta \le \lambda^+ \le 4$. For the values of λ^+ and λ^- as given in the statement of Proposition 2.6, the previous condition is satisfied.

To estimate the norm of $D\mathcal{M}_{\lambda^+,\kappa}(f)$ in $\mathcal{B}_{\mu\nu}$, we proceed as outlined in Section 7.1 and bound each term on the RHS of (7.11). We start with the simple case $h \in \mathcal{B}^{p_r}_{\mu\nu}$. The property $p_s \subseteq p_r$ implies $B_{p_s} h = 0$, so that $Mh = h$. Hence,

$$\begin{aligned}\|D\mathcal{M}_{\lambda^+,\kappa}(f)h\|_{\mu\nu} &\le \|M\|\, \|S_\kappa D\mathcal{N}_{\lambda^+}(f)h\|_{\mu\nu} \\ &\le \|M\|\, \|D\mathcal{N}_{\lambda^+}(f)h\|_{(\kappa\mu)(\nu/\kappa)} \\ &\le \|M\|\, \|D\mathcal{N}_{\lambda^+}(f)h\|_{\mu(\nu/\delta)},\end{aligned} \tag{7.29}$$

for all $\kappa \in [\delta, 1]$. An upper bound on $\|M\|$ was described in the previous section. Representing f by the standard set in $\mathrm{std}(\mathcal{B}_{\mu\nu})^u$ whose affine part is the singleton $\{f^0_{\lambda+}\}$ and whose general term has norm r, the bounds of Section 6 yield 0.85 as an upper bound on the RHS of (7.29) for all h in the unit ball of $\mathcal{L}^{x_0}_\mu$, $\mathcal{C}_{p_r}$, and $\mathcal{R}^{x_n}_\nu$.

Next we consider the more delicate case of $h \in \mathcal{V}^{p_r}$. According to (7.11), one has to estimate the n_r quantities $\|D\mathcal{M}_{\lambda^+,\kappa}(f)\eta_i\|_{\mu\nu}$ where η_i is the normalized characteristic function of the ith interval I_i in the partition p_r. Recalling that $M = 1 + B_{p_s}$, we get from (7.28)

$$D\mathcal{M}_{\lambda^+,\kappa}(f)\eta_i = S_\kappa D\mathcal{N}_{\lambda^+}(f)\eta_i + B_{p_s}(S_\kappa D\mathcal{N}_{\lambda^+}(f) - 1)\eta_i. \tag{7.30}$$

The bound on the map $D\mathcal{N}_\lambda(f) : \mathcal{X} \to \mathcal{B}_{\alpha\gamma}$ previously constructed yields the function $D\mathcal{N}_{\lambda^+}(f)\eta_i$ represented by a standard set in $\mathrm{std}(\mathcal{A})$ and a general term. We denote the former by ρ_i and the latter by g_i, i.e.,

$$D\mathcal{N}_{\lambda^+}(f)\eta_i = \rho_i + g_i,$$

and rewrite (7.30) as

$$\begin{aligned} D\mathcal{M}_{\lambda^+,\kappa}(f)\eta_i &= S_\kappa(\rho_i + g_i) + B_{p_s}(S_\kappa(\rho_i + g_i) - \eta_i) \\ &= MS_\kappa g_i + S_\kappa\rho_i + B_{p_s}(S_\kappa\rho_i - \eta_i). \end{aligned} \tag{7.31}$$

The norm of the first term is bounded as before for all $\kappa \in [\delta, 1]$ by

$$\|MS_\kappa g_i\|_{\mu\nu} \le \|M\|\,\|g_i\|_{\mu(\nu/\delta)}. \tag{7.32}$$

To treat the remaining terms in (7.31), we express them as

$$\begin{aligned} S_\kappa\rho_i + B_{p_s}(S_\kappa\rho_i - \eta_i) = S_\kappa&(\rho_i + B_{p_s}(\rho_i - \eta_i)) \\ &+ B_{p_s}(S_\kappa - 1)\rho_i \\ &+ (1 - S_\kappa)B_{p_s}(\rho_i - \eta_i). \end{aligned} \tag{7.33}$$

As we shall see below, the last two terms are of order $1 - \kappa$, and the first term is small due to the cancelations arising by construction from the action of M. This term, therefore, needs to be computed explicitly.

Let us begin with this term first. One starts by factorizing the action of S_κ by using again

$$\|S_\kappa(\rho_i + B_{p_s}(\rho_i - \eta_i))\|_{\mu\nu} \le \|\rho_i + B_{p_s}(\rho_i - \eta_i)\|_{\mu(\nu/\delta)}, \tag{7.34}$$

which is valid for all $\kappa \in [\delta, 1]$. Next, since $p_s \subseteq p_r$, it is enough to construct a bound on the map

$$\mathcal{A} \times \mathcal{X} \ni (\rho, \chi_I) \mapsto \|\rho + \mathcal{J}_p C\mathcal{I}_p(\rho - \chi_I)\|_{\alpha\gamma}, \tag{7.35}$$

for $p \in \mathcal{P}_n$ and C an $n \times n$ matrix, restricted to intervals I satisfying $I \subseteq I_i$ for some interval I_i in the partition p. A bound on $\rho \mapsto \mathcal{I}_p\rho$ has already been discussed in the previous section. Denoting by I_i the ith interval in the partition p, one has $(\mathcal{I}_p\chi_I)_i = 0$ if $I \cap I_i = \emptyset$, and otherwise

$$(\mathcal{I}_p\chi_I)_i = |I|/|I_i|. \tag{7.36}$$

A bound on the map $C : \mathbb{R}^n \to \mathbb{R}^n$ is readily implemented with interval analysis, and it only remains to consider the map $(\rho, v) \mapsto \|\rho + \mathcal{J}_p v\|_{\alpha\gamma}$, $(\rho, v) \in \mathcal{A} \times \mathbb{R}^n$. Let us denote $\pi(\rho) = (p_\rho, \cdot)$, $\tilde{p} = p \cup p_\rho = \{y_j\}_{j=0}^N$ and $\tilde{\rho}_j = \rho(y_j)$. Imposing the restriction $\operatorname{supp}(p_\rho) \subseteq \operatorname{supp}(p)$, one obtains

$$\|\rho + \mathcal{J}_p v\|_{\alpha\gamma} \leq \sum_{i=1}^{n} \sum_{\substack{j \\ \tilde{I}_j \subseteq I_i}} \sup_{x \in \tilde{I}_j} w_{\alpha\gamma}(x)|\tilde{I}_j| \frac{|\tilde{\rho}_j + v_i| + |\tilde{\rho}_{j-1} + v_i|}{2}, \tag{7.37}$$

where $\tilde{I}_j$ stands for the jth interval in $\tilde{p}$. This finishes the construction of a bound on the map (7.35), which, given standard sets containing ρ_i and η_i, provides an estimate on the RHS of (7.34).

Next, the second term in (7.33) is simply bounded by

$$\|B_{p_s}(S_\kappa - 1)\rho_i\|_{\mu\nu} \leq \|B_{p_s}\| \, \|(S_\kappa - 1)\rho_i\|_{\mu\nu}. \tag{7.38}$$

The operator norm of B_{p_s} has been determined in the previous section, and Lemma 5.1 provides a bound uniform in κ for the second factor, namely

$$\|(S_\kappa - 1)\rho_i\|_{\mu\nu} \leq (1 - \delta)\big(\|\rho_i\|_{\mu\nu} + \|x\rho_i'\|_{\mu(\nu/\delta)}\big).$$

To treat the last term in (7.33), we use the

Lemma 7.1. *Let $p = \{x_0, \ldots, x_n\} \in \mathcal{P}_n$, C an $n \times n$ matrix with coefficients $\{c_{ij}\}$, and $0 < \kappa \leq 1$. Then the operator norm of $(1 - S_\kappa)C_p$ in $\mathcal{B}_{\alpha\beta}$ satisfies*

$$\|(1 - S_\kappa)C_p\| \leq R_p^{\alpha\beta}(\kappa) N_p^{\alpha\beta}(C), \tag{7.39}$$

where $N_p^{\alpha\beta}(C)$ is given by (7.27) and

$$R_p^{\alpha\beta}(\kappa) = (1 - \kappa) + \max_{i=1,\ldots,n} \frac{1}{\int_{I_i} w_{\alpha\beta}(x)dx} \Big(\int_{x_{i-1}}^{x_{i-1}/\kappa} w_{\alpha\beta}(x)\, dx + \int_{x_i}^{x_i/\kappa} w_{\alpha\beta}(x)\, dx \Big).$$

The only dependence on κ in (7.39) is in the factor $R_p^{\alpha\beta}(\kappa)$. Furthermore, $R_p^{\alpha\beta}(\kappa)$ is decreasing in κ. Hence, one obtains

$$\|(1 - S_\kappa)B_{p_s}(\rho_i - \eta_i)\|_{\mu\nu} \leq R_{p_s}^{\mu\nu}(\delta) N_{p_s}^{\mu\nu}(B)(\|\rho_i\|_{\mu\nu} + 1), \tag{7.40}$$

for all $\kappa \in [\delta, 1]$.

Finally, a bound on $\|D\mathcal{M}_{\lambda+,\kappa}(f)\eta_i\|_{\mu\nu}$ follows from (7.32), (7.34), (7.38) and (7.40). As mentioned earlier, computing this bound for the $n_r = 5050$ basis vectors η_i of $\mathcal{V}^{p_r}$ accounts for most of the computation time. In the terms involving explicitly η_i, one can, using the linearity, factorize the value of η_i, that is $(\int_{I_i} w_{\mu\nu})^{-1}$. Therefore, one only needs to compute an upper bound on this quantity. Furthermore, by proceeding like this one can take advantage of the fact that the value of χ_{I_i} can be represented by the standard set containing only the representable number one. This leads to a standard set containing $\rho_i + g_i$ which is more localized and improves the quality of the final bound.

We end this section with the

Proof of Lemma 7.1. For $f \in \mathcal{B}_{\alpha\beta}$, one has

$$\begin{aligned}\|(1 - S_\kappa)C_p f\|_{\alpha\beta} &= \int_0^\infty w_{\alpha\beta}(x)\Big|(1 - S_\kappa)\sum_{i=1}^n\sum_{j=1}^n c_{ij}(\mathcal{I}_p f)_j \chi_{I_i}(x)\Big|\, dx \\ &\leq \sum_{j=1}^n |(\mathcal{I}_p f)_j| \sum_{i=1}^n |c_{ij}| \int_0^\infty w_{\alpha\beta}(x)|(1 - S_\kappa)\chi_{I_i}(x)|\, dx. \end{aligned} \tag{7.41}$$

Furthermore, one has

$$\int_0^\infty w_{\alpha\beta}|(1 - S_\kappa)\chi_{I_i}| \leq \int_{x_{i-1}}^{x_{i-1}/\kappa} w_{\alpha\beta} + (1 - \kappa)\int_{I_i} w_{\alpha\beta} + \kappa \int_{x_i}^{x_i/\kappa} w_{\alpha\beta}.$$

Factorizing $\int_{I_i} w_{\alpha\beta}$ in the previous expression and inserting the bound (7.12) on $|(\mathcal{I}_p f)_i|$ into (7.41) finally leads to (7.39).

■

Remark. A bound on the map $(\rho, v) \mapsto \|\rho + \mathcal{J}_{p_s} v\|_{\alpha\gamma}$ is implemented in the procedure `snorm_add`, and the product Cv is implemented in `linear_app`. A uniform bound on the norm of $(1 - S_\kappa)B_{p_s}$ is computed in `compute_norm_of_1mSB`. Given $i \in \{1, \ldots, n_r\}$, the subroutine `init_chi` returns both a standard set in $\mathrm{std}(\mathcal{X})$ containing χ_{I_i} and the value of η_i, whereas the subroutine `fDM_chi` computes a bound on $\|D\mathcal{M}_{\lambda+,\kappa}(f)\eta_i\|_{\mu\nu}$. Finally, for all $f \in B_r(f^0_{\lambda+})$ and $\kappa \in [\delta, 1]$, a uniform bound on the norm of the tangent maps $D\mathcal{M}_{\lambda+,\kappa}(f)$ is implemented according to (7.11) in `compute_norm_of_DM`.

Acknowledgments

J.W. would like to thank Jean–Pierre Eckmann, Peter Wittwer, and the Department of Theoretical Physics at the University of Geneva for their warm hospitality while part of this work was carried out. A.S. and P.W. would like to thank Jan Wehr and the Department of Mathematics at the University of Arizona at Tucson for their warm hospitality and for providing us generously with computer resources while part of this work was carried out.

Appendix

Proof of Proposition 2.3.

If for some fixed $\lambda \in (1,4)$ and $\alpha, \beta > 0$, f_λ is a fixed point of $\mathcal{N}_\lambda$ and belongs to $\mathcal{B}_{\alpha\beta}\backslash\mathcal{H}$, then Remark 1.2 and Proposition 2.2 imply that $f_\lambda \in \mathcal{B}$. We now prove that, in addition, f_λ is at least once differentiable, with $f'_\lambda \in \mathcal{B}$. The regularization properties of the convolution imply then immediately that f_λ is of class $\mathcal{C}^\infty(\mathbb{R}_+)$. For $\zeta, \eta \geq 0$, let $\mathcal{B}^1_{\zeta\eta}$ denote the Sobolev space of functions in $\mathcal{B}_{\zeta\eta}$ with one (distributional) derivative in $\mathcal{B}_{\zeta\eta}$, i.e.,

$$\mathcal{B}^1_{\zeta\eta} = \{f \in \mathcal{B}_{\zeta\eta} \mid f' \in \mathcal{B}_{\zeta\eta}\},$$

with the norm

$$\|f\|^1_{\zeta\eta} = \|f\|_{\zeta\eta} + \|f'\|_{\zeta\eta}\,.$$

In the sequel, we adopt the shorter notation $\mathcal{B}_\zeta = \mathcal{B}_{\zeta\zeta}$ and $\mathcal{B}^1_\zeta = \mathcal{B}^1_{\zeta\zeta}$. One shows that the fixed point f_λ belongs to $\mathcal{B}^1_\zeta$ for all $\zeta > 0$ by the following argument. One exhibits an $h \in \mathcal{B}^1_\zeta$ and two sequences $\{f_n\}_{n\geq 0}$ and $\{g_n\}_{n\geq 0}$ satisfying $f_\lambda = h + f_n + g_n$ for all $n \geq 0$, such that $\{f_n\}_{n\geq 0}$ is Cauchy in $\mathcal{B}^1_\zeta$ and $\{g_n\}_{n\geq 0}$ converges to zero in $\mathcal{B}_\zeta$. Hence, f_λ is equal in $\mathcal{B}_\zeta$ to a function belonging to $\mathcal{B}^1_\zeta$. Since $\mathcal{N}_\lambda$ preserves the regularity, this function is also a fixed point of $\mathcal{N}_\lambda$. Therefore, it is equal to f_λ in $\mathcal{B}^1_\zeta$.

We first construct recursively the sequences $\{f_n\}_{n\geq 0}$ and $\{g_n\}_{n\geq 0}$. Since f_λ belongs to $\mathcal{B}_\zeta$ for all $\zeta > 0$, and since $\mathcal{C}^\infty_0(\mathbb{R}_+)$ is dense in $\mathcal{B}_\zeta$, there exist for every $\delta_0 > 0$ an $h \in \mathcal{B}^1_\zeta$ and a $g_0 \in \mathcal{B}_\zeta$ satisfying

$$f_\lambda = h + g_0, \tag{A.1}$$

with

$$\|g_0\|_\zeta \leq \delta_0. \tag{A.2}$$

Moreover, one defines

$$f_0 \equiv 0. \tag{A.3}$$

Denoting $c_\lambda(f_\lambda) = \bar{c}_\lambda$ and $\overline{\mathcal{N}}_\lambda = \bar{c}_\lambda \mathcal{N}^1_\lambda + c_2 \mathcal{N}^2_\lambda$, we now define for all $n \geq 0$,

$$\begin{aligned} f_{n+1} &= \overline{\mathcal{N}}_\lambda(h + f_n) - h + \mathcal{C}_\lambda(f_n, g_n), \\ g_{n+1} &= \overline{\mathcal{N}}_\lambda(g_n), \end{aligned} \tag{A.4}$$

where

$$\mathcal{C}_\lambda(f,g) = \overline{\mathcal{N}}_\lambda(h+f+g) - \overline{\mathcal{N}}_\lambda(h+f) - \overline{\mathcal{N}}_\lambda(g).$$

Note that $\mathcal{C}_\lambda(f,g)$ contains only cross terms between $h+f$ and g. We now check that the sequences $\{f_n\}_{n\geq 0}$ and $\{g_n\}_{n\geq 0}$ have the desired properties, i.e., $f_\lambda = h + f_n + g_n$, $\{g_n\}_{n\geq 0}$ converges to zero in $\mathcal{B}_\zeta$, and $\{f_n\}_{n\geq 0}$ is Cauchy in $\mathcal{B}^1_\zeta$. Since f_λ is a fixed point of $\overline{\mathcal{N}}_\lambda$, it first follows from (A.1) and (A.4) that

$$f_\lambda = h + f_n + g_n,$$

for all $n \geq 0$. Furthermore, $\zeta > 0$ and $\lambda \in (1,4)$ together with Proposition 2.2 imply that $\{g_n\}_{n\geq 0}$ converges to zero in $\mathcal{B}_\zeta$. Indeed, $\overline{\mathcal{N}}_\lambda$ is well defined as a map from $\mathcal{B}_\zeta$ to $\mathcal{B}_\zeta$, and the bounds obtained in the proof of Proposition 2.2 lead to

$$\|g_n\|_\zeta \leq \bar{c}_\lambda \|g_{n-1}\|_\zeta^2 + c_2\|g_{n-1}\|_\zeta^4.$$

Applying this inequality recursively and using $\|g_0\|_\zeta \leq \delta_0 < 1$, one gets for all $n \geq 1$

$$\|g_n\|_\zeta \leq \delta^{2^n}, \tag{A.5}$$

where

$$\delta = (\bar{c}_\lambda + c_2)\delta_0 < 1$$

for δ_0 small enough. Note that (A.1), (A.2) and (A.5) imply, for δ_0 small enough, the uniform bound

$$\begin{aligned}\|f_n\|_\zeta &\leq \|f_\lambda - h\|_\zeta + \|g_n\|_\zeta \\ &\leq 2\delta_0.\end{aligned} \tag{A.6}$$

Next, in order to show that $f_n \in \mathcal{B}^1_\zeta$ for all $n \geq 0$, one proceeds as in Proposition 2.2 and studies the maps which enter the definition of $\overline{\mathcal{N}}_\lambda$ and $\mathcal{C}_\lambda$, i.e., S_λ, T and the convolution operator. From (2.8) and $(f*g)' = f' * g$, it follows that

$$\|f*g\|^1_{(4\sigma)\tau} \leq \|f\|^1_{\sigma\tau}\|g\|_{\sigma\tau}, \tag{A.7}$$

whereas (2.9) together with $\lambda > 1$ and $(S_\lambda f)' = \lambda S_\lambda f'$ leads to

$$\begin{aligned}\|S_\lambda f\|^1_{(\sigma/\lambda)(\lambda\tau)} &\leq \|f\|_{\sigma\tau} + \lambda\|f'\|_{\sigma\tau} \\ &\leq \lambda\,\|f\|^1_{\sigma\tau}.\end{aligned} \tag{A.8}$$

(A.7) and (A.8) imply in particular

$$\|S_\lambda(f*g)\|^1_{(4\zeta/\lambda)(\lambda\zeta)} \leq \lambda\|f\|^1_\zeta\|g\|_\zeta. \tag{A.9}$$

We now show that for all $\tau > \tau'$, T is a bounded operator from $\mathcal{B}^1_{\sigma\tau}$ to $\mathcal{B}^1_{\tau'\sigma}$. One has $\|Tf\|_{\tau'\sigma} = \|f\|_{\sigma\tau'} \leq \|f\|_{\sigma\tau}$, and using

$$(Tf)'(x) = -\frac{1}{x^2}\Big(2x(Tf)(x) + (Tf')(x)\Big),$$

one gets

$$\begin{aligned}
\|(Tf)'\|_{\tau'\sigma} &\leq 2\int_0^\infty \frac{1}{x}\, w_{\tau'\sigma}(x)\, |Tf|(x)\, dx + \int_0^\infty \frac{1}{x^2}\, w_{\tau'\sigma}(x)\, |Tf'|(x)\, dx \\
&= 2\int_0^\infty x\, w_{\sigma\tau'}(x)\, |f|(x)\, dx + \int_0^\infty x^2\, w_{\sigma\tau'}(x)\, |f'|(x)\, dx \\
&\leq 2 \sup_{x>0} \frac{(1+x^2) w_{\sigma\tau'}(x)}{w_{\sigma\tau}(x)} \|f\|^1_{\sigma\tau} \\
&\leq C_{\tau\tau'} \|f\|^1_{\sigma\tau},
\end{aligned} \tag{A.10}$$

where $C_{\tau\tau'}$ is finite as long as $\tau > \tau'$. In particular, since $\lambda > 1$, (A.9) and (A.10) imply

$$\|TS_\lambda(f * g)\|^1_{\zeta(4\zeta/\lambda)} \leq C\|f\|^1_\zeta \|g\|_\zeta,$$

which in turn, with $\lambda < 4$, leads to

$$\|T\big(TS_\lambda(f * g) * TS_\lambda(\bar{g} * \bar{\bar{g}})\big)\|^1_\zeta \leq C\|f\|^1_\zeta \|g\|_\zeta \|\bar{g}\|_\zeta \|\bar{\bar{g}}\|_\zeta. \tag{A.11}$$

Therefore, $\overline{\mathcal{N}}_\lambda$ is well defined as a map from $\mathcal{B}^1_\zeta$ to $\mathcal{B}^1_\zeta$ for $\zeta > 0$ and $\lambda \in (1,4)$. Assume now that $f_n \in \mathcal{B}^1_\zeta$ and that δ_0 is small enough. Then, (A.6), (A.9), and (A.11) lead to

$$\begin{aligned}
\|\overline{\mathcal{N}}_\lambda(h + f_n)\|^1_\zeta &\leq \|\overline{\mathcal{N}}_\lambda(h)\|^1_\zeta + C\|f_n\|_\zeta(\|h\|^1_\zeta + \|f_n\|^1_\zeta) \\
&\leq C_1 + \frac{1}{2}\bar{\delta}\|f_n\|^1_\zeta,
\end{aligned} \tag{A.12}$$

for some positive $\bar{\delta} < 1$. Similarly, using (A.5) and (A.6), one gets

$$\|\mathcal{C}_\lambda(f_n, g_n)\|^1_\zeta \leq C\|g_n\|_\zeta\big(\|f_n\|^1_\zeta + \|h\|^1_\zeta\big). \tag{A.13}$$

Therefore, (A.12) and (A.13) lead, together with (A.5), to

$$\|f_{n+1}\|^1_\zeta \leq C_2 + \bar{\delta}\|f_n\|^1_\zeta.$$

Using this bound recursively, one obtains $f_n \in \mathcal{B}^1_\zeta$ for all $n \geq 0$, together with the uniform estimate

$$\|f_n\|^1_\zeta \leq C_2 \sum_{k=0}^n \bar{\delta}^k \leq K. \tag{A.14}$$

Finally, we check that the sequence $\{f_n\}_{n\geq 0}$ is Cauchy in $\mathcal{B}^1_\zeta$. Since (A.5), (A.13) and (A.14) imply that

$$\lim_{n\to\infty} \|\mathcal{C}_\lambda(f_n, g_n)\|^1_\zeta = 0,$$

it only remains to show that $\{\overline{\mathcal{N}}_\lambda(h+f_n)\}_{n\geq 0}$ is Cauchy in $\mathcal{B}^1_\zeta$. We first verify that the sequence $\{h_n\}_{n\geq 0}$, with

$$h_n \equiv \mathcal{N}^1_\lambda(h+f_n),$$

is Cauchy in $\mathcal{B}^1_{(4\zeta/\lambda)(\lambda\zeta)}$. Since $\lambda \in (1,4)$, this implies in particular the convergence of $\{\mathcal{N}^1_\lambda(h+f_n)\}_{n\geq 0}$ in $\mathcal{B}^1_\zeta$. Defining

$$\tilde{\mathcal{N}}^1_\lambda(f,g) = S_\lambda(f * g),$$

one has

$$\begin{aligned}\|h_n - h_m\|^1_{\sigma\tau} &\leq 2\|\tilde{\mathcal{N}}^1_\lambda(h, f_n) - \tilde{\mathcal{N}}^1_\lambda(h, f_m)\|^1_{\sigma\tau} + \|\mathcal{N}^1_\lambda(f_n) - \mathcal{N}^1_\lambda(f_m)\|^1_{\sigma\tau} \\ &\leq 2\|\tilde{\mathcal{N}}^1_\lambda(h, f_n - f_m)\|^1_{\sigma\tau} + \|\tilde{\mathcal{N}}^1_\lambda(f_n, f_n - f_m)\|^1_{\sigma\tau} + \|\tilde{\mathcal{N}}^1_\lambda(f_m, f_m - f_n)\|^1_{\sigma\tau},\end{aligned}$$

which leads, with (A.9) and (A.14), to

$$\begin{aligned}\|h_n - h_m\|^1_{(4\zeta/\lambda)(\lambda\zeta)} &\leq C(2\|h\|^1_\zeta + \|f_n\|^1_\zeta + \|f_m\|^1_\zeta)\|f_n - f_m\|_\zeta \\ &\leq K\|g_n - g_m\|_\zeta.\end{aligned}$$

Therefore, the convergence of $\{h_n\}$ in $\mathcal{B}^1_{(4\zeta/\lambda)(\lambda\zeta)}$ follows from the convergence of $\{g_n\}$ in $\mathcal{B}_\zeta$. Finally, in order to see that $\{\mathcal{N}^2_\lambda(h+f_n)\}$ is Cauchy in $\mathcal{B}^1_\zeta$, one observes that

$$\mathcal{N}^2_\lambda(h+f_n) = F(h_n),$$

where

$$F(f) = T\big(Tf * Tf\big),$$

and that the bounds obtained above imply the continuity of F as a map from $\mathcal{B}^1_{(4\zeta/\lambda)(\lambda\zeta)}$ to $\mathcal{B}^1_\zeta$ for $\zeta > 0$ and $\lambda \in (1,4)$. Hence, the convergence of $\{\mathcal{N}^2_\lambda(h+f_n)\}$ in $\mathcal{B}^1_\zeta$ follows from the convergence of $\{h_n\}$ in $\mathcal{B}^1_{(4\zeta/\lambda)(\lambda\zeta)}$.

■

Proof of Lemma 4.1.

For $x \in I_i$, the functions ρ and $T\tilde{\rho}$ are given by

$$\rho(x) = \frac{1}{\varepsilon}\Big(\rho_i(x - x_{i-1}) + \rho_{i-1}(x_i - x)\Big), \tag{A.15}$$

$$\begin{aligned}\tilde{\rho}(1/x) &= \Big(\frac{1}{x_{i-1}} - \frac{1}{x_i}\Big)^{-1}\Big(x^2_{i-1}\rho_{i-1}\big(\frac{1}{x} - \frac{1}{x_i}\big) + x_i^2\rho_i\big(\frac{1}{x_{i-1}} - \frac{1}{x}\big)\Big) \\ &= \frac{1}{\varepsilon x}\Big(x^3_{i-1}\rho_{i-1}(x_i - x) + x_i^3\rho_i(x - x_{i-1})\Big),\end{aligned} \tag{A.16}$$

and one computes

$$
\begin{aligned}
|\rho(x) - T\tilde{\rho}(x)| &= |\rho(x) - \tilde{\rho}(1/x)/x^2| \\
&= \frac{1}{\varepsilon}\Big|\rho_{i-1}(x_i - x)\big(1 - \frac{x_{i-1}^3}{x^3}\big) - \rho_i(x - x_{i-1})\big(\frac{x_i^3}{x^3} - 1\big)\Big| \\
&= \frac{(x - x_{i-1})(x_i - x)}{\varepsilon x^3}\big|\rho_{i-1}(x - x_{i-1})^2 - \rho_i(x_i - x)^2 - 3x(x_i\rho_i - x_{i-1}\rho_{i-1})\big| \\
&\leq \frac{\varepsilon}{4}\Big(\frac{|\rho_i - \rho_{i-1}|}{x} + \frac{|x_i\rho_i - x_{i-1}\rho_{i-1}|}{x^2} + \frac{|x_i^2\rho_i - x_{i-1}^2\rho_{i-1}|}{x^3}\Big).
\end{aligned}
\tag{A.17}
$$

Integrating the expression on the RHS of (A.17) leads to the stated result.

■

Proof of Lemma 4.2.

By definition of $\mathcal{A}$, it is clear that $(\rho * \sigma)'' \in \mathcal{A}$. Furthermore, because ρ and σ have a uniform partition with identical mesh size ε, $(\rho * \sigma)''$ is also defined on a uniform partition. It is given by $\{z_k\}_{k=0}^{2n}$ where z_k is defined in (4.20). It remains to compute $v_k \equiv (\rho * \sigma)''(z_k)$. With $\rho'(x) = \sum_{i=1}^n \rho_i' \chi_{I_i}(x)$ and $\sigma'(x) = \sum_{j=1}^n \sigma_j' \chi_{J_j}(x)$, where $\rho_i' = (\rho_i - \rho_{i-1})/\varepsilon$ and $\sigma_j' = (\sigma_j - \sigma_{j-1})/\varepsilon$, one gets

$$v_k = \varepsilon \sum_{i+j=k+1} \rho_i' \sigma_j'.$$

Expressing the RHS of the previous equality in terms of the coefficients (4.19) finally leads to the relation (4.21).

■

Proof of Lemma 4.4.

For $k = 2l$, $l = 0, \ldots, n-1$, and $\theta \in (0, 2\varepsilon)$, one has

$$\tilde{\rho}(z_k + \theta) = C_0(k) + \frac{\theta}{2\varepsilon}\Big(C_0(k+2) - C_0(k)\Big).$$

The continuity properties of $\rho * \sigma$ imply

$$
\begin{aligned}
C_0(k+1) &= C_0(k) + \varepsilon C_1(k) + \varepsilon^2 C_2(k) + \varepsilon^3 C_3(k), \\
C_1(k+1) &= C_1(k) + 2\varepsilon C_2(k) + 3\varepsilon^2 C_3(k), \\
C_2(k+1) &= C_2(k) + 3\varepsilon C_3(k),
\end{aligned}
$$

$k = 0, \ldots, 2n-1$. For $\theta \in (0, \varepsilon)$, these relations allow us to write

$$\tilde{\rho}(z_k + \theta) = C_0(k) + \theta C_1(k) + \varepsilon\theta\Big(2C_2(k+1) - \frac{5\varepsilon}{2}C_3(k) + \frac{\varepsilon}{2}C_3(k+1)\Big),$$

and, from (4.23),

$$(\rho * \sigma)(z_k + \theta) = C_0(k) + \theta C_1(k) + \theta^2 \Big(C_2(k+1) - (3\varepsilon - \theta) C_3(k) \Big).$$

Hence,

$$(\tilde{\rho} - \rho * \sigma)(z_k + \theta) = \theta(2\varepsilon - \theta) C_2(k+1) + \theta\Big(\theta(3\varepsilon - \theta) - \frac{5\varepsilon^2}{2}\Big) C_3(k) + \frac{\varepsilon^2 \theta}{2} C_3(k+1),$$

which leads to the estimate

$$\int_{z_k}^{z_{k+1}} |(\tilde{\rho} - \rho * \sigma)(y)| dy \leq \varepsilon^3 \Big(\frac{2}{3} |C_2(k+1)| + \frac{\varepsilon}{2} |C_3(k)| + \frac{\varepsilon}{4} |C_3(k+1)| \Big).$$

For $\theta \in (\varepsilon, 2\varepsilon)$, one proceeds similarly and obtains

$$\int_{z_{k+1}}^{z_{k+2}} |(\tilde{\rho} - \rho * \sigma)(y)| dy \leq \varepsilon^3 \Big(\frac{2}{3} |C_2(k+1)| + \frac{\varepsilon}{4} |C_3(k)| + \frac{\varepsilon}{2} |C_3(k+1)| \Big),$$

and the bound (4.27) follows immediately.

■

References

ANW] Ahlberg, J.H., E.N. Nilson and J.L Walsh: *The Thory of Splines and Their Applications*, Academic Press, New York London, (1967).

[Be1] Bernasconi, J.: Electrical conductivity in disordered systems. Phys. Rev. B **7**, 2252–2260 (1972).

[Be2] Bernasconi, J.: Conduction in anisotropic disordered systems: Effective–medium theory. Phys. Rev. B **9**, 4575–4579 (1974).

[Be3] Bernasconi, J.: Real–space renormalization of bound–disordered conductances lattices. Phys. Rev. B **18**, 2185–2191 (1978).

[Bl] Blumenfeld, R.: Probability densities of homogeneous functions: explicit approximation and applications to percolating networks. J. Phys. A **21**, 815–825 (1988).

[BO] Berker, A.N. and S. Ostlund: Renormalisation–group calculations of finite systems: order parameter and specific heat for epitaxial ordering. J. Phys. C **12**, 4961–4975 (1979).

[BS] Burbanks, A. and A. Stirnemann: Hölder continuous Siegel disc boundary curves. Nonlinearity **8**, 901–920 (1995).

[BSW] Bernasconi, J., W.R. Schneider and H.J. Wiesmann: Some rigorous results for random planar conductance networks. Phys. Rev. B **16**, 5250–5255 (1977).

[BW] Bernasconi, J. and H.J. Wiesmann: Effective–medium theories for site disordered resistances networks. Phys. Rev. B **13**, 1131–1139 (1976).

[C] Celletti, A.: Construction of librational invariant tori in the spin–orbit problem. Journal of Applied Mathematics and Physics (ZAMP) **45**, 61 (1993).

[CC] Celletti, A. and L. Chierchia: Construction of Analytic KAM Surfaces and Effective Stability Bounds. Commun. Math. Phys. **118**, 119–161 (1988).

[dlL] de la Llave, R.: Computer assisted proofs of stability of matter. In: Computer Aided Proofs in Analysis, K. Meyer and D. Schmidt (eds.), The IMA Volumes in Mathematics **28**, 116–126 (1991).

[EB] Essoh, C.D. and J. Bellissard: Resistance and fluctuation of a fractal network of random resistors: a non–linear law of large numbers. J. Phys. A **22**, 4537–4548 (1989).

[EKW1] Eckmann, J.–P., H. Koch and P. Wittwer: Existence of a fixed point of the doubling transformation for area–preserving maps of the plane. Phys. Rev. A **26**, 720–722 (1982).

[EKW2] Eckmann, J.–P., H. Koch and P. Wittwer: A computer–assisted proof of universality for area–preserving maps. Providence, Memoirs of the AMS **47**, 1–121 (1984).

[EW1] Eckmann, J.–P. and P. Wittwer: *Computer Methods and Borel Summability Applied to Feigenbaum's Equation*, Springer–Verlag, Berlin Heidelberg New York Tokyo, Lecture Notes in Physics **227** (1985).

[EW2] Eckmann, J.–P. and P. Wittwer: A Complete Proof of the Feigenbaum Conjectures. J. Stat. Phys. **46**, 455–475 (1987).

[FL] Feffermann, C. and R. de la Llave: Relativistic Stability of Matter. Revista Matemática Iberoamericana **2/1,2**, 119–213 (1986).

[FS] Feffermann, C. and L. Seco: Aperiodicity of the Hamiltonian Flow in the Thomas–Fermi Potential. Revista Matemática Iberoamericana **9/3**, 409–551 (1993).

[G] Grimmet, G.: *Percolation*, Springer, Berlin New York, 2nd ed. (1999).

[K1] Kirkpatrick, S.: Classical Transport in Disordered Media: Scaling and Effective-Medium Theories. Phys. Rev. Lett. **27**, 1722–1725 (1971).

[K2] Kirkpatrick, S: Percolation and Conduction. Rev. Mod. Phys. **45**, 574–588 (1973).

[K3] Kirkpatrick, S.: Percolation thresholds in Ising magnets and conducting mixtures. Phys. Rev. B **15**, 1533–1538 (1977).

[KP] MacKay, R.S. and I.C. Percival: Converse KAM: Theory and Practice. Commun. Math. Phys. **98**, 469–512 (1985).

[KSW] Koch, H., A. Schenkel and P. Wittwer: Computer–Assisted Proofs in Analysis and Programming in Logic: A Case Study. SIAM Review **38**, No. 4, 565–604 (1996).

[KW1] Koch, H. and P. Wittwer: A Non–Gaussian Renormalization Group Fixed Point for Hierarchical Scalar Lattice Field Theories. Commun. Math. Phys. **106**, 495–532 (1986).

[KW2] Koch, H. and P. Wittwer: Rigorous Computer–Assisted Renormalization Group Analysis. In: VIIIth International Congress on Mathematical Physics, M. Mebkhout and R. Sénéor (eds.), World Scientific (1986).

[KW3] Koch, H. and P. Wittwer: Computing Bounds on Critical Indices. In: Nonlinear Evolution and Chaotic Phenomena, G. Gallavotti and P.F. Zweifel (eds.), NATO ASI Series B: Phys. **176**, 269–277 (1987).

[KW4] Koch, H. and P. Wittwer: The Unstable Manifold of a Nontrivial RG Fixed Point. Canadian Mathematical Society, Conference Proceedings **9**, 99–105 (1988).

[KW5] Koch, H. and P. Wittwer: On the Renormalization Group Transformation for Scalar Hierarchical Models. Commun. Math. Phys. **138**, 537–568 (1991).

[KW6] Koch, H. and P. Wittwer: A Nontrivial Renormalization Group Fixed Point for the Dyson–Baker Hierarchical Model. Commun. Math. Phys. **164**, 627–647 (1994).

[KW7] Koch, H. and P. Wittwer: Bounds on the Zeros of a Renormalization Group Fixed Point. Math. Phys. EJ **1**, No 6, 24pp. (1995).

[L1] Lanford III, O.E.: A computer–assisted proof of the Feigenbaum conjectures. Bull. of the AMS **6**, 427–434 (1982).

[L2] Lanford III, O.E.: Computer–Assisted Proofs in Analysis. Physica **124 A**, 465–470 (1984).

[L3] Lanford III, O.E.: A Shorter Proof of the Existence of the Feigenbaum Fixed Point. Commun. Math. Phys. **96**, 521–538 (1984).

[LR1] de la Llave, R. and D. Rana: Accurate Bounds in K.A.M. Theory. In: VIIIth International Congress on Mathematical Physics, M. Mebkhout and R. Sénéor (eds.), World Scientific (1986).

[LR2] de la Llave, R. and D. Rana: Accurate Strategies for Small Divisor Problems. Bull. of the AMS **22**, 85–90 (1990).

[LR3] de la Llave, R. and D. Rana: Accurate strategies in K.A.M. problems and their implementation. In: Computer Aided Proofs in Analysis, K. Meyer and D. Schmidt (eds.), The IMA Volumes in Mathematics **28**, 127–146 (1991).

[M] Mestel, B.D.: A computer assisted proof of universality for cubic critical maps of the circle with golden mean rotation number. Ph.D. Thesis, Math. Dept., University of Warwick (1985).

[N] Nürnberger, G.: *Approximation by spline functions*, Springer–Verlag, Berlin (1989).

[PFTV] Press, W.H., B.P. Flannery, S.A. Teukolsky and W.T Vetterling: *Numerical Recipes in Fortran: The Art of Scientific Computing*, Cambridge University Press, Cambridge, 2nd ed. (1992).

[R] Rana, D.: Proof of accurate upper and lower bounds to stability domains in small denominator problems. Ph.D. Thesis, Princeton University (1987).

[Se1] Seco, L.: Lower bounds for the ground state energy of atoms. Ph.D. Thesis, Princeton University (1989).

[Se2] Seco, L.: Computer Assisted Lower Bounds for Atomic Energies. In: Computer Aided Proofs in Analysis, K. Meyer and D. Schmidt (eds.), The IMA Volumes in Mathematics **28**, 241–251 (1991).

[Sh] Shneiberg, I.: Hierarchical Sequences of Random Variables. Theory Probab. Appl. **31**, 137–141 (1987).

[St] Stirnemann, A.: Existence of the Siegel disc renormalization fixed point. Nonlinearity **7**, 959–974 (1994).

[SS] Schlösser, T. and H. Spohn: Sample–to–Sample Fluctuations in the Conductivity of a Disordered Medium. J. Stat. Phys. **69**, 955–967 (1992).

[SW] Stinchcombe, R.B. and P.B. Watson: Renormalization group approach for percolation conductivity. J. Phys. C: Solid State Phys. **9**, 3221–3247 (1976).

[W1] Wehr, J.: A strong law of large numbers for iterated functions of independent random variables. J. Statist. Phys. **86**, no 5–6, 1373–1384 (1997).

[W2] Wehr, J.: A lower bound on the variance of conductance in random resistor networks. J. Statist. Phys. **86**, no 5–6, 1359–1365 (1997).

[WW] Wehr, J. and J.-M. Woo: A central limit theorem for nonlinear hierarchical sequences of random variables. Annals of Probability (to appear).

[Z] Ziman, J.M.: The localization of electrons in ordered and disordered systems: I. Percolation of classical particles. J. Phys. C **1**, 1532–1538 (1968).

MPEJ

MATHEMATICAL PHYSICS ELECTRONIC JOURNAL

ISSN 1086-6655
Volume 6, 2000

Paper 4
Received: Jun 17, 2000, Revised: Jul 17, 2000, Accepted: Aug 14, 2000
Editor: J. Avron

Degenerate space-time paths and the non-locality of quantum mechanics in a Clifford substructure of space-time

Kaare Borchsenius *

Abstract

The quantized canonical space-time coordinates of a relativistic point particle are expressed in terms of the elements of a complex Clifford algebra which combines the complex properties of $SL(2.C)$ and quantum mechanics. When the quantum measurement principle is adapted to the generating space of the Clifford algebra we find that the transition probabilities for twofold degenerate paths in space-time equal the transition amplitudes for the underlying paths in Clifford space. This property is used to show that the apparent non-locality of quantum mechanics in a double slit experiment and in an EPR type of measurement is resolved when analyzed in terms of the full paths in the underlying Clifford space. We comment on the relationship of this model to the time symmetric formulation of quantum mechanics and to the Wheeler-Feynman model.

*Bollerisvej 8, 3782 Klemensker, Denmark, e-mail: bdge@post5.tele.dk

1 Substructure of the canonical space-time coordinates

The fact that half-integer spin representations of the Lorentz group are realized in nature casts doubt on the assumption that space-time is a primary space. More specifically, as pointed out by Penrose [1], the fact that different spatial directions of a spin-one-half particle correspond to different complex linear combinations of the two quantum states suggests that there is a direct connection between the structure of space and the need for complex state vectors in quantum mechanics. Taken together, considerations like these point to the existence of a substructure of space-time which combines the complex properties of the Lorentz group and quantum mechanics. Substructures of space-time have been discussed in Schwartz and Van Nieuwenhuizen [2] and in Borchsenius [3, 4, 5].

To determine the nature of such a complex substructure of space-time we shall use the canonical quantization of a relativistic point particle as a model. We shall adopt Dirac's method in which space and time are treated on an equal footing, both being regarded as functions of a parameter-time τ. Reparametrization invariance imposes a constraint which can be used to define a Hamiltonian together with a set of canonical variables. The quantization results in a set of hermitian canonical space-time coordinates, the components of which satisfy

$$X^{\mu *}_{ab}(\tau) = X^{\mu}_{ba}(\tau) \tag{1}$$

These components transform under a Lorentz transformation in the index μ and under a unitary change of basis in Hilbert space in the indices a and b. To bring out the complex properties of the Lorentz group, we make use of the connection between a real four-vector and a second-rank hermitian spinor

$$V^{\mu} = \frac{1}{2}\sigma^{\mu}_{A\dot{B}} V^{A\dot{B}}, \qquad V^{A\dot{B}} = \sigma^{A\dot{B}}_{\mu} V^{\mu} \tag{2}$$

where σ_{μ} are the four hermitian Pauli matrices. The spinor form of the canonical space-time coordinates

$$X^{A\dot{B}}_{ab} \stackrel{\text{def}}{=} \sigma^{A\dot{B}}_{\mu} X^{\mu}_{ab} \tag{3}$$

exhibits two hermitian properties, one related to $SL(2.C)$ and the other to the unitary group in Hilbert space. To find a substructure of X corresponding to these two groups, we observe that the components (3) form a hermitian matrix in the combined indices (A, a) and (B, b)

$$\left(X^{A\dot{B}}_{ab}\right)^{*} = X^{B\dot{A}}_{ba} \tag{4}$$

As shown in the appendix, any hermitian matrix can be expressed in terms of the elements of a complex Clifford algebra according to (66). For the canonical space-time coordinates (4) this implies that there exists a complex Clifford algebra with elements C^{A}_{a} so that

$$X^{A\dot{B}}_{ab} = \{C^{A}_{a}, C^{*\dot{B}}_{b}\}, \qquad \{C^{A}_{a}, C^{B}_{b}\} = 0 \tag{5}$$

The complex linear space which generates the Clifford algebra, and to which the C's belong, we shall call Clifford space, and we shall refer to its elements as Clifford coordinates, borrowing from space-time terminology. To write (5) in abstract form we shall adopt the following notation. The components C_a^A which transform like a right-handed two-component spinor in the index A and as a ket vector in the index a shall be written as $\overset{>}{C}{}^A$ where the ket on top is used to distinguish it from a quantum operator and an ordinary eigenvector. Likewise $C_b^{*\dot{B}}$ will be written as the bra vector $\overset{<}{C}{}^{\dot{B}} = (\overset{>}{C}{}^B)^\dagger$ where $\dagger$ performs both the complex involution of the Clifford algebra and the quantum conjugation in Hilbert space. The commutator between a ket vector $\overset{>}{\chi}$ and a bra vector $\overset{<}{\psi}$ shall be defined as

$$\{\overset{>}{\chi}, \overset{<}{\psi}\}_{ab} \overset{\text{def}}{=} \{\chi_a, \psi_b\}, \qquad \{\overset{<}{\psi}, \overset{>}{\chi}\} \overset{\text{def}}{=} \{\psi_a, \chi_a\} \tag{6}$$

that is, we adopt the convention that the order of the ket and bra vectors in the first term in the commutator determines whether *both* terms are direct products or contractions. With this notation, (5) can be written in the abstract form

$$X^{A\dot{B}} = \{\overset{>}{C}{}^A, \overset{<}{C}{}^{\dot{B}}\}, \qquad \{\overset{>}{C}{}^A, \overset{>}{C}{}^B\} = 0 \tag{7}$$

X and $\overset{>}{C}$ can be expressed in terms of a complete set of eigenstates $|x_r\rangle$ and their eigenvalues

$$X^\mu = |x_r^\mu\rangle x_r^\mu \langle x_r^\mu| \tag{8}$$

$$\overset{>}{C}{}^A = |x_r^\mu\rangle c_r^A, \qquad c_r^A \overset{\text{def}}{=} \langle x_r^\mu|\, \overset{>}{C}{}^A \tag{9}$$

When these expressions are inserted into (7) we obtain

$$\{c_r^A, c_s^{*\dot{B}}\} = \delta_{rs} x_s^{A\dot{B}}, \qquad \{c_r^A, c_s^B\} = 0 \tag{10}$$

Hence the eigenvalues of X are determined by a set of mutually orthogonal elements c_r^A of the Clifford algebra. To make our discussion more transparent we shall refer to these elements as 'eigenvalues' and write the eigenstates $|x_r\rangle$ as $|c_r\rangle$. By use of (7) we obtain the expression for the expectation value of X in the state $|s\rangle$

$$\langle s|X^{A\dot{B}}|s\rangle = \langle s|\{\overset{>}{C}{}^A, \overset{<}{C}{}^{\dot{B}}\}|s\rangle = \{\bar{c}^A, \bar{c}^{*\dot{B}}\} \tag{11}$$

$$\bar{c}^A \overset{\text{def}}{=} \langle s|\, \overset{>}{C}{}^A \tag{12}$$

(12) are the Clifford coordinates corresponding to the expectation value of the space-time coordinates. Applying (9) they become

$$\bar{c}^A = \langle s|x_r\rangle c_r^A \tag{13}$$

The relationship of this equation to the expression for the expectation value of the space-time coordinates

$$\bar{x}^{\mu} = |\langle s|x_r\rangle|^2 x_r^{\mu} \tag{14}$$

can be described as a linear extraction of the quantum amplitudes as a complex substructure of the probabilities, and of the Clifford coordinates as a complex substructure of the space-time coordinates. If, conversely, we had sought a substructure of space-time which had the quantum amplitudes as a linear space of weights as in (13), we would have been led to something of the nature of the orthogonality relations (10).

In the continuum limit X has a Continuous spectrum and in the coordinate representation (9) and (10) become

$$\overset{>}{C^A} = \int |x\rangle c^A(x)\, dx \tag{15}$$

$$\{c^A(x), c^{*\dot{B}}(x')\} = x^{A\dot{B}}\delta(x-x'), \qquad \{c^A(x), c^B(x')\} = 0 \tag{16}$$

(16) generates an infinite dimensional Clifford Algebra of a type well known from the Algebra of creation and annihilation operators for a Fermi field.

The stability of Clifford space under $SL(2.C)$ implies that there are at least two values c and $-c$ of the Clifford coordinates,which correspond to the same space-time coordinates x. The well known degeneracy of $SO(1.3)$ transformations with respect to $SL(2.C)$ transformations is hereby extended to space-time itself. As we shall see in sections 3 and 4, this has the consequence that the path of a particle in Clifford space has a starting point in physical time. This is the only physical consequence of the Clifford model which differs from those of conventional quantum mechanics. To reconcile these starting times with experience in a satisfactory way, the Clifford substructure would presumably have to be applied to objects more fundamental than point particles (e.g. fields or strings) and examined in the context of a cosmological model. The viability of our model rests upon the possible success of such a program.

2 Canonical equations

We consider the action:

$$\int L(c(\tau), \dot{c}(\tau))\, d\tau \tag{17}$$

Since the Lagrangian is real-valued it is natural to assume that the Clifford variables c and $\dot{c}$ occur within anticommutators. In this case the variation of L can be expressed as

$$\delta L = \{\frac{\partial L}{\partial c^A}, \delta c^A\} + c.c. + \{\frac{\partial L}{\partial \dot{c}^A}, \delta \dot{c}^A\} + c.c. \tag{18}$$

which defines the derivatives with respect to c and $\dot{c}$ up to terms which anticommute with δc. The conjugate to c is defined as

$$d_A^* = \frac{\partial L}{\partial \dot{c}^A} \tag{19}$$

If $\dot{c}$ can be eliminated in favour of d^* the Hamiltonian becomes

$$H(c,d) = \{\dot{c}^A, d^*_A\} + c.c. - L(c,\dot{c}) \tag{20}$$

with the equations of motion

$$\dot{c}^A = \frac{\partial H}{\partial d^*_A}, \quad \dot{d}^*_A = -\frac{\partial H}{\partial c^A} \tag{21}$$

In case the action (17) has local symmetries the Hamiltonian is found by the methods of constrained dynamics.

We shall only consider Hamiltonians which can be expressed in the form

$$H(c,d) = H(x,p), \quad x^{A\dot{B}} = \{c^A, c^{*\dot{B}}\}, \quad p_{A\dot{B}} = \{d^*_A, d_{\dot{B}}\} \tag{22}$$

The system corresponding to the action (17) cannot be quantized in the usual way through Poisson brackets because c and d^* become vectors $\overset{>}{C}$ and $\overset{<}{D}$ and not operators in Hilbert space. Instead we shall determine the conditions which have to be imposed on $\overset{>}{C}$ and $\overset{<}{D}$ in order to obtain the usual canonical quantization of the system (22) with p as the momenta conjugate to x. For the Hamiltonian (22) the equations of motion (21) become

$$\dot{c}^A = \frac{\partial H}{\partial p_{A\dot{E}}} d_{\dot{E}}, \quad \dot{d}^*_A = -c^{*\dot{E}} \frac{\partial H}{\partial x^{A\dot{E}}} \tag{23}$$

The quantized form of these equations will be

$$\frac{d}{d\tau}\overset{>}{C}{}^A = -\frac{1}{2i\hbar}[H, X^{A\dot{E}}]\, \overset{>}{D}_{\dot{E}}, \quad \frac{d}{d\tau}\overset{<}{D}_A = -\frac{1}{2i\hbar}\, \overset{<}{C}{}^{\dot{E}}\, [H, P_{A\dot{E}}] \tag{24}$$

$$X^{A\dot{B}} = \{\overset{>}{C}{}^A, \overset{<}{C}{}^{\dot{B}}\}, \quad P_{A\dot{B}} = \{\overset{>}{D}_{\dot{B}}, \overset{<}{D}_A\} \tag{25}$$

Applying the equations of motion (24) to (25) gives

$$\begin{aligned} \frac{d}{d\tau}X^{A\dot{B}} &= -\frac{1}{2i\hbar}[H, X^{A\dot{E}}]\{\overset{>}{D}_{\dot{E}}, \overset{<}{C}{}^{\dot{B}}\} - \frac{1}{2i\hbar}\{\overset{>}{C}{}^A, \overset{<}{D}_E\}[H, X^{\dot{B}E}] \\ \frac{d}{d\tau}P_{A\dot{B}} &= -\frac{1}{2i\hbar}[H, P_{E\dot{B}}]\{\overset{>}{C}{}^E, \overset{<}{D}_A\} - \frac{1}{2i\hbar}\{\overset{>}{D}_{\dot{B}}, \overset{<}{C}{}^{\dot{E}}\}[H, P_{A\dot{E}}] \end{aligned} \tag{26}$$

For these equations to reduce to the usual space-time canonical equations of motion we must impose the commutation relations

$$\{\overset{>}{C}{}^A, \overset{<}{D}_B\} = \delta^A_B \mu(\tau) \tag{27}$$

where $\mu(\tau)$ is a real scalar function of τ. Then (26) becomes

$$\frac{d}{d\tau}X^{A\dot{B}} = -\frac{\mu(\tau)}{i\hbar}[H, X^{A\dot{B}}], \quad \frac{d}{d\tau}P_{A\dot{B}} = -\frac{\mu(\tau)}{i\hbar}[H, P_{A\dot{B}}] \tag{28}$$

or in reparametrized form

$$\frac{d}{d\bar{\tau}}X^{A\dot{B}} = -\frac{1}{i\hbar}[H, X^{A\dot{B}}], \quad \frac{d}{d\bar{\tau}}P_{A\dot{B}} = -\frac{1}{i\hbar}[H, P_{A\dot{B}}], \quad \frac{d\bar{\tau}}{d\tau} = \mu(\tau) \tag{29}$$

Though we obtain the standard space-time canonical equations of motion, they are subject to the (as we shall see) important restriction that the parameter $\bar{\tau}$ is only well defined for $\mu(\tau) \neq 0$.

Normally the compatibility of the commutation relations with the equations of motion is ensured by the Poisson brackets. This also applies in the present case to the space-time commutation relations

$$[X^\mu, X^\nu] = 0, \quad [P_\mu, P_\nu] = 0, \quad [X^\mu, P_\nu] = i\hbar\delta^\mu_\nu \tag{30}$$

which are compatible with the equations of motion (28). These equations, however, assume the validity of the Clifford commutation relations (27) which are not related to any poisson brackets. We shall prove the compatibility of these commutation relations with the equations of motion in the classical case where they reduce to

$$\{c_A, d^*_B\} = -\epsilon_{AB}\,\mu(\tau) \tag{31}$$

Since all skewsymmetric second rank spinors are proportional to ϵ_{AB}, (31) is equivalent to the vanishing of the symmetric part of the commutator

$$\{c_{(A}, d^*_{B)}\} = 0 \tag{32}$$

The equations of motion (21) subject to the constraint (32) can be obtained from

$$\int \{\dot{c}^A, d^*_A\} + c.c. - H(c,d) + \lambda^{AB}(\tau)\{c_{(A}, d^*_{B)}\} + c.c.\, d\tau, \quad \lambda^{AB} = \lambda^{BA} \tag{33}$$

by independent variation of c and d where λ^{AB} are six Lagrange multipliers. A local $SL(2.C)$ transformation

$$c_A = S_A{}^E(\tau)\bar{c}_E, \quad d^*_A = S_A{}^E(\tau)\bar{d}^*_E \tag{34}$$

turns (33) into

$$\int \{\dot{\bar{c}}^A, \bar{d}^*_A\} + c.c. - H(\bar{c}, \bar{d}) + (\dot{S}^{EA}S_E{}^B + \bar{\lambda}^{AB})\{\bar{c}_{(A}, \bar{d}^*_{B)}\} + c.c.\, d\tau \tag{35}$$

The last two terms in (35) can be made to vanish if

$$\dot{S}^{EA}(\tau)S_E{}^B(\tau) = -\bar{\lambda}^{AB}(\tau) \tag{36}$$

Taking λ to be small, the infinitesimal $SL(2.C)$ transformation

$$S^{AB}(\tau) = \epsilon^{AB} + \kappa^{AB}(\tau), \qquad \kappa^{AB} = \kappa^{BA} \tag{37}$$

turns (36) into

$$\dot{\kappa}^{AB} = -\lambda^{AB} \tag{38}$$

which can always be solved for $\kappa^{AB}(\tau)$ in terms of $\lambda^{AB}(\tau)$. The constraint (32) can therefore be absorbed into a local $SL(2.C)$ transformation of the dynamical variables and will accordingly preserve the form of the equations of motion. In section 4 we shall examine a specific model of the relativistic point particle and find that also the quantum form of the Clifford commutation relations leads to a consistent result.

3 Degenerate space-time paths

When $\mu(\tau)$ in (27) has a zero, the parameter $\bar{\tau}$ of the space-time equations of motion is ill-defined. We should therefore be prepared to encounter complete solutions $C(\tau)$,$D(\tau)$ to the equations of motion (24) which generate incomplete solutions $X(\tau)$,$P(\tau)$ to the space-time equations of motion. To understand what happens, let us assume that $\mu(0) = 0$. Then for $\tau = 0$ the commutation relations (27) reduce to

$$\{\overset{>}{C^A}(0), \overset{<}{D_B}(0)\} = 0 \tag{39}$$

Let us expand $C(\tau)$

$$\overset{>}{C}(\tau) = \overset{>}{C}(0) + \cdots + \frac{1}{n!}\overset{>}{C^{(n)}}(0)\tau^n + \cdots \tag{40}$$

The higher order derivatives $C^{(n)}(\tau)$ are obtained by differentiating the equations of motion (24) and reinserting the expressions for $\dot{X}$ and $\dot{P}$ obtained from (26). Because of the commutation relations (39) this can only result in coefficients which contain terms of the form

$$F^A_{\dot{B}}(X(0),P(0))\overset{>}{C^B}(0) \quad \text{or} \quad G^{A\dot{B}}(X(0),P(0))\overset{>}{D_{\dot{B}}}(0) \tag{41}$$

When $C(\tau)$,$D(\tau)$ is a solution to (24), so is $-C(-\tau)$,$D(\tau)$. Thus $C(\tau)$ must be odd under a change of sign of $C(0)$ and τ. It follows that in the expansion (40) all terms of even order must have coefficients of the first type in (41) and all terms of odd order must have coefficients of the second type. When therefore the expansion (40) is inserted into (5) to determine $X(\tau)$ we find that, because of the commutation relations (39), all anti-commutators between terms of odd order and terms of even order vanish. Accordingly $X(\tau)$ can only contain terms of even order and must therefore be an even function of τ:

$$X(-\tau) = X(\tau) \tag{42}$$

This implies that $C(\tau)$ reproduces $X(\tau)$ twice, making it twofold degenerate. Hence there exist complete paths in Clifford space which have either a beginning or an end in physical time. If we assume that it is the first possibility which applies, then the only way to avoid a contradiction with experience is to assume that the 'starting times' of the particles are of cosmological origin. The viability of the Clifford model therefore depends on the construction of a cosmological model which would presumably go beyond the framework of the quantum mechanics of point particles.

The classical paths will, like $X(\tau)$, be even functions of τ. In the quantum regime, however, paths for which $x(\tau) \neq x(-\tau)$ will also contribute to the transition amplitudes. Consequently the Clifford model will seem to be non-local from a space-time point of view. We shall interpret this non-locality in section 6.

4 The relativistic point particle

Since there exists no $SL(2.C)$ invariant hermitian second rank spinor, but only the real skewsymmetric metric ϵ_{AB}, the simplest reparametrization invariant action for a relativistic point particle which only depends on $\dot{c}$ is

$$-2^{\frac{7}{4}}\sqrt{m}\int \sqrt[4]{\{\dot{c}^A, \dot{c}^{*\dot{B}}\}\{\dot{c}_A, \dot{c}^*_{\dot{B}}\}}\, d\tau \tag{43}$$

The conjugate to c is

$$d^*_A = -2^{\frac{3}{4}}\sqrt{m}(\{\dot{c}^E, \dot{c}^{*\dot{F}}\}\{\dot{c}_E, \dot{c}^*_{\dot{F}}\})^{-\frac{3}{4}}\{\dot{c}_A, \dot{c}^*_{\dot{B}}\}\dot{c}^{*\dot{B}} \tag{44}$$

Not unexpectedly the Hamiltonian (20) vanishes because of reparametrization invariance. By use of the relation

$$V_{A\dot{F}}V^{B\dot{F}} = \delta^B_A V_\mu V^\mu \tag{45}$$

for a hermitian second rank spinor, we obtain from (44) the associated constraint

$$\frac{1}{2}\{d^*_A, d_{\dot{B}}\}\{d^{*A}, d^{\dot{B}}\} = m^2 \tag{46}$$

or

$$p_\mu p^\mu = m^2 \tag{47}$$

This is the same constraint as would have been obtained from the usual space-time Lagrangian $m\sqrt{\dot{x}^2}$, but with the important difference that p_μ is no longer a primary dynamical variable. The new Hamiltonian is proportional to the constraint:

$$H(\overset{>}{C}, \overset{<}{D}) = \nu(\tau)(P_\mu P^\mu - m^2) \tag{48}$$

The gauge is fixed by choosing $\nu(\tau) = \frac{1}{2m}$. By use of the space-time commutation relations the equations of motion (24) become

$$\frac{d}{d\tau}\overset{>}{C}{}^A = \frac{1}{2m}P^{A\dot{E}}\overset{>}{D}_{\dot{E}}, \qquad \frac{d}{d\tau}\overset{<}{D}_A = 0 \tag{49}$$

with the solution

$$\overset{>}{C^A}(\tau) = \overset{>}{C^A}(0) + \frac{1}{2m} P^{A\dot{E}}(0) \overset{>}{D}_{\dot{E}}(0)\tau, \qquad \overset{<}{D}_A(\tau) = \overset{<}{D}_A(0) \tag{50}$$

From (50) we obtain

$$\{\overset{>}{C^A}(\tau), \overset{<}{D}_B(\tau)\} = \{\overset{>}{C^A}(0), \overset{<}{D}_B(0)\} + \frac{1}{2m} P_{B\dot{E}}(0) P^{A\dot{E}}(0)\tau \tag{51}$$

Applying (45) and the quantum form of (47) to (51) it becomes

$$\{\overset{>}{C^A}(\tau), \overset{<}{D}_B(\tau)\} = \{\overset{>}{C^A}(0), \overset{<}{D}_B(0)\} + \delta^A_B \frac{m}{2}\tau \tag{52}$$

Accordingly, the Clifford commutation relations (27) are preserved in time by the equations of motion, and with the choice $\tau = 0$ for the zero-point of $\mu(\tau)$ we obtain

$$\mu(\tau) = \frac{m}{2}\tau, \qquad \bar{\tau} = \frac{m}{4}\tau^2 \tag{53}$$

The corresponding space-time solution is

$$X^\mu(\bar{\tau}) = X^\mu(0) + \frac{1}{m} P^\mu(0)\bar{\tau}, \qquad P_\mu(\bar{\tau}) = P_\mu(0) \tag{54}$$

In accordance with the general result in section 3, the complete solutions to the Clifford equations of motion (49) are double coverings of the incomplete solutions $X(\bar{\tau}), \bar{\tau} \geq 0$ to the space-time equations of motion.

5 Measurement principle

The measurement principle in quantum mechanics says that the (abstract) state vector is constant in time as long as no measurement is being performed. After a measurement has been performed the state vector is replaced by the eigenvector of the measured quantity for subsequent times ('state vector reduction'). This measurement principle applies equally well to Dirac's parameter-time formalism when 'time' is taken to be a parameter-time with the same direction as our $\bar{\tau}$ in the foregoing. Recognizing the primary character of Clifford space, we shall instead assume that the reduction of the state vector takes place in the positive direction of τ itself which therefore comes to represent the true direction of causality:

> Measurement Principle. *The state vector of the particle is constant in parameter-time τ as long as no measurement is being performed. When the particle is measured to be in the eigenstate $|x_P\rangle$ of X the state vector is replaced by $|c_P\rangle = |x_P\rangle$ for parameter-times $\tau > \tau_P$ where c_P is an 'eigenvalue' of $\overset{>}{C}(\tau_P)$ and c_P and $\overset{>}{C}(\tau_P)$ satisfy $\{c_P, c_P^*\} = x_P$ and $\{\overset{>}{C}(\tau_P), \overset{<}{C}(\tau_P)\} = X$ respectively*

Using a convenient terminology we shall say that the Clifford position of the particle has been measured to be c_P at $\tau = \tau_P$. The measurement principle respects the fact that since the interaction-Hamiltonians used for measuring space-time positions depend only on C *through* X, the state vector reduction in Clifford space should also be defined trough X and its eigenvalues. In the following we shall examine the consequences of this principle.

Let the space-time position of the particle have been measured to be x_Q. From (42) and (10) it follows that $C(\tau)$ satisfies the criteria in the measurement principle at two parameter-times $\tau = \pm\tau_Q$. Let us call the corresponding 'eigenvalues' for c_{Q+} and c_{Q-}. Hence the state vector will be $|c_{Q-}\rangle$ and $|c_{Q+}\rangle$ (both equal to $|x_Q\rangle$) right after $\tau = -\tau_Q$ and $\tau = \tau_Q$ respectively. If no measurement is being performed between $\tau = -\tau_Q$ and $\tau = \tau_Q$ the particle will arrive at $\tau = \tau_Q$ in the state $|c_{Q-}\rangle$. Since after $\tau = \tau_Q$ the state is $|c_{Q+}\rangle$, the transition amplitude is $\langle c_{Q-}|c_{Q+}\rangle = \langle x_Q|x_Q\rangle = 1$. Therefore the measurement principle is self-consistent as long as no measurement is being performed between $\tau = -\tau_Q$ and $\tau = \tau_Q$. Let us now assume that such a measurement *is* being performed, resulting in the space-time position x_P corresponding to the Clifford positions $c_{P\pm}$ at $\tau = \pm\tau_P$ respectively, where $\tau_P < \tau_Q$. The transition amplitude for the particle to pass from c_{Q-} through c_{P-} and c_{P+} to c_{Q+} is

$$\langle c_{Q-}|c_{P-}\rangle\langle c_{P+}|c_{Q+}\rangle = |\langle x_P|x_Q\rangle|^2 \tag{55}$$

and therefore equals the transition probability for the particle to move from x_P to x_Q. We conclude that the space-time transition probabilities arise as transition amplitudes for the complete paths in Clifford space.

Note that viewed from space-time it appears as if there are two amplitudes, one moving forward in time from x_P to x_Q and the other moving backwards in time from x_Q to x_P. This resembles the situation in the time symmetric formulation of quantum mechanics by Aharonov and Vaidman [6], Costa de Beauregard [7], and Werbos [8]. In the present model the two state vectors of time symmetric quantum mechanics are recognized to be one and the same, propagating along a path which covers the space-time path twice. The use of parameter-time in our model is necessitated by the secondary character of physical time, but it has the added advantage of ensuring manifest Lorentz invariance.

The present model should also be compared to the so-called 'double space-time interpretation of quantum mechanics' Bialynicki-Birula [9], inspired by Schwinger's time loop integrated amplitudes. The main problem in this interpretation is how to join the two space-time sheets at infinity to allow a particle to travel along a single path on the two sheets.

The choice of taking the causal direction of state vector reduction to be in the positive direction of parameter-time τ rather than of 'affine time' $\bar{\tau}$ strongly suggests that the same should apply to the direction of propagation of classical fields. The following heuristic observation shows that this is not necessarily inconsistent with experience. Let the union of all possible particle trajectories for $\tau \leq 0$ and for $\tau \geq 0$ form regions Ω_- and Ω_+ of Clifford space which correspond to the same space-time region. For the field to propagate in the positive direction of τ we should choose the advanced field on Ω_- and the retarded field on Ω_+. The contribution to the electrodynamic action in the proper-time interval $[\bar{\tau}_1; \bar{\tau}_2]$ of a test-particle with charge e traversing this region is

$$\frac{1}{2}\int_{-\tau_2}^{-\tau_1} m\sqrt{\dot{x}^2} + A_{adv}e(-\dot{x})\, d\tau + \frac{1}{2}\int_{\tau_1}^{\tau_2} m\sqrt{\dot{x}^2} + A_{ret}e\dot{x}\, d\tau$$

$$= \int_{\bar{\tau}_1}^{\bar{\tau}_2} m\sqrt{\dot{x}^2} + (\frac{1}{2}A_{adv} + \frac{1}{2}A_{ret})e\dot{x}\, d\bar{\tau} \qquad (56)$$

The test-particle will therefore detect the effective field to be the time symmetric half-advanced plus half-retarded field. Assuming complete absorption and no self-interaction Wheeler and Feynman [10] have shown that this time symmetric field leads to the conventional rules of electrodynamics.

6 Interpretation of non-locality

Consider a particle which travels in space-time from a point P to a point Q and is forced to travel trough two alternative points S_1 and S_2. This corresponds to the double slit experiment with the two slits being opened at given times. As follows from our foregoing discussion the particle can follow four alternative sets of paths in Clifford space corresponding to the the four sequences of positions in Clifford space ordered according to parameter time:

$$c_{Q-},\ c_{S_i-},\ c_{P-},\ c_{P+},\ c_{S_j+},\ c_{Q+},\ i,j = 1,2 \qquad (57)$$

The amplitude for the particle to travel from c_{Q-} to c_{Q+} is the sum of the amplitudes for all four different sets of paths

$$\sum_{i,j=1}^{2} \langle c_{Q-}|c_{S_i-}\rangle\langle c_{S_i-}|c_{P-}\rangle\langle c_{P+}|c_{S_j+}\rangle\langle c_{S_j+}|c_{Q+}\rangle$$

$$= |\langle x_P|x_{S_1}\rangle\langle x_{S_1}|x_Q\rangle + \langle x_P|x_{S_2}\rangle\langle x_{S_2}|x_Q\rangle|^2 \qquad (58)$$

which is the well known probability for the particle to travel from P to Q. The customary interpretation of this transition probability is that there are two alternative paths and that the transition probability is the sum of the probabilities for each path plus two interference terms which seem to signal a non-local influence of one path on the other. From (58) we see that there are really four different sets of paths and that the two interference terms are the amplitudes for the two sets of paths where the particle goes through each slit at opposite parameter times. The apparent non-locality can be entirely attributed to the twofold degeneracy of the space-time paths.

If we measure the position of the particle at one of the slits, say S_1, then according to our measurement principle the particle has to travel through both

c_{S_1-} and c_{S_1+} or neither of them, and this excludes the two sets of paths where the particle passes through both slits. This removes the interference terms in accordance with the space-time view of quantum mechanics. This analysis is readily extended to a many-slit experiment by observing that all interference terms arise from pairs of slits.

As the second example of non-locality we shall consider an EPR type of measurement. Consider a composite system (PQ) consisting of two spin $\frac{1}{2}$ particles P and Q. First the position and the total spin of the composite system is measured to be $x_{(PQ)}$ and 0. After this measurement P and Q become separated by a spacelike distance and their position and spin along some axis are measured to be x_P and $\frac{1}{2}$ and x_Q and $-\frac{1}{2}$ respectively. The last two measurements appear to be correlated despite the spacelike separation of P and Q, giving thereby the impression of 'action at a distance'. However, according to our measurement principle, the position measurements, and together with them the spin measurements, each correspond to two measurements in Clifford space at opposite values of τ . If the measurements of $x_{(PQ)}$, x_P and x_Q correspond to parameter-times $\tau = \pm\tau_{(PQ)}$, $\tau = \pm\tau_P$ and $\tau = \pm\tau_Q$ respectively, then the sequence of events for negative τ can be described as follows. First at parameter-times $\tau = -\tau_P$ and $\tau = -\tau_Q$ the spins along some axis of P and Q are measured to be $\frac{1}{2}$ and $-\frac{1}{2}$ respectively. At the later parameter-time $\tau = -\tau_{PQ} > -\tau_P, -\tau_Q$, P and Q merge into a composite system (PQ) and the total spin is measured to be 0 . We would not object to this last sequence of events because it suggests no correlation between the spacelike separated states of P and Q. Rather, it suggests an obvious correlation between the states of P and Q on the one hand and the state of the composite system (PQ) on the other, which invokes no need for 'action at a distance'. Nevertheless these two sequences of events, corresponding to opposite values of τ, together form a single series of events in the causal direction of state vector reduction in Clifford space and are both the result of the same space-time measurements on a degenerate space-time path. They are therefore on an equal footing and we conclude that it has no absolute meaning to say whether the spin-measurements on P and Q are correlated or independent. Accordingly, the apparent manifestation of 'action at a distance' loses its significance.

A Appendix. Clifford algebras and Hermitian quadratic forms

A real Clifford algebra arises naturally as the 'square root' of a real quadratic form Q on a linear space V:

$$v^2 = Q(v),\ v \in V \tag{59}$$

Q can have any signature (N_+, N_0, N_-) . In case Q is degenerate ($N_0 \neq 0$), the algebra contains Grassmann elements. When v is expanded on an orthogonal basis e_i of V it follows that (59) is satisfied if

$$\frac{1}{2}\{e_i, e_j\} = \delta_{ij} Q(e_i) \tag{60}$$

The basis e_i generates the Clifford algebra. Consider now a quadratic form Q with signature $(2N_+, 2N_0, 2N_-)$. We can rearrange the generators e_i into two sets a_i and b_i, $i = 1, \ldots, N$ which when normalized satisfy

$$\frac{1}{2}\{a_i, a_j\} = \frac{1}{2}\{b_i, b_j\} = \delta_{ij} Q(a_i), \qquad \{a_i, b_j\} = 0 \tag{61}$$

a_i and b_i can be used as 'real' and 'imaginary' parts to define the complex quantities

$$f_j = a_j + i\, b_j \tag{62}$$

where i is the imaginary unit. The elements f_j are seen to satisfy the commutation relations

$$\frac{1}{4}\{f_i, f_j^*\} = \delta_{ij} Q(a_i), \qquad \{f_i, f_j\} = 0 \tag{63}$$

where $*$ is any complex involution induced by a self-involution in the real algebra. The algebra generated by f_i is a complex Clifford algebra.

For any hermitian quadratic form H on a linear space V there exists a complex Clifford algebra generated by V which satisfies

$$\frac{1}{2}\{v, v^*\} = H(v),\ v \in V, \qquad v^2 = 0,\ v \in V \tag{64}$$

The proof follows by expanding v on an orthogonal basis f_i of V. (64) is seen to be satisfied if

$$\frac{1}{2}\{f_i, f_j^*\} = \delta_{ij} H(f_i), \qquad \{f_i, f_j\} = 0 \tag{65}$$

which is recognized to be the generating algebra of a complex Clifford algebra. Expressed in matrix language, (64) implies that any hermitian matrix H_{ij} can be expressed in terms of elements v_i of a complex Clifford algebra:

$$H_{ij} = \{v_i, v_j^*\}, \qquad \{v_i, v_j\} = 0 \tag{66}$$

References

[1] Penrose, R. (1967). In *Battelle Rencontre*, C.M. DeWitt and J.A. Wheeler, eds., Benjamin, New York.

[2] Schwartz, J.H. and Van Nieuwenhuizen, P. (1982). *Lettere al Nuovo Cimento*, 34, 21.

[3] Borchsenius, K. (1987). *General Relativity and Gravitation*, 19, 643.

[4] Borchsenius, K. (1989). *General Relativity and Gravitation*, 21, 959.

[5] Borchsenius, K. (1995). *International Journal of Theoretical Physics*, 34, 1863.

[6] Aharonov, Y. and Vaidman, L. (1990). *Phys.Rev.*, A41, 11.

[7] Costa de Beauregard, O. (1989). In *Bell's theorem, quantum theory, and conceptions of the universe*(ed. M. Kafatos).Kluwer, Dordrecht.

[8] Werbos, P. (1989). In *Bell's theorem, quantum theory, and conceptions of the universe*(ed. M. Kafatos).Kluwer, Dordrecht.

[9] Bialynicki-Birula, I. (1986). In *Quantum Concepts in Space and Time*, R.Penrose and C.J.Isham, eds., Claredon Press, Oxford, p. 226.

[10] Wheeler, J.A. and Feynman, R. (1945) *Reviews of Modern Physics*, 17, 157.

M P E J

Mathematical Physics Electronic Journal

ISSN 1086-6655
Volume 6, 2000

Paper 5
Received: Mar 16, 2000, Revised: Nov 9, 2000, Accepted: Nov 27, 2000
Editor: R. de la Llave

Periodic orbits of renormalisation for the correlations of strange nonchaotic attractors

B. D. Mestel
School of Mathematical Sciences
University of Exeter
Exeter
EX4 4QE, UK

A. H. Osbaldestin
Department of Mathematical Sciences
Loughborough University
Loughborough
LE11 3TU, UK

Abstract

We calculate all piecewise-constant periodic orbits (with values ± 1) of the renormalisation recursion arising in the analysis of correlations of the orbit of a point on a strange nonchaotic attractor. Our results make rigorous and generalise previous numerical results.

1 Introduction

The occurrence and robustness of strange nonchaotic attractors was first noted by Grebogi *et al* in their seminal paper [4]. A strange nonchaotic attractor is an attractor whose geometry is "strange", and on which the dynamics is "nonchaotic" (i.e. for which there is no positive Lyapunov exponent). Grebogi *et al* [4] considered quasiperiodically forced systems of the type

$$x_{n+1} = f(x_n, \theta_n)\,, \tag{1.1}$$

$$\theta_{n+1} = \theta_n + \omega \pmod 1\,, \tag{1.2}$$

in which ω is irrational, the dynamical variable x (and f) may be scalar or higher dimensional, and f satisfies $f(x, \theta+1) = f(x, \theta)$. (Such systems are examples of skew-product systems.) Strange nonchaotic attractors have since been reported in other theoretical and experimental situations. References to such occurrences may be found in [7].

In the scalar example studied in some detail in [4] the function f in equation (1.1) takes the form

$$f(x, \theta) = 2\lambda \tanh(x) \sin(2\pi\theta)\,. \tag{1.3}$$

For $|\lambda| < 1$ the invariant line $x = 0$ is the attractor. When $|\lambda| > 1$ this invariant line is no longer an attractor; however, since orbits are confined to a bounded region of phase space an attractor does exist. This is shown to be strange and nonchaotic in [4].

In [9] the autocorrelation of the orbit on the strange attractor is seen to be self-similar and possess a singular continuous spectrum. As in [2], however, we shall confine our attention to a coarser description of the dynamics. Namely we consider only the sign of the variable x, defining

$$y = -\mathrm{sign}(x)\,. \tag{1.4}$$

For the systems under consideration the dynamics are thereby reduced to the linear circle map (1.2) together with a recording (y) of whether θ is in $[0, 1/2)$ or $(1/2, 1)$.

In the case of golden mean forcing, the autocorrelation function of y is seen to be self-similar with structure determined by the renormalisation recursion relation

$$Q_n(x) = Q_{n-1}(-\omega x) Q_{n-2}(\omega^2 x + \omega)\,, \tag{1.5}$$

where $\omega = (\sqrt{5} - 1)/2$ is the golden mean. For completeness we shall include from [2] the derivation of this equation, in section 2.

In [2] Feudel *et al* numerically found a piecewise-constant period-6 orbit of this recursion. This periodic orbit is shown in figure 1.

In this paper we shall give an explicit construction of this periodic orbit, and moreover analyse all piecewise-constant periodic orbits. These periodic orbits correspond to taking a different coarse-grained description from merely noting in which half of the interval θ lies.

In a different but related work, Kuznetsov *et al* [7] have given an elegant analysis of the birth of a strange nonchaotic attractor. The same recursion is used to explain the occurrence of universal scaling factors. In this case however periodic orbits of (1.5) of a different nature are considered.

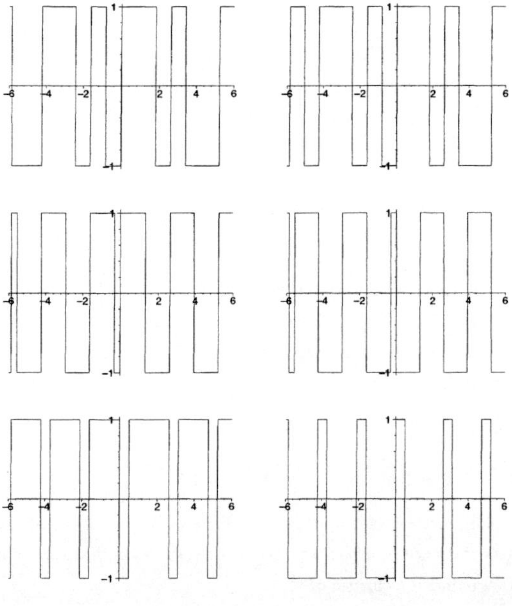

Figure 1: The period-6 orbit discovered by Feudel *et al* ([2])

Remarkably Ketoja and Satija [5] also derive this same equation in their analysis of the self-similar fluctuations of the localized eigenstates of the golden mean Harper equation (also known as the almost Mathieu equation)

$$\psi_{n+1} + \psi_{n-1} + 2\lambda \cos(2\pi(n\omega + \phi))\psi_n = E\psi_n \tag{1.6}$$

in the supercritical regime $\lambda > 1$. This finite difference eigenvalue equation is valuable in the study the localization transition in incommensurate systems. The recursion (1.5) helps explain the universality of the supercritical regime. Note that in [5] the iteration occurs in the form

$$\tilde{Q}_n(x) = -\tilde{Q}_{n-1}(-\omega x)\tilde{Q}_{n-2}(\omega^2 x + \omega)\,, \tag{1.7}$$

but the substitution $\tilde{Q}_n = -Q_n$ renders it equivalent to (1.5). A fixed point of this recursion characterises the universal fluctuations, and this is numerically found in [5]. The same recursion is also used in [5] to analyse a generalised Harper equation describing Bloch electrons on a square lattice with nearest neighbour anisotropy as in (1.6), and the addition of a next-nearest neighbour coupling term. Many periodic orbits are found, and Ketoja and Satija [5] conjecture the existence of a universal strange attractor under the action of the renormalization operator.

In a recent paper [8] we have proved that indeed there is a fixed point of the type numerically found in [5]. (See also [6].) We hope to be able to extend our results on smooth solutions to shed more light on the work of Kuznetsov *et al* [7] on the scenario of the birth of a strange nonchaotic attractor.

In [6] these two seemingly distinct scenarios are linked, and indeed an analogy with the critical dissipative standard map is also drawn.

In this paper we study periodic orbits of (1.5) for which Q_n is piecewise-constant with Q_n taking values ± 1 for all $x \in \mathbb{R}$. By piecewise-constant we mean that for each n the function $Q_n(x) = \pm 1$ has finitely many discontinuities in any bounded interval of $\mathbb{R}$, although Q_n may, and generally will, have infinitely many discontinuities on $\mathbb{R}$. Although this condition might appear somewhat restrictive, we shall see in section 2 that this is the appropriate condition for the renormalisation analysis given by Feudel *et al* in [2] of the correlation function of the sign of orbits in strange nonchaotic attractors. Moreover, as we shall see, the periodic orbit structure for (1.5) is already very rich in this case.

Let us define, for $x \in \mathbb{R}$, the discontinuity function

$$R_n(x) = \frac{Q_n(x+)}{Q_n(x-)}, \tag{1.8}$$

the ratio of the right-hand limit to the left-hand limit of Q_n at x. Then, since every discontinuity of Q_n is isolated, R_n is well defined.

Because we are not primarily interested in the value of Q_n at the discontinuity points, we shall identify any two functions having the same discontinuity points (i.e. those x with $R_n(x) = -1$) and agreeing at all continuity points (i.e. at those x with $R_n(x) = 1$).

We are now in a position to give a summary of the main results of the paper. In what follows we shall show that, if Q_n is a periodic orbit of (1.5) of period p, then R_n is also periodic with period m where $m \mid p$. (Here, and subsequently, the *period* is understood to refer to the minimal period.) Moreover we shall see that $p = m$, $2m$, or $3m$. Reducing the study of periodic orbits of (1.5) on $\mathbb{R}$ to a neighbourhood of the fundamental interval $[-\omega, 1]$, we shall identify the set of discontinuities on $[-\omega, 1]$ for the orbit and show that it is a finite union of periodic orbits of the map $F : [-\omega, 1] \to [-\omega, 1]$ given by

$$F(x) = \begin{cases} -\omega^{-1}x, & x \in [-\omega, \omega^2]; \\ \omega^{-2}x - \omega^{-1}, & x \in [\omega^2, 1]. \end{cases} \tag{1.9}$$

Such periodic orbits are classified by their codes (also called itineraries or kneading sequences) and we shall determine the possible values of m in terms of the codes of these orbits. We shall also identify in detail the cases in which $p = m$, $2m$ and $3m$ can occur. This latter analysis is somewhat complicated and involves some non-intuitive number-theoretic conditions on the codes. A consequence of this analysis is that we shall show *inter alia*

Theorem 1. *For every positive integer $p \geq 1$ there is a periodic orbit Q_n of (1.5) of period p.*

The paper is organised as follows. In the next section, closely following Feudel *et al* [2], we briefly review how the recursion (1.5) arises in the renormalisation analysis of the autocorrelation function for a strange nonchaotic attractor. In section 3 we establish some notation and indicate how an iterated function system and its 'inverse', the function F above (1.9), naturally arise in the recursion. The iterated function system has as invariant set the interval $[-\omega, 1]$, and we show in section 4 that it suffices to consider the recursion (1.5) restricted to this interval. Since we are solely concerned with piecewise-constant functions Q_n taking the values ± 1, much of the nature of the recursion can be understood from a study of the discontinuity function R_n defined above (1.8). This we consider in detail in section 5. However an analysis of the discontinuity function is not in itself sufficient, and in section 6 we relate the periodicity of the discontinuities to that of Q_n itself. This relationship is nontrivial and requires a careful consideration of the orbits of the map F. The results are summarised in section 7. In section 8 we give an analysis of the construction of periodic orbits of (1.5). The period-6 orbit of Feudel *et al* [2] shown in figure 1 is seen to be but one example.

2 Renormalisation analysis of the autocorrelation function

In this section we review the work of Feudel *et al* [2] and show in particular how equation (1.5) arises in a renormalisation analysis of the autocorrelation function for a strange nonchaotic attractor.

In all that follows we shall take $\omega = (\sqrt{5}-1)/2$ and assume that $\lambda > 0$. Recall that the Fibonacci numbers are given by: $F_0 = 0$, $F_1 = 1$, $F_n = F_{n-1} + F_{n-2}$, for $n > 1$.

In terms of the discrete variable y defined above (1.4), our mapping (1.1)–(1.2), with the choice of f given by a function of the form (1.3), is now just

$$y_{n+1} = y_n \Phi(\theta_n)\,, \tag{2.1}$$

$$\theta_{n+1} = \theta_n + \omega \pmod 1\,, \tag{2.2}$$

where the "modulation function"

$$\Phi(\theta) = \begin{cases} -1\,, & 0 \le \theta < 1/2\,; \\ +1\,, & 1/2 \le \theta < 1\,. \end{cases} \tag{2.3}$$

Thus

$$y_n = \prod_{k=0}^{n-1} \Phi(\theta_k)\,, \tag{2.4}$$

$$\theta_n = \theta_0 + n\omega \pmod 1\,, \tag{2.5}$$

where we take $y_0 = 1$. The dynamics of y are nothing other than the recording of the location of iterates of the linear circle map, and depend only on the initial angle θ_0.

The autocorrelation function $C(t)$ of y (which has zero mean and unit variance) is the limit time average

$$C(t) = \lim_{T\to\infty} \frac{1}{T} \sum_{i=0}^{T-1} y_i y_{i+t}\,, \tag{2.6}$$

which in view of (2.4), and the fact that $\Phi = \pm 1$, is

$$C(t) = \lim_{T\to\infty} \frac{1}{T} \sum_{i=0}^{T-1} \prod_{k=0}^{i-1} \Phi(\theta_k) \prod_{k=0}^{i+t-1} \Phi(\theta_k) = \lim_{T\to\infty} \frac{1}{T} \sum_{i=0}^{T-1} \prod_{k=i}^{i+t-1} \Phi(\theta_k)\,. \tag{2.7}$$

Now the ergodicity of the linear circle map allows us to write

$$\lim_{T\to\infty} \frac{1}{T} \sum_{i=0}^{T-1} \prod_{k=i}^{i+t-1} \Phi(\theta_k) = \int_0^1 \prod_{k=i}^{i+t-1} \Phi(\theta_k) d\theta_0\,, \tag{2.8}$$

and, since Φ has unit period, we may also change the integration variable (initial condition θ_0) to $\theta_0 - i\omega$ resulting in

$$C(t) = \int_0^1 \prod_{k=0}^{t-1} \Phi(\theta_k) d\theta = \int_0^1 y_t(\theta) d\theta\,, \tag{2.9}$$

where we explicitly note the dependence of y_t on the (initial) angle θ.

The autocorrelation function is observed to have scaling about Fibonacci times, and so to analyse this we define $S_n(\theta) = y_{F_n}(\theta)$, and have, with $\theta_k = \theta + k\omega \pmod 1$,

$$S_n(\theta) = \prod_{k=0}^{F_n-1} \Phi(\theta_k) \tag{2.10}$$

$$= \prod_{k=0}^{F_{n-1}-1} \Phi(\theta_k) \prod_{k=F_{n-1}}^{F_n-1} \Phi(\theta_k) \tag{2.11}$$

$$= S_{n-1}(\theta) S_{n-2}(\theta + F_{n-1}\omega)\,, \tag{2.12}$$

which, using the fact that $F_{n-1}\omega = F_{n-2} - (-\omega)^{n-1}$, is

$$S_n(\theta) = S_{n-1}(\theta) S_{n-2}(\theta - (-\omega)^{n-1})\,. \tag{2.13}$$

To analyse the scaling, we define $Q_n(x) = S_n((-\omega)^n x)$ giving

$$Q_n(x) = Q_{n-1}(-\omega x) Q_{n-2}(\omega^2 x + \omega)\,, \tag{2.14}$$

which is equation (1.5). As noted in [2]

$$C(F_n) = \int_0^1 y_{F_n}(\theta) d\theta = \int_0^1 S_n(\theta) d\theta = \frac{1}{(-\omega)^{-n}} \int_0^{(-\omega)^{-n}} Q_n(x) dx\,. \tag{2.15}$$

Thus the autocorrelation function for Fibonacci times can be determined from the average of the function Q_n. For n not a multiple of three we have that F_n is odd which gives $C(F_n) = 0$. Indeed, as above, by changing the range of the product we may write

$$C(2m+1) = \int_0^1 \prod_{k=0}^{2m} \Phi(\theta_k) d\theta = \int_0^1 \prod_{k=-m}^{m} \Phi(\theta_k) d\theta = \int_0^1 \Phi(\theta) \prod_{k=1}^{m} \Phi(\theta_k)\Phi(\theta_{-k}) d\theta = 0\,, \tag{2.16}$$

since the integrand is odd about 1/2. When n is a multiple of three it is numerically observed in [2] that the average approaches approximately 0.55 for large n. This is the relative height of the secondary peaks in the autocorrelation function.

The results of this paper explain the periodic behaviour of the functions Q_n in the specific example studied by Feudel *et al*, and also determine the behaviour in the presence of more general modulation than equation (2.3).

3 Iterated function system and the inverse map F

We may write equation (1.5) in the form

$$Q_n(x) = Q_{n-1}(\phi_1(x)) Q_{n-2}(\phi_2(x))\,, \tag{3.1}$$

where

$$\phi_1(x) = -\omega x\,, \qquad \phi_2(x) = \omega^2 x + \omega\,, \tag{3.2}$$

and $\omega = (\sqrt{5} - 1)/2$ is the golden mean satisfying $\omega^2 + \omega = 1$.

Associated with this equation is an iterated function system (IFS) on $\mathbb{R}$ given by the two contractions ϕ_1, ϕ_2 satisfying the following properties:

1. ϕ_1 and ϕ_2 are linear contractions with fixed points 0 and 1 respectively, and with $\phi_1'(x) = -\omega$ and $\phi_2'(x) = \omega^2$.

2. The interval $I = [-\omega, 1]$ is the fixed point set for the IFS. Indeed

$$\phi_1([-\omega, 1]) = [-\omega, \omega^2]\,, \qquad \phi_2([-\omega, 1]) = [\omega^2, 1]\,, \tag{3.3}$$

so that

$$\phi_1(I) \cup \phi_2(I) = I\,. \tag{3.4}$$

We shall henceforth refer to I as the *fundamental interval.*

3. The fundamental interval I is the attractor for the IFS. Indeed given any compact subset $K \subset \mathbb{R}$ and any $\varepsilon > 0$, there exists $N \in \mathbb{N}$ such that for any $k \geq N$ and any choice $i_1, \ldots, i_k \in \{1, 2\}$ we have

$$\phi_{i_1} \circ \cdots \circ \phi_{i_k}(x) \in [-\omega - \varepsilon, 1 + \varepsilon] \tag{3.5}$$

for any $x \in K$. This property will be important when we consider the behaviour of equation (1.5) outside the fundamental interval I.

We refer the reader to the book [1] for the theory of iterated function systems.

On the fundamental interval we may define a unique inverse map to the pair ϕ_1, ϕ_2. Let $F : [-\omega, 1] \to [-\omega, 1]$ be defined by

$$F(x) = \begin{cases} \phi_1^{-1}(x) = -\omega^{-1}x, & x \in [-\omega, \omega^2]\,; \\ \phi_2^{-1}(x) = \omega^{-2}x - \omega^{-1}, & x \in [\omega^2, 1]\,, \end{cases} \tag{3.6}$$

as drawn in figure 2.

We shall see below that periodic points of F correspond to discontinuities of the periodic solutions of (1.5). It is therefore appropriate to study the periodic orbit structure of F, but, before so doing, it is worth noting that for any periodic point $y \in [-\omega, 1]$, precisely one of $\phi_1(y)$, $\phi_2(y)$ is also a periodic point of F. For suppose $F^\ell(y) = y$ for some $\ell \in \mathbb{N}$. Then $F(F^{\ell-1}(y)) = y$, so $\phi_i^{-1}(F^{\ell-1}(y)) = y$ for some $i \in \{1, 2\}$, which depends on whether the periodic point $F^{\ell-1}(y) \in [-\omega, \omega^2]$ (in which case $i = 1$), or $F^{\ell-1}(y) \in [\omega^2, 1]$ (in which case $i = 2$). We have that $F^{\ell-1}(y) \neq \omega^2$, since ω^2 is not periodic under F. Thus one of $\phi_1(y)$, $\phi_2(y)$ equals $F^{\ell-1}(y)$, which is periodic.

Now suppose that both $\phi_1(y)$ and $\phi_2(y)$ are periodic. Then there exist ℓ_1, $\ell_2 \in \mathbb{N}$ such that $F^{\ell_1}(\phi_1(y)) = \phi_1(y)$, $F^{\ell_2}(\phi_2(y)) = \phi_2(y)$. Then $\phi_1(y) = F^{\ell_1\ell_2}(\phi_1(y)) = F^{\ell_2\ell_1}(\phi_2(y)) = \phi_2(y)$, where we have used the fact that $F(\phi_i(x)) = x$, for $i = 1$, 2, a simple consequence of the definition of F. Now the only solution of the equation $\phi_1(y) = \phi_2(y)$ is $y = -\omega$ so we must have $\phi_1(y) = \omega^2 = \phi_2(y)$, which is impossible since ω^2 is not a periodic point of F and the result is proved.

We now consider periodic orbits of the map F.

We may analyse the dynamics of F in terms of the code of a point $x \in I$. It is convenient for our purposes to define the code in terms of the symbols 1 and 2, rather than 0, 1, or +1, −1, as is usually done. Let

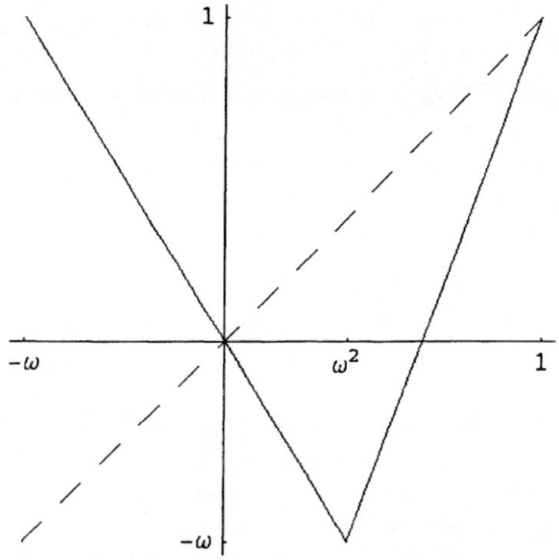

Figure 2: The function F.

the interval $[-\omega, \omega^2)$ be encoded with the symbol 1 and $(\omega^2, 1]$ with the symbol 2. We define the code of $x \in I$ to be the sequence $(a_n)_{n \geq 0}$ in $\{1,2\}^{\mathbb{N}_0}$ given by

$$a_n = \begin{cases} 1, & F^n(x) \in [-\omega, \omega^2)\,; \\ 2, & F^n(x) \in (\omega^2, 1]\,. \end{cases} \tag{3.7}$$

As is usual we ignore the (countable) set of points whose orbits under F include the point ω^2. (Such points are not periodic points of F.) Hence the codes are all infinite sequences. In terms of the code $a_0 a_1 a_2 \ldots$ of a point $x \in [-\omega, 1]$, we have

$$F(x) = (-\omega^{-1})^{a_0} x - (a_0 - 1)\omega^{-1}\,. \tag{3.8}$$

Since F is uniformly expanding ($|F'(x)| \geq \omega^{-1}$) every point $x \in I$ corresponds to a unique code and vice versa. In particular, periodic orbits of F correspond to periodic codes in $\{1,2\}^{\mathbb{N}_0}$ under the shift map σ:

$$\sigma(a_0 a_1 a_2 \ldots) = a_1 a_2 \ldots \tag{3.9}$$

A periodic orbit $y_0, y_1, \ldots, y_{k-1}$ of period k of F is given uniquely by a periodic code

$$a_0 a_1 \ldots a_{k-1} a_0 a_1 \ldots a_{k-1} \cdots, \tag{3.10}$$

which we henceforth denote by $\overline{a_0 a_1 \ldots a_{k-1}}$.

It is straightforward to calculate the periodic orbit $y_0, y_1, \ldots, y_{k-1}$ of F corresponding to a given code $\overline{a_0 a_1 \ldots a_{k-1}}$. For we have $\phi_{a_{k-1}}^{-1} \circ \cdots \circ \phi_{a_0}^{-1}(y_0) = y_0$, or, equivalently, $\phi_{a_0} \circ \cdots \circ \phi_{a_{k-1}}(y_0) = y_0$. The (unique) solution of this equation is readily calculated to be

$$y_0 = \frac{-\sum_{j=0}^{k-1}(a_j - 1)(-\omega)^{1+\sum_{i=0}^{j-1} a_i}}{1 - (-\omega)^{\Sigma a_j}}, \tag{3.11}$$

where empty sums are defined to be zero. The other points of the orbit may be calculated by applying this formula with the code $a_0a_1\dots a_{k-1}$ cyclically permuted.

For example, $F(y)$ has two fixed points: $y = 0$ with code 1, and $y = 1$ with code 2. The period-2 orbit with code 21 is given by $y_0 = 1/2$ and $y_1 = -\omega/2$. It is the fixed point $y = 0$ and this period-2 orbit that are the discontinuity points in the fundamental interval of the period-6 orbit shown in figure 1. As an example of applying the formula (3.11), we calculate the period-4 orbit of F with code 1211: $y_0 = -\omega^2/(1+\omega^5)$, $y_1 = \omega/(1+\omega^5)$, $y_2 = -\omega^4/(1+\omega^5)$, $y_3 = \omega^3/(1+\omega^5)$.

In what follows it will be the code of the periodic orbit that is important, not the orbit itself. Therefore, from now on, we shall principally refer to periodic orbits of F just by their codes.

4 Reduction of Q_n on $\mathbb{R}$ to the fundamental interval

In this section we consider equation (3.1) outside the fundamental interval $[-\omega, 1]$ i.e., on the whole of $\mathbb{R}$. In what follows we restrict to Q_n taking values ± 1.

Because the fundamental interval I attracts points under the IFS (3.2), we have the following lemma:

Lemma 1. *Let Q_0, Q_1 be initial conditions on $\mathbb{R}$ and let $\varepsilon > 0$ be such that $Q_0(x) = Q_1(x) = 1$, for all $x \in [-\omega-\varepsilon, 1+\varepsilon]$, and let Q_n satisfy equation (3.1). Then for each $L > 1$, there exists $N > 0$ (depending only on L) such that $Q_n(x) = 1$ for all $x \in [-L, L]$ and all $n > N$.*

Proof. Let Q_0, Q_1 satisfy the hypotheses of the lemma, and let $L > 0$ be given. Since $\phi_1([-\omega-\varepsilon, 1+\varepsilon])$, $\phi_2([-\omega-\varepsilon, 1+\varepsilon]) \subseteq [-\omega-\varepsilon, 1+\varepsilon]$, we have that $Q_n(x) = 1$ for all $n \geq 0$ and all $x \in [-\omega-\varepsilon, 1+\varepsilon]$.

Now, from the properties of iterated function systems (section 3), it follows that there exists $N_1 \in \mathbb{N}$ such that for any $k \geq N_1$ and any choice $i_1, \dots, i_k \in \{1, 2\}$ we have

$$\phi_{i_1} \circ \cdots \circ \phi_{i_k}(x) \in [-\omega-\varepsilon, 1+\varepsilon] \tag{4.1}$$

for all $x \in [-L, L]$. Iterating (3.1), we see that $Q_n(x)$ may be written as a product

$$Q_n(x) = \prod_{i_1, i_2, \dots, i_k \in \{1,2\}} Q_{n-\sum_{j=1}^k i_j}(\phi_{i_1} \circ \cdots \circ \phi_{i_k}(x)). \tag{4.2}$$

Hence setting $N = 2N_1$, we have for $n > N$ and $x \in [-L, L]$,

$$Q_n(x) = \prod_{i_1, i_2, \dots, i_k \in \{1,2\}} Q_{n-\sum_{j=1}^k i_j}(\phi_{i_1} \circ \cdots \circ \phi_{i_k}(x)) = 1. \tag{4.3}$$

This completes the proof of the lemma. □

From the lemma we may prove the following proposition:

Proposition 1. *Let Q_n be a piecewise-constant periodic solution of (3.1) of period p on $\mathbb{R}$ with $Q_n(1+) = Q_n(1)$. Then Q_n is periodic with period p on the fundamental interval I. Conversely, suppose that Q_n is periodic with period p on I. Then there is a unique extension $\tilde{Q}_n$ of Q_n to $\mathbb{R}$ such that $\tilde{Q}_n$ is periodic on $\mathbb{R}$ with period p.*

Proof. First of all let Q_n be periodic on $\mathbb{R}$ with period p, and with $Q_n(1+) = Q_n(1)$ for all $n \geq 0$. Then, clearly, Q_n is periodic on I with period p' dividing p. Let $\varepsilon > 0$ be chosen so that for all $n \geq 0$ there are no discontinuities of Q_n in the intervals $[-\omega - \varepsilon, -\omega)$ and $(1, 1+\varepsilon]$, and such that ϕ_1 and ϕ_2 map $[-\omega - \varepsilon, -\omega)$ into I. Such an ε exists since the discontinuities are isolated on $\mathbb{R}$, Q_n is periodic, and $\phi_1(-\omega) = \phi_2(-\omega) = \omega^2$, which is not a discontinuity of any Q_n. Furthermore, since Q_n has no discontinuities on $(1, 1+\varepsilon]$, we have that $Q_n(x) = Q_n(1)$ for $x \in [1, 1+\varepsilon]$ and so Q_n is periodic on $[1, 1+\varepsilon]$ with period dividing p'. Now for $x \in [-\omega - \varepsilon, -\omega)$, we have that $\phi_1(x), \phi_2(x) \in I$, so from (3.1) we have that $Q_n(x)$ is periodic with period dividing p'. By the multiplicative property of equation (3.1), the functions $\tilde{Q}_n = Q_{n+p'}/Q_n$ satisfy equation (3.1) and $\tilde{Q}_0$, $\tilde{Q}_1$ are identically 1 on $[-\omega - \varepsilon, 1+\varepsilon]$. Applying lemma 1, we have that $Q_{n+p'}(x) = Q_n(x)$ for all $x \in \mathbb{R}$, since Q_n is periodic on $\mathbb{R}$. Hence $p = p'$ and Q_n is periodic with period p on I.

Conversely, suppose Q_n is periodic with period p on I. Let $\varepsilon > 0$ be given as above such that $\phi_1([-\omega - \varepsilon, -\omega])$ and $\phi_2([-\omega - \varepsilon, -\omega])$ do not contain discontinuities of Q_n for $n \geq 0$. (Such an ε exists since $\omega^2 = \phi_1(-\omega) = \phi_2(-\omega)$ is not a discontinuity of any Q_n, and there are only finitely many discontinuities of the Q_n on I since Q_n is periodic.) Then we may extend Q_0, Q_1 to $[-\omega - \varepsilon, 1+\varepsilon]$ by setting $Q_0(x) = Q_0(-\omega)$, $Q_1(x) = Q_1(-\omega)$ for $x \in [-\omega - \varepsilon, -\omega)$ and $Q_0(x) = Q_0(1)$, $Q_1(x) = Q_1(1)$ for $x \in (1, 1+\varepsilon]$. Moreover, since $\phi_1([-\omega - \varepsilon, 1+\varepsilon])$, $\phi_2([-\omega - \varepsilon, 1+\varepsilon]) \subseteq [-\omega - \varepsilon, 1+\varepsilon]$, for $n \geq 2$ we may define Q_n on $[-\omega - \varepsilon, -\omega)$ and $(1, 1+\varepsilon]$ by equation (3.1) and then Q_n is periodic with period p on $[-\omega - \varepsilon, 1+\varepsilon]$. Consider now, the initial conditions $\hat{Q}_0 = Q_p/Q_0$, $\hat{Q}_1 = Q_{p+1}/Q_1$. Then $\hat{Q}_0$, $\hat{Q}_1$ satisfy lemma 1. Let $x' \in \mathbb{R}$ and let $L > 1$ be such that $x' \in [-L, L]$. Applying lemma 1, and using the multiplicative property of equation (3.1), we have that there exists $N > 0$, depending only on L, such that $Q_{n+p}(x) = Q_n(x)$ for all $x \in [-L, L]$ and all $n > N$. We define $\tilde{Q}_0(x') = Q_{kp}(x')$ and $\tilde{Q}_1(x') = Q_{kp+1}(x')$, where k is an integer such that $kp > N$. (Note that any such k will give the same value for $\tilde{Q}_0(x')$ and $\tilde{Q}_1(x')$.) Now, for $n \geq 2$, we define $\tilde{Q}_n$ by equation (3.1) (with the Q's replaced by the $\tilde{Q}$'s). We observe that for $n \geq 2$, $\tilde{Q}_n(x') = Q_{kp+n}(x')$. This follows by induction: $\tilde{Q}_n(x') = \tilde{Q}_{n-1}(\phi_1(x'))\tilde{Q}_{n-2}(\phi_2(x')) = Q_{kp+n-1}(\phi_1(x'))Q_{kp+n-2}(\phi_2(x')) = Q_{kp+n}(x')$, by equation (3.1). (Here we have used the fact that $\phi_1(x')$, $\phi_2(x') \in [-L, L]$ so the same value of k is applicable for these points.) Thus it follows that $\tilde{Q}_{n+p}(x') = Q_{kp+n+p}(x') = Q_{kp+n}(x') = \tilde{Q}_n(x')$ for all $n \geq 0$. This defines $\tilde{Q}_n$ on $\mathbb{R}$ such that equation (3.1) holds, $\tilde{Q}_n = Q_n$ on I, $\tilde{Q}_0$, $\tilde{Q}_1$ are right continuous at 1, and $\tilde{Q}_n$ is periodic on $\mathbb{R}$ with period dividing p. Since Q_n is periodic on I with period p, it follows that $\tilde{Q}_n$ has period p.

This completes the proof of the proposition. □

In fact the fundamental interval I 'drives' the recurrence (3.1) as we see from the following proposition, which follows from lemma 1.

Proposition 2. *Let Q_0, Q_1 be piecewise-constant initial conditions on $\mathbb{R}$ and let $Q_0(1+) = Q_0(1)$, $Q_1(1+) = Q_1(1)$. Let Q_n satisfy (3.1), and be periodic of period p on the fundamental interval I. Then the sequence Q_n converges to the unique periodic extension $\tilde{Q}_n$ given by proposition 1, i.e., for all integers $r \geq 0$ we have $Q_{r+np}(x) \to \tilde{Q}_r(x)$ as $n \to \infty$.*

From the results of this section we see that, without loss of generality, we may restrict our analysis of the periodic orbits of (3.1) to the fundamental interval I, as we shall henceforth do.

5 Analysis of the discontinuities

In order to study the piecewise-constant periodic orbits of the recurrence (1.5) with $Q_n(x) = \pm 1$ and with initial conditions Q_0, Q_1, it is helpful to consider the dynamics of the discontinuities of Q_n. We may define, for each $x \in \mathbb{R}$ and $n \geq 0$,

$$R_n(x) = \frac{Q_n(x+)}{Q_n(x-)}, \tag{5.1}$$

the ratio of the right-hand limit at x to the left-hand limit at x. Since $Q_n(x) = \pm 1$, we have $R_n(x) = \pm 1$, and it is clear that $R_n(x) = -1$ if and only if Q_n has a discontinuity at x. Since Q_n has at most finitely many discontinuities in any compact interval we have that R_n is well defined. Because of the multiplicative nature of the recurrence (1.5), and because ϕ_1, ϕ_2 are orientation reversing and preserving respectively, we have

$$R_n(x) = \frac{Q_n(x+)}{Q_n(x-)} = \frac{Q_{n-1}(\phi_1(x)-)\,Q_{n-2}(\phi_2(x)+)}{Q_{n-1}(\phi_1(x)+)\,Q_{n-2}(\phi_2(x)-)}, \tag{5.2}$$

so, using $R_n(x) = 1/R_n(x)$, we obtain

$$R_n(x) = R_{n-1}(\phi_1(x))R_{n-2}(\phi_2(x)), \tag{5.3}$$

and R_n satisfies the same recurrence relation as Q_n. However $R_n(x) = 1$ except at points of discontinuity of Q_n, where $R_n(x) = -1$.

We first of all discuss the dynamics of R_n and then relate the dynamics of Q_n to those of R_n. Indeed, it is clear that if Q_n is periodic with period $p \in \mathbb{N}$, then R_n is also periodic with period m dividing p. Our task is to determine the possible periods m of R_n and relate m to p, the period of Q_n.

From now on we assume that Q_n is periodic with period p and that R_n is periodic with period m, and, in view of proposition 1, we only consider the behaviour of Q_n and R_n on the fundamental interval $[-\omega, 1]$. We denote by

$$D = \{x \in [-\omega, 1] : R_n(x) = -1 \text{ for some } n \geq 0\}, \tag{5.4}$$

the *restricted discontinuity set.* Then D is the set of points in the fundamental interval $[-\omega, 1]$ for which Q_n has a discontinuity for at least one $n \geq 0$. One important observation is that since each Q_n is piecewise-constant (and so the set of discontinuities of Q_n on $[-\omega, 1]$ is finite), and since Q_n is periodic, it follows that D is a finite set.

5.1 The restricted discontinuity set and the map F

In this section we show that the restricted discontinuity set D consists of finitely many periodic orbits of the map F. Indeed we have the following result:

Proposition 3. *Let Q_n be a periodic orbit of (1.5) with $Q_n(x) = \pm 1$, and let D be the restricted discontinuity set. Then D consists of a finite collection of periodic orbits of the map F.*

For suppose $y \in D$. Then $R_n(y) = -1$ for some $n \geq 0$. From (5.3) we have that $R_{n-i_1}(\phi_{i_1}(y)) = -1$ for some $i_1 \in \{1, 2\}$. We therefore have $\phi_{i_1}(y) \in D$. Continuing in this way, we obtain a sequence

$i_1, i_2, \ldots \in \{1,2\}$ such that $\phi_{i_k} \circ \cdots \circ \phi_{i_1}(y) \in D$. Since D is finite there exist ℓ, $\ell' \in \mathbb{N}$ with $\ell > \ell'$ and $\phi_{i_{\ell'}} \circ \cdots \circ \phi_{i_1}(y) = \phi_{i_\ell} \circ \cdots \circ \phi_{i_1}(y)$. Applying F^ℓ to this equation gives $F^{\ell-\ell'}(y) = y$, so that y is a periodic point of F of period k dividing $\ell - \ell'$.

Now let $y_0 = y, y_1, \ldots, y_{k-1}$ be the points on the orbit of y_0 under F with $y_{i+1 \pmod k} = F(y_i)$ for $i = 0, 1, \ldots, k-1$, and let $a_0 a_1 \ldots a_{k-1}$ be the code of the orbit. Then for $0 \le i \le k-1$ we have

$$\phi_{a_i}^{-1}(y_i) = y_{i+1} \quad \text{or, equivalently,} \tag{5.5}$$

$$\phi_{a_{i-1}}(y_i) = y_{i-1}\,, \tag{5.6}$$

where here, and in what follows, we assume that expressions relating to the periodic orbit $y_0, y_1, \ldots, y_{k-1}$ are reduced modulo k.

Moreover, by the results of section 3, we have that precisely one of $\phi_1(y_i)$, $\phi_2(y_i)$ is periodic, so that $\phi_2(y_i) \notin D$ if $a_{i-1} = 1$ and $\phi_1(y_i) \notin D$ if $a_{i-1} = 2$. It follows that the recurrence (5.3) becomes

$$R_n(y_i) = \begin{cases} R_{n-1}(y_{i-1}), & a_{i-1} = 1 \\ R_{n-2}(y_{i-1}), & a_{i-1} = 2\,, \end{cases} \tag{5.7}$$

where we have used the facts that $R_{n-2}(\phi_2(y_i)) = 1$ if $a_{i-1} = 1$ and $R_{n-1}(\phi_1(y_i)) = 1$ if $a_{i-1} = 2$. This can be written as

$$R_n(y_i) = R_{n-a_{i-1}}(y_{i-1})\,. \tag{5.8}$$

From this we see that $R_{n+a_0+\cdots+a_{i-1}}(y_i) = R_n(y_0)$, so that if $y_0 \in D$ and $R_n(y_0) = -1$ we have $y_i \in D$, since $R_{n+a_0+\cdots+a_{i-1}}(y_i) = R_n(y_0) = -1$.

We conclude that not only must every point y in D be a periodic point of F, but that every point on the periodic orbit of y also lies in D, so that D consists of complete orbits of F. Since D is finite, proposition 3 now follows.

From (5.8) we see that only one of the factors in the right-hand side of (5.3) is different from $+1$, although which one depends on the code $a_0 a_1 \ldots a_{k-1}$. We also observe that in (5.8) n decreases by a_{i-1}. Now, over the whole of the orbit $y_0, y_1, \ldots, y_{k-1}$ we have that n decreases by

$$\ell = \sum_{i=0}^{k-1} a_i\,, \tag{5.9}$$

i.e.,

$$R_n(y_i) = R_{n-\ell}(y_i)\,, \tag{5.10}$$

for $0 \le i \le k-1$. It follows that we must have $m \mid \ell$. We therefore conclude the following:

Proposition 4. *The period m of the discontinuity function R_n restricted to a periodic orbit $y_0, \ldots, y_{k-1}$ of F divides ℓ, the sum of the code over the orbit of F.*

We now introduce three examples of periodic orbits of F which we shall use to illustrate the theory as it develops.

Example 1. Period-4 orbit of F with code 1122. Then $\ell = 6$.

Example 2. Period-4 orbit of F with code 1211. Then $\ell = 5$.

Example 3. Period-6 orbit of F with code 111222. Then $\ell = 9$.

5.2 The discontinuity matrix

Let $y_0y_1\dots y_{k-1}$ be a periodic orbit of F with code $a_0a_1\dots a_{k-1}$, and with ℓ given by equation (5.9). We shall first consider the dynamics of R_n on this orbit.

It is helpful at this point to introduce an $\ell\times k$ matrix M, the *discontinuity matrix*, with entries ± 1 defined by

$$M_{n,i} = R_n(y_i)\,, \tag{5.11}$$

for $0 \le n \le \ell - 1$, $0 \le i \le k-1$. Then the entry in row n and column i is the value of R_n at the point y_i on the orbit $y_0, y_1, \cdots, y_{k-1}$.

The relation (5.8) above gives a special structure to the matrix M. Indeed (5.8) translates to

$$M_{n,i} = M_{n-a_{i-1},i-1}\,, \tag{5.12}$$

where here, and in what follows, indices referring to the periodicity of R_n are reduced modulo ℓ.

The structure (5.12) can be more easily understood as follows. Column i of the matrix M is simply column $(i-1)$ cyclically permuted downwards by a_{i-1} single cyclic permutations. This observation also holds when $i = 0$, for then (5.12) becomes

$$M_{n,0} = M_{n-a_{k-1},k-1}\,. \tag{5.13}$$

Let us denote the column 0 by $(X_0, X_1, \cdots, X_{\ell-1})$, i.e., $M_{n,0} = X_n$ for $0 \le n \le \ell - 1$. Then the relation (5.12) tells us that

$$M_{n,1} = M_{n-a_0,0} = X_{n-a_0}\,, \tag{5.14}$$

and, in general,

$$M_{n,i} = M_{n-\sum_{j=0}^{i-1} a_j,0} = X_{n-\sum_{j=0}^{i-1} a_j}\,, \tag{5.15}$$

so that the columns of M are simply cyclic permutations of the column 0 of M.

As an illustration consider example 2. Recall that the code is 1211, the period k is 4, and $\ell = 5$. The matrix M is

$$M = \begin{pmatrix} X_0 & X_4 & X_2 & X_1 \\ X_1 & X_0 & X_3 & X_2 \\ X_2 & X_1 & X_4 & X_3 \\ X_3 & X_2 & X_0 & X_4 \\ X_4 & X_3 & X_1 & X_0 \end{pmatrix}. \tag{5.16}$$

Note that the column i is obtained from the column $(i-1)$ by cyclically permuting the column $(i-1)$ downwards by a_{i-1}.

5.3 Periodicity of the discontinuities

Not only does any periodic orbit R_n with discontinuities at a periodic orbit $y_0, y_1, \cdots, y_{k-1}$ of F have a discontinuity matrix M with the structure (5.15), but also, conversely, any matrix M satisfying (5.15) corresponds to a periodic orbit of R_n, by defining $R_n(x) = 1$ except on the points $y_0, \ldots, y_{k-1}$ where we define $R_n(y_i) = M_{n \bmod \ell, i}$ for $n \geq 0$. The period of R_n certainly divides ℓ, but may not actually be equal to ℓ. Indeed, trivially, setting $M_{n,i} = -1$ for all i, n gives a periodic orbit of period 1 for R_n. In fact the period of R_n depends only on the period of column 0 of M viewed as a sequence of ± 1. This is because this column is periodic with period m if and only if it is invariant under m single cyclic permutations and m is the least positive integer for which this is true, i.e., $X_{n+m} = X_n$ and $X_{n+j} \neq X_n$ for all n if $1 \leq j < m$. Now the other columns of M are obtained from column 0 by cyclically permuting and thus they will also have period m. In fact, since any column of M can be obtained from any other one by cyclic permutations it follows that all columns of M have the same period m. Indeed, for $r \in \mathbb{N}$

$$M_{n+r,i} = X_{n+r-\sum_{j=0}^{\ell-1} a_j} = X_{n-\sum_{j=0}^{\ell-1} a_j + r} = M_{n,i} \tag{5.17}$$

if, and only if, $m \mid r$. This is because the X_n have period m. Thus each column of M has period m. We conclude that the period m of R_n is the period of the first column $(X_0, X_1, \ldots, X_{\ell-1})$ of M. It is now clear that $m \mid \ell$ and that for every m dividing ℓ we can find a column $(X_0, X_1, \ldots, X_{\ell-1})$ with period m.

It is worth remarking that the first two rows $n = 0$ and $n = 1$ of M are not independent, so that, although the recursion (1.5) is second order, we cannot choose R_0 and R_1 arbitrarily on $y_0, \ldots, y_{k-1}$ and obtain a periodic orbit.

We have therefore solved the question of the periodic behaviour of the discontinuities R_n for a single periodic orbit $y_0, y_1, \ldots, y_{k-1}$ of F with code $a_0 a_1 \ldots a_{k-1}$. In summary, the period m of R_n corresponds precisely to the period of a single column of M, i.e., $R_n(y_i)$ for any $0 \leq i \leq k-1$. We have m divides $\ell = \sum_{j=0}^{k-1} a_j$, and conversely, for every m dividing ℓ, we can, for suitable choice of the column 0 of M, viz., $(X_0, X_1, \ldots, X_{\ell-1})$, arrange for $(X_0, X_1, \ldots, X_{\ell-1})$, and thus M and R_n, to have period m.

Consider example 2. Here the first column is $(X_0, X_1, X_2, X_3, X_4)$. The only positive integers dividing $\ell = 5$ are 1 and 5, so the only possible periods in this case are $m = 1$ and $m = 5$. Setting $X_0 = X_1 = X_2 = X_3 = X_4 = -1$ gives period 1, whilst any other choice (with at least one -1) gives period 5. Setting $X_0 = X_1 = X_2 = X_3 = X_4 = 1$ gives period 1, but then the orbit of F will not lie in D.

5.4 Multiple periodic orbits in D

Having considered the dynamics of the discontinuity function R_n on a single periodic orbit of F, we now consider the case in which the restricted discontinuity set D consists of more than one periodic orbit of F. To do this, we must establish some notation.

Firstly, let t be the number of periodic orbits of F in D. For $0 \leq s \leq t-1$, we consider the periodic orbit s of F in D. We make the general convention that superscript s refers to the orbit s. Let k^s denote its period and let the points $y_0^s, \ldots, y_{k^s-1}^s$ be the members of the orbit. We denote the code by $a_0^s \ldots a_{k^s-1}^s$. Let

$$\ell^s = \sum_{j=0}^{k^s-1} a_j^s . \tag{5.18}$$

Now, from the multiplicative structure of (5.3), we have that a product of solutions is again a solution of the equation. Moreover, because the periodic orbits in D are distinct, and are never mapped to each other under the two maps ϕ_1, ϕ_2, we have that the dynamics of R_n on each of the periodic orbits in D

are independent. Indeed, we may write

$$R_n(x) = \prod_{s=0}^{t-1} R_n^s(x)\,, \tag{5.19}$$

where R_n^s is the restriction of R_n to the periodic orbit s, i.e.,

$$R_n^s(x) = \begin{cases} R_n(x)\,, & x \in \{y_0^s, \dots, y_{k^s-1}^s\}\,; \\ 1\,, & \text{otherwise}\,. \end{cases} \tag{5.20}$$

We may apply the analysis of the previous subsections to each of the functions R_n^s. This is because $R_n^s(x) = 1$, except when x is one of the points on the periodic orbit $y_0^s \dots y_{k^s-1}^s$ of F. In particular, for each orbit in D we can formulate the $\ell^s \times k^s$ discontinuity matrix M^s, where, for $0 \le n \le \ell^s - 1$ and $0 \le i \le k^s - 1$,

$$M_{n,i}^s = R_n^s(y_i^s)\,. \tag{5.21}$$

We observe that these matrices are independent of each other since the dynamics of R_n on each periodic orbit in D are independent.

The theory for R_n that we discussed above carries over in a straightforward manner to the function R_n^s. To simplify notation, we adopt the convention that, when dealing with periodic orbit s and its matrix, expressions relating to the periodic orbit $y_0^s, \dots, y_{k^s-1}^s$ are reduced modulo k^s whilst those relating to the periodicities of R_n are reduced modulo ℓ^s. Thus, as in (5.12), we have

$$M_{n,i}^s = M_{n-a_{i-1}^s, i-1}^s\,, \tag{5.22}$$

for $0 \le n \le \ell^s - 1$ and $0 \le i \le k^s - 1$, and the matrix M^s is determined by its column 0: $(X_0^s, X_1^s, \dots, X_{\ell^s-1}^s)$. Indeed, as in (5.15),

$$M_{n,i}^s = X_{n-\sum_{j=0}^{i-1} a_j^s}^s\,, \tag{5.23}$$

and the period m^s of the column 0 is precisely the row period of M^s. We also have $m^s \mid \ell^s$. Conversely, let $\ell = \text{lcm}(\ell^0, \dots, \ell^{t-1})$. Then for any $m \mid \ell$ we define $m^s = \gcd(m, \ell^s)$. Then $m^s \mid \ell^s$ and by appropriate choices of $(X_0^s, X_1^s, \dots, X_{\ell^s-1}^s)$ we may construct a matrix M^s with row period any m^s dividing ℓ^s, and, extending periodically to all $n \ge 0$, we have that R_n has period m^s restricted to the orbit $y_0^s, \dots, y_{k^s-1}^s$.

We therefore have the following proposition for piecewise-constant, right-continuous functions taking the values ± 1:

Proposition 5. *Let Q_n be a periodic orbit of (1.5). Then the period m of the discontinuity function R_n is given by*

$$m = lcm(m^0, \dots, m^{t-1})\,, \tag{5.24}$$

where m^s is the period of the function R_n^s and is given by the period of $(X_0^s, X_1^s, \dots, X_{\ell^s-1}^s)$, i.e., column 0 of the discontinuity matrix M^s. Furthermore, m divides

$$\ell = lcm(\ell^0, \dots, \ell^{t-1})\,. \tag{5.25}$$

Moreover, by appropriate choices of $(X_0^s, X_1^s, \dots, X_{\ell^s-1}^s)$, for any m dividing ℓ we may construct a periodic orbit of R_n with period m.

Let us illustrate this result when D is the union of examples 1–3 in subsection 5.1. Then $\ell = \mathrm{lcm}(6,5,9) = 90$. Hence R_n has period dividing 90 and, conversely, for any m dividing 90, we may ensure that R_n has period m.

6 The relationship between Q_n and R_n

Let D be the restricted discontinuity set. We now consider the period of the functions Q_n and relate it to that of the discontinuity functions R_n. We first of all note that R_n does not completely determine Q_n. However, Q_n *is* determined by R_n together with the value $Q_n(x)$ at a single point x. Although any choice of x would be sufficient, for our purposes it is convenient to take $x = 1+$, the right-hand limit at $x = 1$. We write

$$Q_n^{1+} = Q_n(1+)\,. \tag{6.1}$$

Indeed, since Q_n is right-continuous, this is just $Q_n(1)$, but we write Q_n^{1+} to emphasise the fact that it is the right-hand limit. Now, on the fundamental interval, we have

$$Q_n(x) = Q_n(x+) = Q_n^{1+} \prod_{\substack{x<y\leq 1\\ y\in D}} R_n(y) \tag{6.2}$$

$$Q_n(x-) = Q_n^{1+} \prod_{\substack{x\leq y\leq 1\\ y\in D}} R_n(y)\,, \tag{6.3}$$

for $x \in [-\omega, 1]$. It follows that Q_n is periodic with period p if and only if R_n is periodic with period m dividing p and Q_n^{1+} is periodic with period p, or R_n is periodic with period $p = m$ and Q_n^{1+} is periodic with period dividing p. We can therefore reduce the problem of the periodicity of Q_n to that of Q_n^{1+} and of R_n on $[-\omega, 1]$.

To simplify the notation in what follows we introduce the quantities

$$D_n = \prod_{y\in D} R_n(y)\,, \qquad D_n^s = \prod_{i=0}^{k^s-1} R_n^s(y_i^s)\,. \tag{6.4}$$

We now evaluate (1.5) at $x = 1+$ to obtain

$$\begin{aligned} Q_n^{1+} = Q_n(1+) &= Q_{n-1}(-\omega-)Q_{n-2}(1+) && (6.5)\\ &= Q_{n-1}^{1+}Q_{n-2}^{1+}D_{n-1} && (6.6)\\ &= (Q_{n-2}^{1+})^2 Q_{n-3}^{1+} D_{n-1}D_{n-2} && (6.7)\\ &= Q_{n-3}^{1+} D_{n-1}D_{n-2}\,, && (6.8)\end{aligned}$$

where we have used the fact that $(Q_{n-2}^{1+})^2 = 1$ since $Q_{n-2}^{1+} = \pm 1$.

Now each of the products D_{n-1}, D_{n-2} in (6.8) is a product of entries in the matrices M^s for $0 \leq s \leq t-1$. Indeed for each n in the range $0 \leq n \leq m-1$ we have

$$D_n = \prod_{s=0}^{t-1} D_n^s = \prod_{s=0}^{t-1}\prod_{i=0}^{k^s-1} M_{n,i}^s\,, \tag{6.9}$$

so in (6.8) we have expressed Q_n^{1+} in terms of Q_{n-3}^{1+} and a product of entries in the matrices M^s, $0 \le s \le t-1$, and hence of the X_n^s.

Now an orbit of the second order recurrence (1.5) is periodic with period p if and only if $Q_0 = Q_p$, and $Q_1 = Q_{p+1}$, where p is the least such positive integer. We know that p is a multiple of m, the period of R_n. To obtain the relationship between p and m, we investigate Q_m^{1+}/Q_0^{1+} and Q_{m+1}^{1+}/Q_1^{1+}. We note that $p = m$ if and only if both of these ratios have value 1, i.e.,

$$\frac{Q_m^{1+}}{Q_0^{1+}} = \frac{Q_{m+1}^{1+}}{Q_1^{1+}} = 1\,. \tag{6.10}$$

In what follows we shall need to evaluate products of the form

$$\prod_{\substack{n \not\equiv r \bmod 3 \\ r \le n < m+r}} D_n \tag{6.11}$$

where $m \ge 1$ and $r \in \{0, 1, 2\}$. We therefore prove the following lemma, which we shall use in our subsequent work.

Lemma 2. *Let a' be a positive integer and let $D_n = \pm 1$ have period dividing a', i.e., $D_{n+a'} = D_n$ for all n. Let b' be a positive integer with $b' \equiv 0 \bmod 3$ and $a' \mid b'$, and let $r \in \{0, 1, 2\}$. Then:*

1. *if $a' \equiv 0 \bmod 3$ then*

$$\prod_{\substack{n \not\equiv r \bmod 3 \\ r \le n < b'+r}} D_n = \Big(\prod_{\substack{n \not\equiv r \bmod 3 \\ r \le n < a'+r}} D_n \Big)^{(b'/a')}\,; \tag{6.12}$$

2. *if $a' \equiv 0 \bmod 3$ then*

$$\prod_{\substack{n \not\equiv r \bmod 3 \\ r \le n < 2a'+r}} D_n = 1\,; \tag{6.13}$$

3. *if $a' \not\equiv 0 \bmod 3$ then*

$$\prod_{\substack{n \not\equiv r \bmod 3 \\ r \le n < 3a'+r}} D_n = 1\,; \tag{6.14}$$

4. *if $a' \not\equiv 0 \bmod 3$ then*

$$\prod_{\substack{n \not\equiv r \bmod 3 \\ r \le n < b'+r}} D_n = 1\,. \tag{6.15}$$

Proof. Suppose that $a' \equiv 0 \bmod 3$. Then

$$\prod_{\substack{n \not\equiv r \bmod 3 \\ r \le n < b'+r}} D_n = \prod_{j=0}^{b'/a'-1} \prod_{\substack{n \not\equiv r \bmod 3 \\ ja'+r \le n < (j+1)a'+r}} D_n \tag{6.16}$$

$$= \prod_{j=0}^{b'/a'-1} \prod_{\substack{n-ja' \not\equiv r \bmod 3 \\ r \le n-ja' < a'+r}} D_{n-ja'} \tag{6.17}$$

$$= \Big(\prod_{\substack{n \not\equiv r \bmod 3 \\ r \le n < a'+r}} D_n \Big)^{(b'/a')}. \tag{6.18}$$

In the above calculation we have used the fact that $3 \mid a' \mid b'$ and $D_{n+a'} = D_n$. This proves assertion 1. Assertion 2 follows immediately from 1, with $b'/a' = 2$.

Now suppose $a' \not\equiv 0 \bmod 3$ and $r \leq n < a' + r$. Then precisely one of n, $n + a'$, $n + 2a'$ is congruent to $r \bmod 3$. Furthermore, because $D_n = D_{n+a'} = D_{n+2a'}$ each factor occurs precisely twice in the product

$$\prod_{\substack{n \not\equiv r \bmod 3 \\ r \leq n < 3a'+r}} D_n = \prod_{\substack{n \not\equiv r \bmod 3 \\ r \leq n < a'+r}} D_n \prod_{\substack{n \not\equiv r \bmod 3 \\ a'+r \leq n < 2a'+r}} D_n \prod_{\substack{n \not\equiv r \bmod 3 \\ 2a'+r \leq n < 3a'+r}} D_n \tag{6.19}$$

and since $D_n = \pm 1$ each factor cancels itself out. Thus assertion 3 follows, and assertion 4 follows easily from 1 and 3. □

We now return to evaluating the ratios Q_m^{1+}/Q_0^{1+} and Q_{m+1}^{1+}/Q_1^{1+}. They can be obtained by iterating equation (6.8); their values depend on the residue of m modulo 3. Accordingly, we divide into three cases.

6.1 $m \equiv 1 \bmod 3$

In this case we have, iterating (6.8), and using (6.6),

$$\frac{Q_m^{1+}}{Q_0^{1+}} = \frac{Q_1^{1+}}{Q_0^{1+}} \prod_{\substack{n \not\equiv 1 \bmod 3 \\ 1 \leq n < m}} D_n \tag{6.20}$$

$$\frac{Q_{m+1}^{1+}}{Q_1^{1+}} = \frac{Q_2^{1+}}{Q_1^{1+}} \prod_{\substack{n \not\equiv 2 \bmod 3 \\ 2 \leq n < m+1}} D_n = Q_0^{1+} \prod_{\substack{n \not\equiv 2 \bmod 3 \\ 1 \leq n < m+1}} D_n \,. \tag{6.21}$$

We see that whether (6.10) holds, or not, depends both on the product of the D_n (which themselves are products of entries from the matrices M^s) and on Q_0^{1+}, Q_1^{1+}. In the case (6.10) we have that Q_n has the same periodicity as R_n, i.e., $p = m$. Otherwise we have $p > m$. From (6.8) we have

$$\frac{Q_{3m}^{1+}}{Q_0^{1+}} = \prod_{\substack{n \not\equiv 0 \bmod 3 \\ 0 \leq n < 3m}} D_n \,, \qquad \frac{Q_{3m+1}^{1+}}{Q_1^{1+}} = \prod_{\substack{n \not\equiv 1 \bmod 3 \\ 1 \leq n < 3m+1}} D_n \,. \tag{6.22}$$

Now from lemma 2 (3), with $a' = m$ and $r = 0, 1$, we have $Q_{3m}^{1+}/Q_0^{1+} = Q_{3m+1}^{1+}/Q_1^{1+} = 1$, and hence we have $p = 3m$. (We cannot have $p = 2m$ for this would imply $p = m$, since 2 and 3 have greatest common divisor 1.)

6.2 $m \equiv 2 \bmod 3$

The analysis in the case $m \equiv 2 \bmod 3$ is similar, and again leads to the conclusion that either $p = m$ or $p = 3m$. Indeed, we have

$$\frac{Q_m^{1+}}{Q_0^{1+}} = Q_1^{+} \prod_{\substack{n \not\equiv 2 \bmod 3 \\ 1 \leq n < m}} D_n \tag{6.23}$$

$$\frac{Q_{m+1}^{1+}}{Q_1^{1+}} = \frac{Q_0^{1+}}{Q_1^{1+}} \prod_{\substack{n \not\equiv 0 \bmod 3 \\ 1 \leq n < m+1}} D_n \,, \tag{6.24}$$

and a similar calculation to the above gives $Q_{3m}^{1+}/Q_0^{1+} = Q_{3m+1}^{1+}/Q_1^{1+} = 1$, even when (6.10) does not hold.

6.3 $m \equiv 0 \bmod 3$

We now consider the case $m \equiv 0 \bmod 3$. By iterating equation (6.8), we have

$$\frac{Q_m^{1+}}{Q_0^{1+}} = \prod_{\substack{n \not\equiv 0 \bmod 3 \\ 0 \le n < m}} D_n \,, \qquad \frac{Q_{m+1}^{1+}}{Q_1^{1+}} = \prod_{\substack{n \not\equiv 1 \bmod 3 \\ 1 \le n < m+1}} D_n \,. \tag{6.25}$$

We observe that, in this case, whether equation (6.10) holds or not is independent of Q_0^{1+} and Q_1^{1+}. Now, if (6.10) does not hold, then

$$\frac{Q_{2m}^{1+}}{Q_0^{1+}} = \prod_{\substack{n \not\equiv 0 \bmod 3 \\ 0 \le n < 2m}} D_n = 1 \,, \qquad \frac{Q_{2m+1}^{1+}}{Q_1^{1+}} = \prod_{\substack{n \not\equiv 1 \bmod 3 \\ 1 \le n < 2m+1}} D_n = 1 \,, \tag{6.26}$$

which follows from lemma 2 (2). We therefore conclude that, in this case, either $p = m$ or $p = 2m$.

We may therefore sum up these results as follows.

Proposition 6. *Let Q_n be periodic with period p and let R_n have period m. Then if $m \not\equiv 0 \bmod 3$, then either $p = m$ or $p = 3m$. Otherwise, if $m \equiv 0 \bmod 3$, then either $p = m$ or $p = 2m$.*

7 Theorem 2

We now compile the results of the previous sections into the following theorem.

Theorem 2. *Let Q_n, $n \ge 0$, be a periodic orbit of period p of (1.5) with $Q_n(x) = \pm 1$ for all x, Q_n right-continuous, and such that the restricted discontinuity set D is finite. Let m be the period of the discontinuity function R_n given by (5.1). Then*

1. *D is a finite set of t periodic orbits (y_i^s), $0 \le s \le t-1$, $0 \le i \le k^s - 1$ of F with codes $a_0^s \ldots a_{k^s-1}^s$;*
2. *the period m of R_n divides $\ell = lcm(\ell^0, \ldots, \ell^{t-1})$ where $\ell^s = \sum_{j=0}^{k^s-1} a_j^s$;*
3. *the period p of Q_n is either m, $2m$ or $3m$. If $m \not\equiv 0 \bmod 3$ then $p = m$ or $p = 3m$ depending on the values of R_n and Q_0^{1+}, Q_1^{1+}. However if $m \equiv 0 \bmod 3$ then either $p = m$ or $p = 2m$ and this depends only on the values of R_n.*

Theorem 2 gives only a partial classification of the periodic orbit structure of (1.5). It remains to determine what periods p for Q_n can actually be achieved for a given choice of restricted discontinuity set D. It is this question that we study in the rest of the paper.

8 The construction of periodic orbits

In this section we consider how, by an appropriate choice of the X_n^s, and Q_0^{1+} and Q_1^{1+}, we may construct periodic orbits with a given restricted discontinuity set D.

Let D be a finite collection of t periodic orbits of F. We adopt the notation of subsection 5.4 for the orbits y_i^s, viz., ℓ^s, ℓ etc. We know (by theorem 2) that the period m of R_n must divide ℓ. Now suppose that m is any positive integer dividing ℓ. Let $0 \leq s \leq t-1$ and let $m^s = \gcd(m, \ell^s)$. Then by choosing column 0 of M^s, i.e., $(X_0^s, X_1^s, \ldots, X_{\ell^s-1}^s)$, to have period m^s, we may ensure that R_n^s has period m^s. Thereby we may ensure that R_n will have period $\mathrm{lcm}(m^0, \ldots, m^{t-1})$ which is equal to m, and that the restricted discontinuity set is D.

However, in order to determine the possible values of p, the period of Q_n, we must be more careful in our choices, at least in the case $m \equiv 0 \bmod 3$. We now consider the two cases $m \not\equiv 0 \bmod 3$ and $m \equiv 0 \bmod 3$ separately.

Suppose first that $m \not\equiv 0 \bmod 3$. From subsections 6.1 and 6.2, and in particular from equations (6.20) and (6.23), we see that, if R_n is chosen to have period m, then we may choose Q_0^{1+} and Q_1^{1+} so that either equation (6.10) holds (in which case $p = m$) or else it does not hold (in which case $p = 3m$). We are therefore able to conclude the following result.

Proposition 7. *Let D be a finite collection of t periodic orbits of F with the notation of subsection 5.4 and let m divide ℓ with $m \not\equiv 0 \bmod 3$. Then for both $p = m$ and for $p = 3m$, there exists a periodic orbit Q_n of (1.5) of period p and with restricted discontinuity set D.*

8.1 $m \equiv 0 \bmod 3$

The case $m \equiv 0 \bmod 3$ is more delicate, since in that case we see from equation (6.25) that p is unaffected by the choice of Q_0^{1+} and Q_1^{1+}; it depends only on R_n, and, in particular, the X_n^s. From subsection 6.3, we see that $p = m$ if and only if the products

$$P_0 = \prod_{\substack{n \not\equiv 0 \bmod 3 \\ 0 \leq n < m}} D_n, \qquad P_1 = \prod_{\substack{n \not\equiv 1 \bmod 3 \\ 1 \leq n < m+1}} D_n \tag{8.1}$$

are both equal to 1; otherwise $p = 2m$. We shall have to be more careful in the way the X_n^s are chosen. In particular, we must examine in more detail the products P_0 and P_1.

Writing

$$P_0^s = \prod_{\substack{n \not\equiv 0 \bmod 3 \\ 0 \leq n < m}} D_n^s, \qquad P_1^s = \prod_{\substack{n \not\equiv 1 \bmod 3 \\ 1 \leq n < m+1}} D_n^s, \tag{8.2}$$

we have that

$$P_0 = \prod_{s=0}^{t-1} P_0^s, \qquad P_1 = \prod_{s=0}^{t-1} P_1^s. \tag{8.3}$$

We now study P_0^s and P_1^s. Recall that $m^s = \gcd(m, \ell^s)$. Then we have $m^s \mid \ell^s$ and $m^s \mid m$. If $m^s \not\equiv 0 \bmod 3$ then $(m/m^s) \equiv 0 \bmod 3$. Therefore, from lemma 2 (using (1) with $b' = m$ and $a' = 3m^s$, and then (3) with $a' = m^s$), we have that

$$P_0^s = \Big(\prod_{\substack{n \not\equiv 0 \bmod 3 \\ 0 \leq n < 3m^s}} D_n^s\Big)^{m/(3m^s)} = 1, \qquad P_1^s = \Big(\prod_{\substack{n \not\equiv 1 \bmod 3 \\ 1 \leq n < 3m^s+1}} D_n^s\Big)^{m/(3m^s)} = 1. \tag{8.4}$$

We therefore conclude that, unless $m^s \equiv 0 \bmod 3$, we have $P_0^s = P_1^s = 1$.

Now suppose $m^s \equiv 0 \bmod 3$. Then from lemma 2 (1) (with $b' = m$ and $a' = m^s$), we have

$$P_0^s = (\prod_{\substack{n \not\equiv 0 \bmod 3 \\ 0 \le n < m^s}} D_n^s)^{(m/m^s)}, \qquad P_1^s = (\prod_{\substack{n \not\equiv 1 \bmod 3 \\ 1 \le n < m^s+1}} D_n^s)^{(m/m^s)}. \tag{8.5}$$

We observe that we shall again have $P_0^s = P_1^s = 1$ unless m/m^s is odd. Hence we have:

Proposition 8. *Let* $m \equiv 0 \bmod 3$. *Then*

$$P_0 = \prod_{\substack{s=0 \\ 3|m^s,\, 2\nmid(m/m^s)}}^{t-1} P_0^s, \qquad P_1 = \prod_{\substack{s=0 \\ 3|m^s,\, 2\nmid(m/m^s)}}^{t-1} P_1^s. \tag{8.6}$$

(Here empty products are defined to equal 1.*)*

We now introduce a condition on the orbit s of F which is necessary for $P_0^s = P_1^s = 1$ not to hold. We call orbits satisfying this condition *active.*

8.2 Active periodic orbits of F

In view of the previous proposition, we make the following definition.

Definition. Let m be a positive integer dividing ℓ and divisible by 3. The periodic orbit $y_0^s, y_1^s, \ldots, y_{k^s-1}^s$ of F with code $a_0^s a_1^s \ldots a_{k^s-1}^s$ is said to be *active with respect to* m if

1. $3 \mid \ell^s = \sum_{j=0}^{k^s-1} a_j^s$ (so that $3 \mid m^s = \gcd(m, \ell^s)$);

2. m/m^s is odd; and

3. in the sequence $a_0^s, a_0^s + a_1^s, \ldots, a_0^s + a_1^s + \cdots + a_{k^s-1}^s$ the three residue classes modulo 3 do not all occur with the same parity, i.e., there is at least one residue class that occurs an even number of times and another which occurs an odd number of times.

Note that whether a given code is active or not with respect to m does not depend on the restricted discontinuity set D (and hence ℓ) directly, but only on the choice of integer m dividing ℓ.

This definition is not at all intuitive, and we shall illustrate it with reference to examples 1 – 3 and with $m = 30$.

Example 1: The code is 1122 and the orbit is active with respect to m, since $3 \mid \ell^s = 6$, $m^s = \gcd(30, 6) = 6$ so $m/m^s = 5$ is odd. Furthermore $a_0^s, a_0^s + a_1^s, \ldots, a_0^s + a_1^s + \cdots + a_{k^s-1}^s = 1, 2, 4, 6 \equiv 1, 2, 1, 0 \bmod 3$, so the residue class 1 occurs an even number of times and the residue classes 0, 2 occur an odd number of times.

Example 2: The code is 1211 and the orbit is not active with respect to m, since $3 \nmid \ell^s = 5$.

Example 3: The code is 111222 and the orbit is not active with respect to m. Indeed it fails on two counts, since $3 \mid \ell^s = 9$, and $m^s = \gcd(30, 9) = 3$, so $m/m^s = 10$ is even. Moreover we have $a_0^s, a_0^s + a_1^s, \ldots, a_0^s + a_1^s + \cdots + a_{k^s-1}^s = 1, 2, 3, 5, 7, 9 \equiv 1, 2, 0, 2, 1, 0 \bmod 3$, so each residue class occurs an even number of times.

Now let $m^s \equiv 0 \bmod 3$, let m/m^s be odd, and suppose that $(X_0^s, X_1^s, \ldots, X_{m^s-1}^s)$ is periodic with period m^s. Then the rows of the matrix M^s are also periodic with period m^s, and all entries are one of $X_0^s, X_1^s, \ldots, X_{m^s-1}^s$ which we regard as unknowns taking values ± 1. We are interested in evaluating P_0^s and P_1^s as functions of the unknowns $X_0^s, X_1^s, \ldots, X_{m^s-1}^s$. Now each entry of M^s is one of $X_0^s, X_1^s, \ldots, X_{m^s-1}^s$ and so P_0^s and P_1^s are each products of the unknowns $X_0^s, X_1^s, \ldots, X_{m^s-1}^s$. Since each of these is ± 1, the important ones are those that occur to an odd power in the product.

In what follows, by the *parity* of a set, we mean the number reduced modulo 2 of elements in the set.

We may identify these using the following lemma:

Lemma 3. *Let the orbit s in D satisfy $m^s \equiv 0 \bmod 3$ and let m/m^s be odd. Then for $r = 1, 2$, we have*

$$P_r^s = \prod_{0 \le n' < m^s} (X_{n'}^s)^{\Gamma(n',r)} \tag{8.7}$$

where $\Gamma(n', r)$ is the parity of the set

$$\{i \mid 0 \le i < k^s,\, n' + \sum_{j=0}^{i-1} a_j^s \not\equiv r \bmod 3\}\,. \tag{8.8}$$

Proof. Let $r = 0, 1$. From (8.5) we have

$$P_r^s = \prod_{\substack{n \not\equiv r \bmod 3 \\ r \le n < r+m^s}} D_n^s = \prod_{i=0}^{k^s-1} \prod_{\substack{n \not\equiv r \bmod 3 \\ r \le n < r+m^s}} M_{n,i}^s = \prod_{\substack{n \not\equiv r \bmod 3 \\ r \le n < r+m^s}} \prod_{i=0}^{k^s-1} X^s_{n - \sum_{j=0}^{i-1} a_j^s}\,, \tag{8.9}$$

where the subscript is reduced modulo m^s. Writing $n' = n - \sum_{j=0}^{i-1} a_j^s \bmod m^s$, we obtain

$$P_r^s = \prod_{0 \le n' < m^s} \prod_{\substack{i=0 \\ n' + \sum_{j=0}^{i-1} a_j^s \not\equiv r \bmod 3}}^{k^s-1} X_{n'}^s = \prod_{0 \le n' < m^s} (X_{n'}^s)^{\Gamma(n',r)}\,, \tag{8.10}$$

since $X_{n'}^s = \pm 1$. This completes the proof of the lemma. □

From this lemma it follows immediately that those of the X_n^s occurring to an odd power in P_0^s come in residue classes modulo 3, and similarly for P_1^s.

We let $\mathbb{Z}_3$ denote the additive group of residue classes modulo 3, and $\mathbb{Z}_2$ denote the residue classes modulo 2. Let $\tilde{\gamma} : \mathbb{Z}_3 \to \mathbb{Z}_2$ be the number modulo 2 of residue classes modulo 3 occurring in $a_0^s, a_0^s + a_1^s, \ldots, a_0^s + a_1^s + \cdots + a_{k-1}^s$. Then, from the above definition, the orbit is active if and only if $\tilde{\gamma}$ is surjective, i.e., there exists at least one residue class u with $\tilde{\gamma}(u) = 0$, and another, u', with $\tilde{\gamma}(u') = 1$.

Let $\gamma(n', r)$ be the parity of the set $\{i \mid 0 \le i < k^s,\, n' + \sum_{j=0}^{i-1} a_j^s \equiv r \bmod 3\}$. Then, since $\ell^s \equiv 0 \bmod 3$, $\{i \mid 0 \le i < k^s,\, n' + \sum_{j=0}^{i-1} a_j^s \equiv r \bmod 3\} = \{i \mid 1 \le i \le k^s,\, n' + \sum_{j=0}^{i-1} a_j^s \equiv r \bmod 3\}$, and it is clear that $\gamma(n', r) = \tilde{\gamma}(r - n')$ and

$$\Gamma(n', r) = \gamma(n', r+1) + \gamma(n', r+2) = \tilde{\gamma}(r + 1 - n') + \tilde{\gamma}(r + 2 - n')\,, \tag{8.11}$$

where, of course, here, and in what follows, we reduce the sums modulo 2. Furthermore, $\Gamma(n', r) = \Gamma(n' \bmod 3, r)$. This means that we need only examine the three residue classes modulo 3. Moreover we have $\Gamma(n', r) = \Gamma(n' + 1, r + 1)$.

For an inactive orbit we have $\Gamma(n', r) = 0$, since $\tilde{\gamma}(u) + \tilde{\gamma}(u') = 0$ for all choices of u, $u' \in \mathbb{Z}_3$. (This follows because $\tilde{\gamma} \equiv 0$ or $\tilde{\gamma} \equiv 1$.) However if the orbit is active then as n' runs from 0 to 2 precisely two

$\Gamma(n',r)$ will be 1 and the third 0. We show the possibilities for $r=0$ (i.e., for P_0^s) in the following table. We have $\Gamma(n',0)=\tilde{\gamma}(1-n')+\tilde{\gamma}(2-n')$, so that

$$\Gamma(0,0)=\tilde{\gamma}(1)+\tilde{\gamma}(2) \tag{8.12}$$

$$\Gamma(1,0)=\tilde{\gamma}(0)+\tilde{\gamma}(1) \tag{8.13}$$

$$\Gamma(2,0)=\tilde{\gamma}(2)+\tilde{\gamma}(0)\,. \tag{8.14}$$

$\tilde{\gamma}(0)$	$\tilde{\gamma}(1)$	$\tilde{\gamma}(2)$	active	$\Gamma(0,0)$	$\Gamma(1,0)$	$\Gamma(2,0)$
0	0	0	no	0	0	0
1	0	0	yes	0	1	1
0	1	0	yes	1	1	0
0	0	1	yes	1	0	1
1	1	0	yes	1	0	1
0	1	1	yes	0	1	1
1	0	1	yes	1	1	0
1	1	1	no	0	0	0

The parity of the residue class n' in P_1^s is

$$\Gamma(n',1)=\Gamma(n'-1,0)\,. \tag{8.15}$$

Thus the residue classes occurring an odd number of times in P_1^s are those occurring an odd number of times in P_0^s increased by 1 modulo 3. It follows that one residue class occurs in both P_0^s and P_1^s and the other two in one each. We establish the following notation. Let c_0^s, c_1^s be the residue classes occurring in P_0^s and let c_1^s, c_2^s the residue classes occurring in P_1^s. Then

$$P_0^s=\prod_{\substack{n\equiv c_0^s \bmod 3\\ 0\le n<m^s}} X_n^s \prod_{\substack{n\equiv c_1^s \bmod 3\\ 0\le n<m^s}} X_n^s \tag{8.16}$$

$$P_1^s=\prod_{\substack{n\equiv c_1^s \bmod 3\\ 0\le n<m^s}} X_n^s \prod_{\substack{n\equiv c_2^s \bmod 3\\ 0\le n<m^s}} X_n^s\,, \tag{8.17}$$

a product of two residue classes modulo 3.

We now return to example 1, which is active with respect to $m=30$. Recall that the code is 1122, so $\ell^s=6$ and $m^s=6$. For convenience we drop the index s. The matrix M for this orbit is

$$M=\begin{pmatrix} X_0 & X_5 & X_4 & X_2\\ X_1 & X_0 & X_5 & X_3\\ X_2 & X_1 & X_0 & X_4\\ X_3 & X_2 & X_1 & X_5\\ X_4 & X_3 & X_2 & X_0\\ X_5 & X_4 & X_3 & X_1 \end{pmatrix}. \tag{8.18}$$

Then, in view of lemma 2 (1),

$$P_0 = \prod_{\substack{n \not\equiv 0 \bmod 3 \\ 0 \le n < 6}} D_n \tag{8.19}$$

$$= \prod_{n=1,2,4,5} \prod_{i=0}^{3} M_{n,i} \tag{8.20}$$

$$= (X_1X_0X_5X_3)(X_2X_1X_0X_4)(X_4X_3X_2X_0)(X_5X_4X_3X_1) \tag{8.21}$$

$$= X_0X_1X_3X_4\,. \tag{8.22}$$

Similarly $P_1 = X_1X_2X_4X_5$. Now $a_0, a_0+a_1, a_0+a_1+a_2, a_0+a_1+a_2+a_3 = 1,2,4,6 \equiv 1,2,1,0 \bmod 3$. Hence $\tilde{\gamma}(0) = 1$, $\tilde{\gamma}(1) = 0$, $\tilde{\gamma}(2) = 1$, and so $\Gamma(0,0) = \tilde{\gamma}(1) + \tilde{\gamma}(2) = 1$, $\Gamma(1,0) = \tilde{\gamma}(0) + \tilde{\gamma}(1) = 1$, $\Gamma(2,0) = \tilde{\gamma}(2) + \tilde{\gamma}(0) = 0$. Thus $c_0 = 0$, $c_1 = 1$, $c_2 = 2$. We have therefore that

$$P_0 = \prod_{\substack{n \equiv 0 \bmod 3 \\ 0 \le n < 6}} X_n \prod_{\substack{n \equiv 1 \bmod 3 \\ 0 \le n < 6}} X_n \tag{8.23}$$

$$= (X_0X_3)(X_1X_4)\,, \tag{8.24}$$

and

$$P_1 = \prod_{\substack{n \equiv 1 \bmod 3 \\ 0 \le n < 6}} X_n \prod_{\substack{n \equiv 2 \bmod 3 \\ 0 \le n < 6}} X_n \tag{8.25}$$

$$= (X_1X_4)(X_2X_5)\,, \tag{8.26}$$

which indeed agree with the direct calculation above.

We next consider example 3, which is inactive with respect to $m = 30$. Since $m = 30$, and $\ell^s = 9$, we have $m^s = 3$. We have (again dropping the index s) that the matrix M for this orbit is

$$M = \begin{pmatrix} X_0 & X_2 & X_1 \\ X_1 & X_0 & X_2 \\ X_2 & X_1 & X_0 \end{pmatrix}. \tag{8.27}$$

We have

$$\prod_{\substack{n \equiv 1 \bmod 3 \\ 0 \le n < 3}} M_{n,i} \prod_{\substack{n \equiv 2 \bmod 3 \\ 0 \le n < 3}} M_{n,i} = 1\,, \tag{8.28}$$

as is easily seen (using $X_n^2 = 1$ for all n). Hence $P_0 = 1$ and similarly $P_1 = 1$.

In order to simplify the presentation of the next section, we introduce a relabeling of the X_n^s for active orbits. Specifically, we define, for $n \geq 0$

$$Y_n^s = X_{n+c_0^s}^s \tag{8.29}$$

where we reduce modulo m^s. Then $Y_0^s = X_{c_0^s}^s$, $Y_1^s = X_{c_1^s}^s$ and $Y_2^s = X_{c_2^s}^s$, and the column $(Y_0^s, \dots, Y_{m^s-1}^s)$ is simply $(X_0^s, \dots, X_{m^s-1}^s)$ relabeled, and, in particular, the two columns have the same period as sequences of ± 1. Then, in terms of the Y_n^s we have

$$P_0^s = \prod_{\substack{n\equiv 0 \bmod 3\\ 0\le n<m^s}} Y_n^s \prod_{\substack{n\equiv 1 \bmod 3\\ 0\le n<m^s}} Y_n^s \tag{8.30}$$

$$P_1^s = \prod_{\substack{n\equiv 1 \bmod 3\\ 0\le n<m^s}} Y_n^s \prod_{\substack{n\equiv 2 \bmod 3\\ 0\le n<m^s}} Y_n^s . \tag{8.31}$$

In summary, we have the following proposition.

Proposition 9. *Let $m \equiv 0 \bmod 3$. Then the products P_0^s, P_1^s for the orbit s satisfy $P_0^s = P_1^s = 1$ irrespective of the values of $X_0^s, \ldots, X_{m^s-1}^s$ if and only if the orbit s is not active with respect to m. If, on the other hand, the orbit s is active with respect to m, then P_0^s and P_1^s are given by (8.16) and (8.17) where $\Gamma(c_0^s, 0) = \Gamma(c_1^s, 0) = 1$ and $\Gamma(c_1^s, 1) = \Gamma(c_2^s, 1) = 1$. In terms of the relabeled variables $Y_0^s, \ldots, Y_{m^s-1}^s$, P_0^s and P_1^s are given by (8.30) and (8.31).*

In summary, for orbits active with respect to m, we have a systematic method of calculating the products P_0^s, P_1^s given by (8.16–8.17) in terms of the unknowns $X_0^s, X_1^s, \ldots, X_{m^s-1}^s$ or, equivalently, $Y_0^s, Y_1^s, \ldots, Y_{m^s-1}^s$.

8.3 Realisation of the possible values of p for $m \equiv 0 \bmod 3$

We know from theorem 2 that either $p = m$ or $p = 2m$ when $m \equiv 0 \bmod 3$. We now wish to see which of these cases can occur. We recall that if $m \not\equiv 0 \bmod 3$ then both $p = m$ and $p = 3m$ can occur for suitable choices of Q_0^{1+}, Q_1^{1+}. In the case $m \equiv 0 \bmod 3$ these quantities do not determine Q_m^{1+}/Q_0^{1+}, Q_{m+1}^{1+}/Q_1^{1+} and the period p is determined completely by the columns of discontinuities $(X_0^s, \ldots, X_{k^s-1}^s)$.

However, unless the restricted discontinuity set D contains some periodic orbits that are active with respect to m, then the period $p = m$, since we have $P_0 = P_1 = 1$ in this case.

Now suppose that there are some periodic orbits in D which are active with respect to m. We shall now study whether, by appropriate choice of the X_n^s for the active orbits, we may ensure both that R_n has period $m^s = \gcd(\ell^s, m)$ on the orbit s and that either $P_0 = P_1 = 1$ (in which case $p = m$) or not (in which case $p = 2m$). We divide the analysis into three distinct subcases which we treat separately. Our constructions serve only as examples; they are by no means unique. Other choices of the X_n^s can be made to achieve the same result.

8.3.1 Case (i): D contains at least two periodic orbits active with respect to m

Let there be v such orbits labeled $s_0, \ldots, s_{v-1}$ for $1 \le v \le t-1$. Referring to subsection 8.2, we recall that (with respect to the $Y_n^{s_i}$), 0 and 1 are the residue classes occurring to an odd power in $P_0(s_i)$, and 1 and 2 are the residue classes occurring to an odd power in $P_1(s_i)$. Then, by assigning Y_n^s appropriately, we may ensure that that $p = m$ or $p = 2m$.

We first of all assign those X_n^s for $s \ne s_0, \ldots, s_{v-1}$ so that the period of the column $(X_0^s, \ldots, X_{m^s-1}^s)$ is m^s, but otherwise arbitrarily.

Consider the assignments given by the following table:

	s_0	s_1	s_2	$\ldots$	s_{v-1}
Y_0^s	± 1	1	1	$\ldots$	1
Y_1^s	-1	-1	-1	$\ldots$	-1
Y_2^s	1	± 1	1	$\ldots$	1
Y_3^s	1	1	1	$\ldots$	1
$\vdots$	$\vdots$	$\vdots$	$\vdots$	$\ldots$	$\vdots$
$Y_{m^s-1}^s$	1	1	1	$\ldots$	1

For each of $s = s_0, \ldots, s_{v-1}$ we set $Y_n^s = 1$ for $n \neq 0, 1, 2$, so that

$$P_0 = \prod_{i=0}^{v-1} Y_0^{s_i} \prod_{i=0}^{v-1} Y_1^{s_i}, \qquad P_1 = \prod_{i=0}^{v-1} Y_1^{s_i} \prod_{i=0}^{v-1} Y_2^{s_i}. \tag{8.32}$$

Then from (8.32) we have that $P_0 = (\pm 1)(-1)^v$, $P_1 = (-1)^v(\pm 1)$.

The two ± 1 entries can be chosen to obtain either $p = m$ or $p = 2m$, as required. To achieve $p = m$ we choose both the signs to be $(-1)^v$, and to achieve $p = 2m$ we choose (at least one of) the signs to be $-(-1)^v$.

8.3.2 Case (ii): D contains only one periodic orbit active with respect to m, but $m^s \neq 3$

Let the active periodic orbit be s_0. Firstly, the Y_n^s for $s \neq s_0$ are set to have period m^s but are otherwise arbitrary. We assign $Y_n^{s_0}$ according to the following table.

	s_0
Y_0^s	± 1
Y_1^s	-1
Y_2^s	± 1
Y_3^s	1
$\vdots$	$\vdots$
$Y_{m^s-1}^s$	1

Then $P_0 = P_1 = \mp 1$ and so the two ± 1 entries can be chosen to obtain either $p = m$ or $p = 2m$, as required. We note that since $m^{s_0} \geq 6$ we have the column $(Y_0^{s_0}, \ldots, Y_{m^{s_0}-1}^{s_0})$ has period m^s, even when $Y_0^{s_0} = Y_1^{s_0} = Y_2^{s_0} = -1$.

8.3.3 Case (iii): D contains only one periodic orbit active with respect to m, and $m^s = 3$

Let the active periodic orbit be s_0. Then $m^{s_0} = 3$, and $P_0 = Y_0^{s_0} Y_1^{s_0}$, $P_1 = Y_1^{s_0} Y_2^{s_0}$. It is not possible to choose $Y_0^{s_0}$, $Y_1^{s_0}$, $Y_2^{s_0}$ so that $P_0 = P_1 = 1$ and $Y_0^{s_0}$, $Y_1^{s_0}$, $Y_2^{s_0}$ are not all equal. (Note that if they are all equal then the period is 1 not 3.) Thus it is not possible to obtain $p = m$ in this case and $p = 2m$ for any choice of $Y_0^{s_0}$, $Y_1^{s_0}$, $Y_2^{s_0}$ of period 3.

8.4 Theorem 3

In summary, combining the results of section 8, we have proved the following theorem:

Theorem 3. *Let D be a set of t periodic orbits (y_i^s), $0 \leq s \leq t-1$, $0 \leq i \leq k^s - 1$ of F, and let ℓ^s, ℓ be as in section 5.4. Then, for any $m \mid \ell$, setting $m^s = gcd(m, \ell^s)$, the variables $X_0^s, \ldots, X_{m^s-1}^s$ for $0 \leq s \leq t-1$ may be chosen so that:*

1. *R_n is periodic with period m^s on y_i^s;*
2. *R_n is periodic with period m;*
3. *the restricted discontinuity set of Q_n is D.*

The period p of Q_n is related to m as follows:

4. *if $m \not\equiv 0 \bmod 3$ then there are choices of Q_0^{1+}, Q_1^{1+} so that $p = m$ or $p = 3m$;*
5. *if $m \equiv 0 \bmod 3$ and there are no periodic orbits in D active with respect to m, then $p = m$;*
6. *if $m \equiv 0 \bmod 3$ and there are at least two periodic orbits in D active with respect to m then there are choices of X_i^s such that $p = m$ or $p = 2m$;*
7. *similarly, if $m \equiv 0 \bmod 3$ and there is only one periodic orbit in D active with respect to m, and $m^s > 3$, then there are choices of X_i^s such that $p = m$ or $p = 2m$;*
8. *finally, if $m \equiv 0 \bmod 3$ and there is exactly one periodic orbit in D active with respect to m, and, for that orbit, $m^s = 3$, then $p = 2m$ for all choices of X_i^s of period m^s.*

The theorem has an important corollary (theorem 1).

Corollary 1. *For every $p \geq 1$ there is a periodic orbit Q_n of (1.5) of period p.*

Proof. Let $p \in \mathbb{N}$ and let D consist of the orbit with code equal to p copies of 111222. Then $\ell = 9p$ and the orbit is inactive (as can be easily checked). Let $m = p$. We may use part 4 or part 5 of theorem 3 to obtain an orbit of Q_n of period p depending on whether $p \not\equiv 0 \bmod 3$ or $p \equiv 0 \bmod 3$. □

As an illustration of the theory, we refer back to examples 1 – 3 with reference to theorem 3. Let D consist of the union of the three periodic orbits in examples 1 – 3. Recall that $\ell = 90$ in this case. Let us choose $m = 30$. Then we are in case 7 of theorem 3, since we have one active orbit with respect to m and $m^s = \gcd(30, 6) = 6$. We therefore are able to choose the X_n^s so that $p = 30$ or $p = 60$ in this case. Finally, let us return to figure 1 and the periodic orbit found by Feudel *et al* ([2]). In this case D consists of the fixed point with code 1 and the period 2 orbit with code 21. Then $\ell = \text{lcm}(1, 3) = 3$ and, choosing $m = 3$, we have that the period 2 orbit is active with respect to $m = 3$. We are therefore in case 8 of theorem 3 so that we have $p = 2m = 6$ in this case. In fact we have $P_0 = 1$ and $P_1 = -1$, as can be easily seen from figure 1.

9 Conclusion

Orbits of the renormalisation recursion (1.5) arise in the analysis of self-similarity in a variety of phenomena. The recursion, despite being multiplicative in nature, is also nontrivial from a mathematical point of view.

In this paper we have given the complete solution in the case of piecewise-constant functions taking values ± 1. A particular instance of such a solution was numerically calculated by Feudel *et al* [2] in their analysis of the autocorrelation function for a strange nonchaotic attractor.

In a previous paper [8] we have considered analytic solutions of the fixed point equation corresponding to (1.5). This solution helps explain the universality of the supercritical regime of the Harper equation, and is also directly of importance in the study of the onset of a strange nonchaotic attractor [6], [7]. We hope to be able to combine the ideas on periodic orbit structure developed in this paper with the analysis of our previous work [8] to understand the universal strange attractor found in a generalised Harper equation [6].

We remark that an additive version of the renormalisation recursion (1.5) is derived in [3] in the analysis of the self-similarity of the autocorrelation of a quasiperiodically forced two-level system. Again, the piecewise-constant periodic orbits are important in determining the precise nature of the autocorrelation. Much of our work in this paper is also applicable to this additive case, but there are also some subtle differences which we shall explore in the near future.

Finally we remark that the fact that ω in (1.1) is the golden mean is essential for our analysis. It seems likely that a similar study could be undertaken for other quadratic irrationals (which have eventually periodic continued fraction expansions). However it is not clear how to extend our work to more general rotation numbers.

Acknowledgement

The authors would like to thank Peter Vámos and Robin Chapman for helpful discussions, and the referees for suggesting improvements to the paper and correcting errors.

References

[1] K. J. Falconer, *Fractal geometry: mathematical foundations and applications*, John Wiley & Sons, 1990.

[2] U. Feudel, A. Pikovsky, and A. Politi, *Renormalization of correlations and spectra of a strange non-chaotic attractor*, J. Phys. A **29** (1996), 5297–5311.

[3] U. Feudel, A.S. Pikovsky, and M. A. Zaks, *Correlation properties of a quasiperiodically forced two-level system*, Phys. Rev. E **51** (1995), 1762–1769.

[4] C. Grebogi, E. Ott, S. Pelikan, and J. A. Yorke, *Strange attractors that are not chaotic*, Physica D **13** (1984), 261–268.

[5] J. A. Ketoja and I. I. Satija, *Self-similarity and localization*, Phys. Rev. Lett. **75** (1995), 2762–2765.

[6] ———, *Harper equation, the dissipative standard map and strange nonchaotic attractors: Relationship between an eigenvalue problem and iterated maps*, Physica D **109** (1997), 70–80.

[7] S. P. Kuznetsov, A. S. Pikovsky, and U. Feudel, *Birth of a strange nonchaotic attractor: A renormalization group analysis*, Phys. Rev. E **51** (1995), R1629–R1632.

[8] B. D. Mestel, A. H. Osbaldestin, and B. Winn, *Golden mean renormalisation for the Harper equation: the strong coupling fixed point*, to appear in J. Math. Phys. 2000.

[9] A. S. Pikovsky and U. Feudel, *Correlations and spectra of strange non-chaotic attractors*, J. Phys. A **27** (1994), 5209–5219.

MPEJ

MATHEMATICAL PHYSICS ELECTRONIC JOURNAL

ISSN 1086-6655
Volume 6, 2000

Paper 6
Received: May 8, 2000, Revised: Nov 29, 2000, Accepted: Dec 6, 2000
Editor: R. de la Llave

CIRCLE PACKING IN THE HYPERBOLIC PLANE

LEWIS BOWEN

Department of Mathematics
The University of Texas at Austin
Austin, TX 78712-1082

ABSTRACT. We consider circle packings in the hyperbolic plane, by finitely many congruent circles, which maximize the number of touching pairs. We show that such a packing has all of its centers located on the vertices of a triangulation of the hyperbolic plane by congruent equilateral triangles, provided the diameter d of the circles is such that an equilateral triangle in the hyperbolic plane of side length d has each of its angles is equal to $2\pi/N$ for some $N > 6$.

Research supported in part by Texas ARP Grants 152 and 158.

By a *circle packing* we will mean a collection of circles, all of some fixed diameter d, in either the hyperbolic or Euclidean plane such that no two circles overlap except possibly in their boundaries.

For the hyperbolic plane there are well known difficulties in making sense of *optimally dense* circle packings; see for instance [1-4]. In this paper we analyze a class of packings with similar properties using an alternative to density. We will call a *finite* packing *optimal* if the number of tangencies between its circles is not less than the number of tangencies for any other circle packing with the same number of circles and the same radius. Heitmann and Radin [5] proved that any optimal packing of the Euclidean plane is such that the set of its centers is contained in an equilateral triangular lattice. In this paper, we prove a similar result in the hyperbolic plane. If the diameter d of an optimal circle packing is such that an equilateral triangle of side length d has angle $\frac{2\pi}{N}$ for some integer $N \geq 7$, then the set of centers of the packing is contained in the vertices of a tessellation of the hyperbolic plane by equilateral triangles (of side length d).

In both the Euclidean and hyperbolic settings these results can be interpreted physically as the determination of the internal structure and Wulff shape for the energy ground state of a model of matter composed of hard disks with contact attraction. A notable difference between the settings is that in hyperbolic space surface tension is not in fact a surface effect but is comparable in magnitude to the bulk energy.

I. Notation and Statement of Results.

From now on, fix an integer $N \geq 7$. Let d be such that an equilateral triangle of side length d has an interior angle equal to $\frac{2\pi}{N}$. Let ρ be the distance function in the hyperbolic plane, $\mathbb{H}^2$. For any subsets X and Y in the plane, let $\rho(X, Y)$ be the infimum over $\rho(x, y)$ where x is an element of X and y is an element of Y.

An *admissible graph* G is a finite geodesic graph in the hyperbolic plane that satisfies these conditions: (a) every pair of distinct vertices of G are at least a distance d apart, and (b) an edge exists between vertices v and w if and only if the distance between v and w is d.

An *optimal graph* is an admissible graph that has at least as many edges as any admissible graph with the same number of vertices.

There is a natural bijection from the set of finite circle packings of diameter d to the set of admissible graphs which restricts to a bijection between optimal packings and optimal graphs. The bijection is given by considering the centers of a circle packing to be the vertices of an admissible graph. Notice that the tangencies of the packing then correspond to the edges of the graph.

For any admissible graph G, let $V(G)$ denote its vertex set, $E(G)$ its edge set, $F(G)$ its face set, and ∂G its boundary, i.e. the subgraph of G that is contained in the closure of the unbounded component of the complement of G in the plane. For any set S, let $|S|$ denote the cardinality of S. For any face f of G, let $A(f)$ be the area of f. For any vertex v in the boundary of G, let $a(v)$ be the angle subtended by v, i.e. the angle interior to the polygon whose boundary is the boundary of G.

We now define the *spiral* graph on n vertices, S_n, as follows. Let T be the graph formed from the tiling of the plane by equilateral triangles with interior angles equal to $2\pi/N$. Let e be an edge of T and let v_1 and v_2 be its endpoints. We now proceed inductively. Assuming v_j has been chosen for $1 < j \leq i$ we let v_{i+1} be the unique vertex of T defined as follows. Order *all* the previously defined vertices v_j adjacent to v_i in the form v_{b_k}, $1 \leq k \leq t$, with $v_{b_1} = v_{i-1}$, v_{b_j} adjacent to $v_{b_{j+1}}$, and the triangle $v_{b_j} v_{b_{j+1}} v_i$ positively oriented. Then let v_{i+1} be the vertex adjacent to v_i and v_{b_t} such that the triangle $v_{b_t} v_{i+1} v_i$ is positively oriented. We define the spiral graph of order n, S_n, to be the admissible graph whose vertex set is $\{v_1, \ldots, v_n\}$. Note that this uniquely determines S_n up to congruence.

S_n is easily seen to have the following properties.

a) All the faces of S_n are triangles.

b) There is an m such that the vertex set of the boundary of S_n is $\{v_m, v_{m+1}, \ldots, v_n\}$.

c) $4\pi/N \leq a(v_m) \leq 2\pi - 4\pi/N$.

d) If $m < i < n$, then $4\pi/N \leq a(v_i) \leq 6\pi/N$.

e) $2\pi/N \leq a(v_n) \leq 4\pi/N$ and $a(v_m) = 4\pi/N$ if and only $a(v_m) = 2\pi - 4\pi/N$.

The main result follows immediately from

Theorem 1. *Using circles whose radius r is such that an equilateral triangle with side length $2r$ has an interior angle of $2\pi/N$ for some integer $N \geq 7$, then for this radius all spiral graphs are optimal and every face of any optimal graph is triangular.*

II. Proofs.

Before proving Theorem 1 we require two technical lemmas.

Lemma 1. *Let G be an admissable graph. Let $\ell_G = \min\{\rho(e,v) \mid e \in E(G),\ v \in V(G)$ and $v \notin e\}$. Let $\ell = \inf\{\ell_G \mid G$ is an admissable graph$\}$. Then ℓ is the length of the arc depicted in Figure 1:*

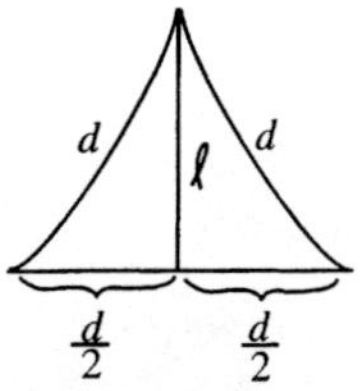

FIGURE 1

Proof. Let G be an admissible graph. Suppose that there exists $(v,e) \in V(G) \times E(G)$ such that $v \notin e$ and $\rho(v,e) < \ell$. Let v_1, v_2 be the endpoints of e. Without loss of generality, assume $\rho(v_1,v) \leq \rho(v_2,v)$. Let $p \in e$ such that $\rho(p,v) \leq \rho(x,v)$ for all $x \in e$. Then $\rho(p,v) < \ell < d < \rho(v,v_2)$. So there exists at point $v_2' \in e$ such that p is in the arc from v_1 to v_2' and $\rho(v,v_2') = \rho(v,v_1)$. See Figure 2. Let m be the midpoint of the arc from v_1 to v_2'.

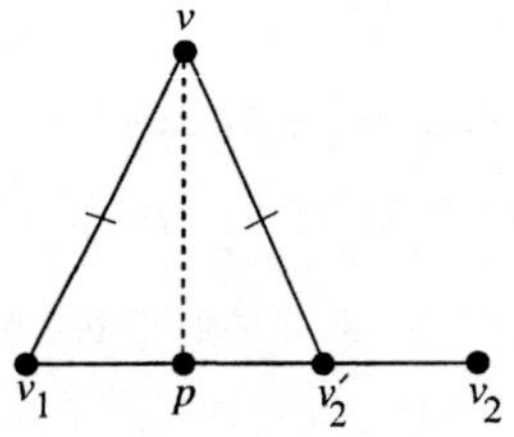

FIGURE 2

Without loss of generality assume p is in the arc from v_1 to m. By symmetry there exists a point p' on the arc from m to v_2' such that $\rho(v,p) = \rho(v,p')$. See Figure 3.

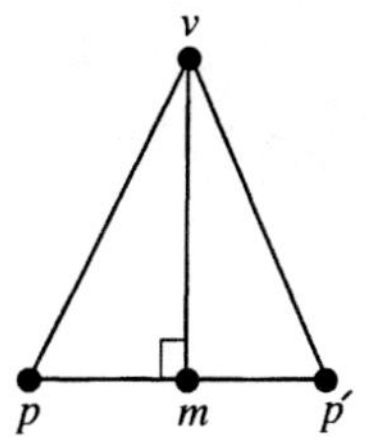

FIGURE 3

By the law of sines,

$$(*)\qquad \begin{aligned} &\frac{\sinh \rho(m,v)}{\sin a(vpm)} = \frac{\sinh \rho(v,p)}{1} \\ \Longrightarrow\quad &\frac{\sinh \rho(m,v)}{\sinh \rho(v,p)} = \sin a(vpm) \le 1 \\ \Longrightarrow\quad &\rho(m,v) \le \rho(v,p) \Longrightarrow m = p\,. \end{aligned}$$

By the law of cosines,

$$\begin{aligned} 0 = \cos\frac{\pi}{2} &= \frac{\cosh \rho(v,p)\cosh \rho(p,v_2') - \cosh \rho(v,v_2')}{\sinh \rho(v,p)\sinh \rho(p,v_2')} \\ \Rightarrow \cosh \rho(v,p) &= \frac{\cosh \rho(v,v_2')}{\cosh \rho(p,v_2')}\,. \end{aligned}$$

So $\rho(v,p)$ is minimized when $\rho(p,v_2')$ is maximized and $\rho(v,v_2')$ is minimized. This occurs precisely when $\rho(p,v_2') = d/2$ and $\rho(v,v_2') = d$, i.e., when $v_2 = v_2'$, vv_1v_2 form an equilateral triangle and $\rho(v,p) = \ell$. □

Lemma 2. *Let G be an admissible graph. Let $f \in F(G)$. If f has n sides then $A(f) \ge (\frac{n}{3})(\pi - 3\alpha)$.*

Proof. Let $e, f \in E(G)$ be different edges with endpoints v_1, v_2 and w_1, w_2 respectively. Let $v_3, w_3 \in \mathbb{H}^2$ such that the triangles $v_1v_2v_3$, $w_1w_2w_3$ have interior angles at v_1, v_2, w_1 and w_2 equal to $\alpha/2$.

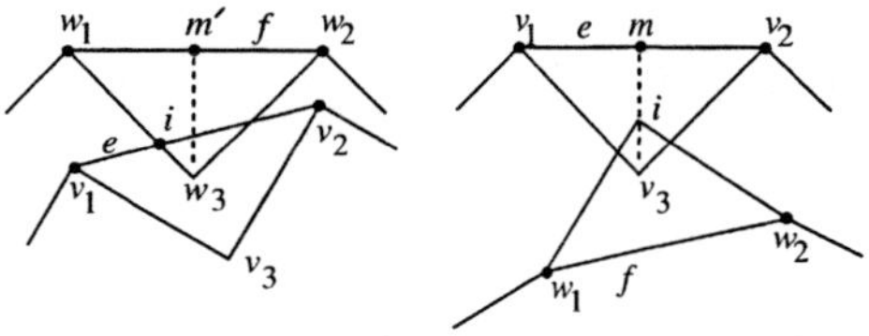

FIGURE 4

Suppose for a contradiction that the triangles $v_1v_2v_3$ and $w_1w_2w_3$ intersect in their interiors. See Figure 4. Since Int $e \cap$ Int $f = \emptyset$ this implies that the triangle $v_1v_2v_3$ intersects $w_1w_2w_3$ in the arcs w_1w_3 and w_2w_3. Let m, m' be the midpoints of e and f respectively. Then $v_1v_2v_3$ intersects the arc $m'w_3$. If v_1v_2 intersects $m'w_3$ then without loss of generality $v_1v_2 \cap w_1w_3 \neq \emptyset$. Let $\{i\} = v_1v_2 \cap w_1w_3$. Then $\rho(w_1, e) \leq \rho(w_1, i) < \rho(w_1, w_3) < \ell$. This contradicts the previous lemma. So $v_1v_2 \cap m'w_3 = \emptyset$. So either v_1v_3 or v_2v_3 intersects $m'w_3$. Without loss of generality v_1v_3 intersects $m'w_3$. Let $\{i\} = v_1v_3 \cap m'w_3$. Then $\rho(v_1, f) \leq \rho(v_1, i) + \rho(i, m') < \rho(v_1, v_3) + \rho(w_3, m') = \ell$ contradicting the previous lemma. Thus $v_1v_2v_3$ and $w_1w_2w_3$ do not intersect in their interiors.

For each edge e of f form the triangle $v_1v_2v_3$ as above so that this triangle lies in the interior of f. Since none of these triangles overlap and each has area $= \frac{1}{3}(\pi - 3\alpha)$, the lemma is proved.

Proof of Theorem 1. If G is a graph of one vertex, then G is spiral and each face of G is a triangle. Let $n \geq 1$. Assume for induction that if G is an admissible graph such that $|V(G)| \leq n$ then if G is spiral then G is optimal and if G is optimal then each face of G is a triangle.

Let O be an optimal graph and let S be a spiral graph such that $|V(O)| = |V(S)| = n+1$. Define F_k as the number of k-gon faces of O. Then $\sum_{k=3}^{\infty} kF_k = 2(|E(O)| - |E(\partial O)|) + |E(\partial O)|$. Combined with the Euler characteristic formula $V - E + F = 1$, this gives:

$$(1) \qquad 3|V(O)| - 3 - \left(|E(\partial O) + \sum_{k=4}^{\infty}(k-3)F_k\right) = |E(O)| \geq |E(S)| = 3|V(S)| - 3 - |E(\partial S)|\,.$$

The last equality holds because all faces of S are triangles. Thus $|E(\partial O)| \le |E(\partial S)|$ with equality if and only if S is optimal and every face of O is a triangle. Assume for a contradiction that $|E(\partial O)| < |E(\partial S)|$.

Let O' be the admissible graph with vertex set $V(O') = V(O) - V(\partial O)$. If O' is empty then by (1) the theorem is true. So assume O' is not empty. Order the vertices of S, $(v_1, \dots, v_n)$ as in the definition of spiral. Let S' be the admissible graph with $V(S') = \{v_1, \dots, v_s\}$ and $s = |V(O')|$. Then S' is spiral. By induction, S' is optimal so $|E(O')| \le |E(S')|$.

By Gauss-Bonnet,

$$A(O) - A(S) = \sum_{v \in \partial O} \pi - a(v) - \sum_{w \in \partial S} \pi - a(w) \tag{2}$$

$$\sum_{v \in \partial O} \pi - a(v) \le |E(\partial O)|\pi - (|E(O)| - |E(O')|)\alpha \tag{3}$$

because there are at least $|E(O)| - |E(O')|$ pairs of adjacent edges touching ∂O and each pair contributes at least α to the sum.

$$\sum_{w \in \partial S} \pi - a(w) = \sum_{w \in \partial S - \partial S'} \pi - a(w) + \sum_{w \in \partial S \cap \partial S'} \pi - a(w) \tag{4}$$

$$= \pi|E(\partial O)| - (|E(S)| - |E(S')| - 1)\alpha + \sum_{w \in \partial S \cap \partial S'} \pi - a(w) \tag{5}$$

because $|V(\partial S) - V(\partial S')| = |E(\partial O)|$, there are exactly $(|E(S)| - |E(S')| - 1)$ pairs of adjacent edges radiating from the vertices in $V(\partial S) - V(\partial S')$ (since if $a(v_1) = \alpha$ for $v_i \in \partial S$ then $i = n$ and the edge connecting v_{n-1} to v_m is not counted twice), and each pair contributes exactly α to the sum since every face of S is a triangle.

$$\sum_{w \in \partial S \cap \partial S'} \pi - a(w) \ge \pi - (2\pi - 2\alpha) + (\pi - 3\alpha)\Big[|E(\partial S)| - |E(\partial O)| - 1\Big] \tag{6}$$

since $a(v_m) \le (2\pi - 2\alpha)$ and $a(v_i) \le 3\alpha$ for $m < i < n$, and

$$|V(\partial S) \cap V(\partial S')| = |E(\partial S)| - |E(\partial O)|.$$

By lemma 2,

$$A(O) - A(S) = \sum_{f\in F(O)} A(f) - \sum_{f\in F(S)} A(f) \tag{7}$$
$$\geq (|F(O)| - |F(S)|)(\pi - 3\alpha)$$
$$= (|E(O)| - |E(S)|)(\pi - 3\alpha)$$

and equality holds if and only if all faces of O are triangles. The last equality holds from the Euler characteristic formula.

Equations (2)–(7) and $\alpha = 2\pi/N$ imply

$$|E(O)| - |E(S)| + |E(S')| - |E(O')| \tag{8}$$
$$\leq N - 6 - \left(\frac{N-6}{2}\right)(|E(\partial S)| - |E(\partial O)|) - \left(\frac{N-6}{2}\right)(|E(O)| - |E(S)|)$$

with equality holding only if all faces of O are triangles. Hence $|E(\partial S)| - |E(\partial O)| \leq 2$ with equality holding only if $|E(O)| = |E(S)|$ and all faces of O are triangles. This contradicts (1). So we may assume $|E(\partial S)| - |E(\partial O)| = 1$.

By (8),

$$(|E(O)| - |E(S)|)\left(\frac{N-4}{2}\right) \leq \frac{N-6}{2} + |E(O')| - |E(S')| \,. \tag{9}$$

since $|E(O')| \leq |E(S')|$, $|E(O)| = |E(S)|$, i.e., S is optimal.

By (1), there exists exactly one nontriangular face, f of O and it is a 4-gon. Suppose, for a contradiction, that there exists a vertex v in f which is not in ∂O. Let k be the number of triangular faces of O containing v. Since v is in the interior of O,

$$f(v) = 2\pi - k\alpha = (N-k)\alpha \tag{10}$$

where $f(v) =$ the interior angle of f at v. But this implies

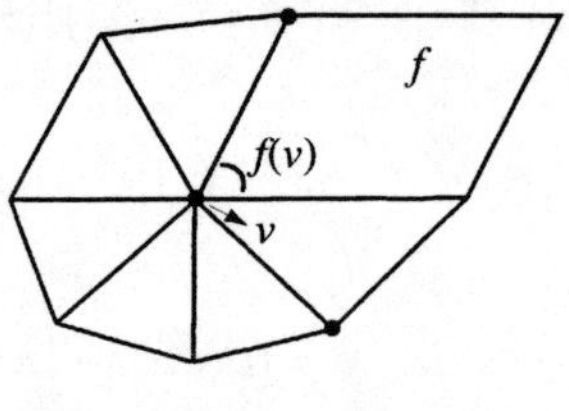

that two opposite vertices of f are distance less than or equal to d apart, contradicting that f is a face of O. Thus all vertices of f are in ∂O. Thus any vertex in the interior of O is contained in N edges of O. For any admissible graph G, let $\hat{E}(G) = \{e \in E(G) | e \cap V(\partial G) \neq \emptyset\}$. Then,

$$\begin{aligned}
(11) \qquad N(|V(O) - V(\partial O)|) &= 2(|E(O) - \hat{E}(O)|) + |\hat{E}(O) - E(\partial O)| \\
&= 2|E(O)| - |\hat{E}(O)| - |E(\partial O)| \\
&= 2|E(S)| - |\hat{E}(O)| - (|E(\partial S)| - 1) \\
(11a) \qquad N(|V(O)| - |V(\partial O)|) &= N(|V(S)| - (|V(\partial S)| - 1)) \\
&= 2|E(S)| - |\hat{E}(S)| - |E(\partial S)| + N \,.
\end{aligned}$$

So,

$$(12) \qquad |\hat{E}(O)| = |\hat{E}(S)| - N + 1 \,.$$

Every edge of $\hat{E}(S)$ which is in $E(S')$ is connected to the unique vertex $v \in V(\partial S) \cap V(\partial S')$. Since $v \in \partial S$, at least 2 edges radiating from v touch other vertices on ∂S and are therefore not in S'. Hence,

$$(13) \qquad |\hat{E}(S)| \leq |E(S)| - |E(S')| + N - 3 \,.$$

(12) and (13) imply

$$|E(O')| = |E(O)| - |\hat{E}(O)| = |E(S)| - |\hat{E}(S)| + N - 1 \geq |E(S')| + 2 \,.$$

This contradicts that S' is optimal. Thus $|E(\partial O)| = |E(\partial S)|$ and by (1), the theorem is proved. □

III. Conclusion.

For some specific diameters (other than those we have just considered) it is easy to guess what the optimal packings look like. For example, let P be a 6-gon such that one angle of P is equal to $\pi/2$ and all other angles are equal to $2\pi/3$. Suppose also that the sides of P that make the $\pi/2$ angle have equal length and the other four sides all have equal length as well. Then P is determined up to congruence and P admits a unique tessellation of the plane for which any copy of P is a fundamental domain. With respect to the diameter of the incircle of P, we believe that any optimal packing will have its center set contained in the center set of a tessellation by P.

For most diameters d, we conjecture that any limit of optimal packings (for d) does not fill space well; in fact, it appears that they may be very "narrow". Imagine placing one circle in the plane after another in such a way as to maximize the number of tangencies at each step. It is easy to increase the number of tangencies by two in a single step but the opportunity to increase the tangencies by three or more is rare. By construction, if d is smaller than the diameter of the incircle of P then the maximum number of tangencies in a circle packing of diameter d with n circles is at least $2n-3+[\frac{n}{12}]+[\frac{n+3}{12}]$. This is the best bound we have so far.

Acknowledgements

I would like to thank Charles Radin for suggesting this problem and for many helpful conversations.

References

1. K. Boroczky, *Sphere packing in spaces of constant curvature (I) (in Hungarian)*, Mat. Lapok. **25** (1974), 265–306.
2. H. S. M. Coxeter, *Regular honeycombs in hyperbolic space*, Proceedings of the International Congress of Mathematicians of 1954, North-Holland, Amsterdam, 1956.
3. L. Fejes Toth, *Regular Figures*, Macmillan, New York 1964.
4. G. Fejes Toth and W. Kuperberg, *Packing and covering with convex sets*, Handbook of convex geometry, 799–860, North-Holland, Amsterdam, 1993.
5. R. Heitmann and C. Radin, *The Ground State for Sticky Disks*, J. Stat. Phys. **22** (1980), 281–287.